SAVAGE KINGDOM

Also by Benjamin Woolley

Heal Thyself: Nicholas Culpeper and the Seventeenth-Century Struggle to Bring Medicine to the People

The Queen's Conjurer: The Science and Magic of Dr John Dee, Adviser to Queen Elizabeth I

Bride of Science: Romance, Reason and Byron's Daughter

Virtual Worlds: A Journey in Hype and Hyperreality

SAVAGE KINGDOM

The True Story of Jamestown, 1607, and the Settlement of America

BENJAMIN WOOLLEY

HarperCollins*Publishers*

HarperCollins books may be purchased for educational, business, or sales promotional use. For information, please write: Special Markets Department, HarperCollins Publishers, 10 East 53rd Street, New York, NY 10022.

Originally published as *Savage Kingdom: Virginia and the Founding of English America* in the United Kingdom in 2007 by HarperPress, an imprint of HarperCollins Publishers.

FIRST EDITION

Maps by John Gilkes

Library of Congress Cataloging-in-Publication Data is available upon request.

ISBN: 978-0-06-009056-2
ISBN-10: 0-06-009056-1

07 08 09 10 11 UK/RRD 10 9 8 7 6 5 4 3 2 1

CONTENTS

LIST OF ILLUSTRATIONS

Plates

Maps

AUTHOR'S NOTE

Spellings have been modernised, and in the case of the names of people and places, standardised. Dates are generally Old Style, but with the new year starting on 1 January.

I am grateful to the British Library, Library of Congress, Library of Virginia, University of Virginia Library, Bodleian Library, Folger Shakespeare Library and Huntington Library for help in consulting their collections. My thanks to: Gregory H. Stoner, Virginia Historical Society; Terry Tuey, McPherson Library, University of Victoria; Nettie Stoppelenburg, Archives Service in Utrecht; Theo Pollemans, Oudewater, Netherlands; Julian Roberts; Nicholas Lee, University of Bristol; Rick Beyer, Plate of Peas Productions; David Sacks, Reed College; Stanley Wells, Shakespeare Birthplace Trust; Jocelyn Wingfield; Paul Embler; John McCavitt; Robert Massey, Royal Observatory; Catherine Beaumont, Otley Hall; Bly Straube, Historic Jamestowne; Patrick Martin, Herbert Law Center, Louisiana State University; Crandall Shifflett, Virginia Tech, creator of Virtual Jamestown (www.virtualjamestown.org) and Richard Rabinowitz, American History Workshop.

Personal thanks to Jack Russell, Barbara Briggs, Sir Roger and Lady Bannister, Arabella Pike, Anthony Sheil, Asha Joseph and Matthew Woolley.

PROLOGUE

The Great White Fleet

IN AUGUST 1907, Marine Guard Paul Elias Embler found himself aboard the USS *Alabama*, coasting New England. Together with eight more ironclads steaming alongside, his eleven-and-a-half-thousand-ton battleship was bound for Virginia to take part in one of the great naval events of history, the rendezvous of what came to be known as the 'Great White Fleet'. On the orders of President Theodore Roosevelt, sixteen battleships of the US navy were to assemble in Chesapeake Bay, to commemorate the three hundredth anniversary of the founding of Jamestown, or, as the President put it, 'the birthday of this nation'.

Embler was an ordinary seaman, but distinguished himself by keeping a diary, which he scribbled during intervals between meals, watch and drill, dangling in his hammock or sitting on a duty bag next to Morrison the tobacco chewer. 'Reveille at 5 AM. Breakfast at 7. Uniform of the day white. Quarters at 9:15' began his first entry, dated August 27, 1907. Those that followed were similarly succinct, details generally being reserved for the meals:

> *Ship is off Newport RI. Dinner roast beef, gravy, spuds with jackets on, butter, bread and blackberry jam and coffee . . .*
>
> *Supper clam chowder, butter, tea. Hammock call 7:30. Tattoo at 9:15.*
>
> *Clear night.*[1]

On September 26, the routine of life at sea was interrupted by the *Alabama* putting in at the Brooklyn Navy Yard for a refit, offering the prospect of ten weeks shore leave in the world's most dynamic city, New York. Embler withdrew a fresh uniform from the ship's laundry, carefully marked each item of clothing with his name, stuffed a plump $13 pay packet in a pocket, and set off for Manhattan.

Crossing the Brooklyn Bridge, he plunged into a world that was quite alien to a farm boy from Asheville, North Carolina. It was a world that swept people up in a 'great torrent of spending', as another visitor put it, and overwhelmed them with the 'luminous epilepsy' of electrified streetlights and billboards. Embler's terse entries for the following weeks provide stroboscopic flashes of a city hurtling recklessly into the twentieth century, twitching with energy, ravenous to consume. There are burlesque shows, brawls with Bowery toughs, 'friendly ladies' asking 25 cents 'to buy them a drink', soaring skyscrapers, squalid flophouses and, for Embler, most dazzling of all, the two gold front teeth of 'a beautiful girl, age 17', one Viola Savage.

At the eye of this maelstrom of modernity, Embler came across the body of a crewmate, Marine Private Conklin. It had been found floating in the East River on November 11. Conklin had apparently been drugged, robbed, thrown off a pier and left to drown. Attempts to track down his family came to nothing, as it transpired he had signed up under an assumed name. So he was buried in the grounds of the Naval Hospital, his anonymous grave decorated with a bouquet of 'very large white chrysanthemums and American Beauties' worth $6, a parting gift from his comrades.[2]

On December 6, Embler left this all behind as the *Alabama* slipped back into the East River, her hull now brilliant white, the gilded scrollwork of her prow glittering like Viola Savage's teeth. Together with her four similarly spick-and-span consorts, she steamed off into the quiet ocean at midnight, bound for Virginia.

A day later, the five ships reached the yawning mouth of Chesapeake Bay, where the rest of the fleet awaited them. To the south, Embler could just make out the feeble winking of Cape Henry Lighthouse, marking the point first sighted by the English adventurers in 1607. Just beyond lay Sewell's Point, the site of the Jamestown Exposition, opened by President Roosevelt back in April to commemorate the three hundredth anniversary of the settlers' arrival. In a stirring inaugural speech, the President had reminded the crowds of the hardships faced by their predecessors when they arrived on those shores. 'Famine and pestilence and war menaced the little band of daring men who had planted themselves alone on the edge of a frowning continent'. But their example had shown that, 'if the average of character in the individual citizen is sufficiently high . . . there is literally no height of triumph unattainable in this vast experiment of government by, of, and for a free people'. With these words, the President was supposed to press a gold button to set in motion the marvelous engines of industry and invention being exhibited. But, due to bureaucratic bungles and technical hitches, the wires leading from the button did not connect to anything. The presidential Oz would have to wait to work his wizardry.[3]

Approximately 3 million visitors had come to Virginia since the exposition's opening, drawn by such attractions as a huge relief map of the Panama Canal, daily re-enactments of the battles of the *Merrimac* and *Monitor*, an exhibition of incubators featuring 'live' premature babies, a simulation of the 1906 San Francisco earthquake and a reconstruction of the Klondike Gold Mine.[4]

Now, on this cold December morning, President Roosevelt was back again, aboard his official yacht, the *Mayflower*—its name an irksome reminder to his hosts of the greater fame still enjoyed by the English exiles who arrived in New England a full thirteen years after the Jamestown settlers came to Virginia. The President had come to initiate the climactic event of the festivities: the sailing of the Great White Fleet now riding at anchor in the bay.

The fleet's rendezvous had been the President's pet project for the Jamestown anniversary. No event, he felt, could better advertise to the host of visiting dignitaries from the Great Powers of the old world how far the upstart United States had come. The 'splendid little' Spanish American War of 1898, in which Roosevelt had made his name with his volunteer force of 'Rough Riders', had toppled the colonial governments of Cuba and the Philippines, dispensing with the last vestiges of the Spanish Empire. This had left the US the dominant power in her hemisphere, with a global network of naval bases in Guantanamo, Manila, Puerto Rico and Guam, soon to be joined by Pearl Harbor. The great fleet now sitting upon the waters of the Chesapeake, ready to sail on the President's orders, was cast-iron proof of the nation's triumphant rise.[5]

Following 'many pathetic scenes of farewell', at 10 am sharp four bells were rung to get 223,836 tons of armament under way—the greatest quantity ever to sail under a single flag, according to the *New York Times*, carrying 12,793 officers and men, provisioned with a million pounds of flour, vegetables and frozen beef.

The *Mayflower* started to make her way towards the open ocean, dragging the great chain of ironclads in her wake, through waters studded with 'shipping of every description', past shores 'black with people'.[6] As the armada approached Cape Henry, the *Mayflower* pulled aside, and beneath a canopy of coal and gun smoke, reviewed the ships processing by, each letting off a thunderous 21-gun salute as it passed: the flagship *Connecticut*, the *Kansas*, the *Vermont*, the *Louisiana*, the *Georgia*, the *New Jersey*, the *Rhode Island*, the *Virginia*, the *Minnesota*, the *Maine*, the *Missouri*, the *Ohio* and, bringing up the rear, Embler's asthmatic *Alabama* (despite the repairs in New York, she suffered from chronic engine problems which would later force her to retire from the fleet), followed by the *Illinois*, the *Kearsage* and the *Kentucky*. A great moving map of the United States entered the ocean and spread across the waters.

A decade later, the site of the Jamestown Exposition was handed over to the Navy and turned into the largest naval base in the world, home to the United States' Atlantic Fleet. Over the coming century an armada of swifter, heavier, deadlier vessels came and went, including the nuclear-powered aircraft carrier USS *Theodore Roosevelt*, commissioned in 1986. It alone displaced nearly half the tonnage of the entire Great White Fleet.[7]

The *Susan Constant*, the *Godspeed* and the *Discovery*, the three English ships that had first slipped through those waters three centuries before, together weighed less than the Great White Fleet's consignment of canned fruit. Crammed inside were a hundred men and boys, a few chickens and some rats, and none but the rats were in a fit state to face the ordeals that awaited them.

One of the great questions of history, which it is the purpose of this book to consider, is how such a morsel of colonial ambition, popped into the mouth of the Chesapeake in 1607, could, a few centuries later, produce a nation capable of disgorging the most powerful navy ever to sail the high seas? Was there anything that those ill-conditioned creatures brought, other than infectious disease, that can account for a transformation never before matched in the annals of humanity?

Part of the answer must lie in where these men had come from, as much as their destination. It transpires that their mission had been shaped not by the prospect of a colonial adventure in North America so much as by convulsive events back home. Powerful forces unleashed by the Protestant Reformation, and the spread of the Muslim Ottoman Empire to the east, had shaken the very foundations of European civilization. This more than anything else was what had propelled them out of London and deposited them in America so ill prepared, badly equipped and poorly financed.

And then there were the 'Indians'—misnamed, maltreated, displaced. While the Pilgrim Fathers' relationship with local tribes has been portrayed as one of happy coexistence, culminating with the ritual of Thanksgiving, the Jamestown colonists encountered, and recorded, a very different experience. Admiration vied with condescension, curiosity with prejudice. Though the Indians were called 'savages' by the English, and their homeland a 'savage kingdom', the interlopers well knew, and half acknowledged, that it was their own savagery, horribly exposed in the religious wars of Europe and carried like an infection across the ocean, that was the greater threat.[8]

Shakespeare's beguiling play *The Tempest* was written during, and partly inspired by, events surrounding the Jamestown adventure. It concerns a group of courtiers who undertake a perilous sea voyage. Caught in the eponymous storm, they are stripped of their rank and dignities and 'belched' by the 'never-surfeited sea' onto a strange, enchanted land. There, their destinies are transformed by mysterious happenings, secret histories and lost souls. Visions of utopia vie with glimpses of hell; love blossoms alongside rape; the amnesia of a sweet deliverance is spoiled by flashbacks of barbarity. It begins with the breakdown of order, and ends with the play's most famous line, the promise of a 'brave new world'.

The story of England's attempt to colonize America is much closer to that of Prospero and Caliban than the pious Pilgrims. Its master narrative is not the Biblical tale of the elected saints finding a promised land—though some at the time hoped it might be. It is about flawed, dispossessed, desperate people trying to reinvent themselves. It is about being caught in a dirty struggle to survive, haunted by failure, hungering for escape, dreaming of riches and hoping for redemption.

SAVAGE
KINGDOM

Part One

ONE

A Feast of Flowers and Blood

ON THE MORNING of 20 September 1565, the sixty-year-old carpenter Nicolas le Challeux awoke to the sound of rain pelting down on the palm-leaf thatch overhead. It had not stopped for days, and a muddy morass awaited him outside.

When he had arrived in Florida the previous month, a sunnier prospect had beckoned. He had left the terrors of his native France far behind, and come to a place where he could practise his craft and religion in peace. Its very name suggested renewal, the Spanish explorer Juan Ponce de León calling it Florida after the season in which he first sighted its shores: Easter Week, or *Pascua Florida*, 'the feast of flowers'.

Florida could furnish all that a man could wish on earth, Challeux had been told. It had received a particular favour from heaven, suffering neither the snow nor raw frost of the North, nor the drying, burning heat of the South. The soil was so fertile, the forest so full of wild animals, the honest and gentle natives could live off the land without having to cultivate it. There were even reports of unicorns, and of veins of gold in a great mountain range to the north called the 'Appalatcy'. It was 'impossible that a man could not find there great pleasure and delight,' Challeux was assured.[1]

The contrast with the state of his homeland was stark. Europe was in turmoil. To the south, the Catholic Spanish and Holy Roman empires, offshoots of a single dynasty, domineered. In the north, Queen Elizabeth reigned over Europe's upstart Protestant monarchy England, while her subjects egged on their co-religionists in the Low

Countries (modern Netherlands and Belgium), who were fighting for independence from their Spanish overlords. To the east stretched the Islamic Ottoman Empire, Suleiman the Magnificent resting an elbow upon the Balkans, a heel upon Basra. And in the middle lay France, a Catholic country penetrated by a powerful Protestant or 'Huguenot' minority. Exposed to so many religious and political tensions, it threatened to disintegrate, and in 1562, a series of civil wars erupted across the kingdom that were so brutal, they gave the word *massacre*, French for a butcher's block, its modern meaning.

It was from the midst of this maelstrom that Gaspard de Coligny, leader of the Huguenots, had dispatched a fleet under the command of his kinsman René de Laudonnière to found a Protestant refuge in Florida. To the eyes of Coligny's Catholic enemies, this was a provocative move. Though its coastline was still only hazily charted, and some even doubted it was a single land mass, all of North America was claimed by the Spanish under a famous 'bull' or edict issued by Pope Alexander VI shortly after Christopher Columbus's historic expedition of 1492. This had donated all the 'remote and unknown mainlands and islands' in the Atlantic to the Iberian kingdoms of Spain and Portugal, so they could bring the native populations 'to the worship of our Redeemer and profession of the Catholic faith'. By sending his men to Florida, which was within convenient reach of Spanish possessions in Cuba and Mexico, Coligny was clearly challenging not only the Spanish claim, but the religious authority underpinning it.[2]

However, Coligny's exiles had found Florida untouched by the Spanish, and settled themselves on the banks of the River of May (now called St John's River, near modern Jacksonville), on a 'pleasant open space covered with various kinds of grasses and plants'. They called their new home Fort Caroline, after France's Catholic King Charles IX, in the hope of forestalling reprisals. Old Challeux had arrived the following year with another consignment of refugees, on a supply ship captained by Jean Ribault, a prominent Huguenot, as well as one of France's most accomplished seamen.[3]

Conditions for the newcomers turned out to be less Elysian than advertised. The hundred or so settlers who had been there a year had run out of supplies, and were living off wild fruits, berries, the occasional crocodile, and goods stolen from the local Indians. There were also reports that the Spanish had been tipped off about Coligny's project, and had sent a fleet which was even now roving the coast.

Over the coming weeks, Challeux joined a team of workmen who, under the direction of John de Hais, master carpenter, tried to reinforce La Caroline's fragile palisade. The state of the fort's defences was pitiful. The triangular layout was breached in two places, along the western side, and the long southerly wall facing the river, where the foundations for a 'grange' to store the settlement's artillery and munitions lay partially built.

The weather hampered the workmen's efforts. Daily deluges washed away the embankment supporting the palisade wall, and intervals of baking sunshine were too fleeting to allow the damage to be repaired. Meanwhile, the surrounding landscape became more and more saturated. Rivers burst their banks, meadows became marshland.

And so, on this September morning, Challeux faced another day of hard labour in the remorseless rain. Nevertheless, he managed to rouse himself, put on a damp and rotting cloak, and gather his tools.

A few hundred yards away, beyond the curtain of incessant rain, Don Pedro de Menéndez de Avilés lay in wait at the head of a column of five hundred soaking, disgruntled but well-armed Spanish troops. Menéndez was a Spanish noble and naval commander. He had arrived in Florida with a fleet of Spanish galleons a few days before Ribault, with orders to exterminate the 'Lutherans' and establish himself as *Adelantado* or governor of Florida, which King Philip II of Spain declared extended all the way from the keys on the southernmost tip of the Florida peninsula to Newfoundland.[4]

Menéndez had anchored his ships in the River of Dolphins (modern Matanzas River), about thirty-five miles south of Fort Caroline. There, on 28 August, he had set about building a military base, which he called St Augustine, in honour of the feast day upon which construction work had begun. After several weeks gathering intelligence about Fort Caroline from local Indians, and harrying Ribault's fleet, he decided to mount a land attack on the French settlement.

The journey from St Augustine to Fort Caroline had proved dangerous and tiring. Though led by Francis Jean, a French defector who had lived in Fort Caroline, Menéndez's troops got lost. They had to make their way through 'morasses and desert paths never yet trod', often up to their armpits in water, holding heavy knapsacks, ladders (for scaling the fort's palisade) and harquebuses (a heavy forerunner to the musket) above their heads. They had finally reached a 'little rise in the ground' overlooking Fort Caroline in the early morning of 20 September, 'the eve of the day of the Blessed Apostle and Evangelist St Matthew'.[5]

Menéndez commanded his men to stay hidden in the woodland while the camp master, Don Pedro de Valdes, and Francis Jean were sent ahead. They made their way through the thick undergrowth until they found their path blocked by a fallen tree. Turning back, a French sentinel who was patrolling nearby glimpsed them through the thicket. '*Qui va la?* [Who goes there?]' he called. '*Un Français* [a Frenchman],' the Spanish captain replied, in a convincing accent. The sentinel approached, and as soon as he was within reach, Valdes drew his sword and slashed him across the face. The sentinel fell backwards, giving a shout.

The cry carried through the thicket of trees, across the waterlogged clearing and turf embankment surrounding the French fort, through the wide gaps of its palisade, reaching the ears of Challeux as he was leaving his quarters. Guards were summoned, and they rushed out to see what had happened, leaving the fort's gates open behind them.

At the same moment, Menéndez, thinking the cry was from his own camp master, gave the order to attack. Spanish troops burst from the cover of trees and poured through the fort's still-open gates, firing and slashing at the panicking French settlers, who emerged from their quarters in their nightshirts to see what was going on.

Challeux saw a Spanish pikeman charging at him, and ran to a ladder or scaffold leaning up against one section of the fort's rampart. 'Nothing but the grace of God enabled me to double my effort,' he recalled. 'Old grey-haired man that I am, I nevertheless jumped over the ramparts, which if I thought about it, I could not have done, for they were eight or nine feet in height.'[6]

Not until he reached the cover of trees did he dare look back. He beheld 'the horrible slaughter of our people'. Some managed to reach boats moored along the nearby shore, and row out to Ribault's fleet, which was anchored in the river. The rest were killed where they stood, the Spanish vying 'with each other as to who could best cut the throats of our men'. They even attacked the ill, the old, women and children, Challeux claimed, 'in such a way that it is impossible to conceive of a massacre which could be equal to this one in cruelty and barbarity'.[7]

Once the savagery had subsided, the corpses were dragged out of the fort and dumped on the river bank. The Spanish soldiers, frustrated that so many had escaped, 'vented their wrath and bloody cruelty on the dead', scooping out their victims' eyeballs, sticking them on the points of their daggers, and hurling them into the river 'with shrieks of abuse and ribald laughter'.

Challeux fled into the woods, and made his way to the coast. He was picked up by one of Ribault's ships the next day, and reached France a few weeks later, where he published a sensational account of his experiences. Others, including Jean Ribault, were less fortunate. Caught in a storm, they were shipwrecked back on the Florida coast. Exhausted and starving, they eventually surrendered to the Spanish, who bound them two by two, took them behind a sand dune, and slaughtered them. Only seven were spared, four who claimed to be Catholics, together with a drummer, a fife player and a carpenter, required by Menéndez to supplement his own troops.[8]

In a report of his actions sent to King Philip, Menéndez boasted,

'I made war with fire and blood as Governor and Captain General of these Provinces upon all who might have come here to settle and to plant this evil Lutheran sect'. Having 'come to this coast to burn and hang' the hated Huguenots, he pledged to continue pursuing them 'by sea and by land' until they were annihilated.

To prevent any other adherents of that 'wicked Lutheran sect' returning to North America – particularly the English, who he noted had nominated Ribault 'Captain General of all [their] fleet' – Menéndez now urged the King to secure the entire eastern coast of North America. In particular, the Spanish should build a fort at the 'Bahia de Santa Maria' at thirty-seven degrees of northerly latitude, 'the key to all the fortifications in this land'. This was the region that would later come to be dubbed by the English 'Virginia'.[9]

Even in Spain, Menéndez was denounced for 'inhumanely condemning so many souls to hell for ever' at Fort Caroline. But he claimed that such brutality had been necessary. He had prevented the weed of Protestantism which had spread so quickly across the northern parts of Europe, from taking hold in the fertile soil of the New World. The massacre was a godly act, born of 'divine inspiration, rather than from any dictate of human understanding'. Some even suggested that Menéndez had been merciful, 'since he nobly and honourably put them to the sword, when by every right he could have burnt them alive', the preferred punishment for heresy.

King Philip, who saw himself and his mighty empire as custodians of the true faith, reinforced this view. 'The justice you meted out to the Lutheran corsairs who attempted to occupy and fortify Florida in order to sow the seeds of their wicked sect,' he informed Menéndez, was 'fully justified'. In the great battle for the soul of Europe and the future of mankind, North America had become the new front line.[10]

TWO

Machiavelli

ON A COLD JANUARY DAY IN 1606, a messenger walked inconspicuously across the cobbles of London's Strand, carrying a parcel for delivery to an address opposite the Savoy. It was a journey of only a few hundred yards, but one that crossed from one era of history to another. Just beyond Charing Cross lay York House, where, in 1599, the Earl of Essex, 'a man of great designs', had been imprisoned for his attempts to rouse Queen Elizabeth's hesitant government to a heroic war against the 'tyranny' of Catholic Spain. A few yards further on was Durham House, once the home of Sir Walter Raleigh, the swashbuckling adventurer who had tried, and failed, to make North America an English colony, and a beachhead for attacking the Spanish empire. Essex had been executed for his rebellious behaviour in 1601. Raleigh was arrested for treason two years later, and now languished in the Tower of London, his bold schemes mere echoes in the empty halls of his great residence.

Past these mausoleums to old follies and glories lay the messenger's destination, a brand-new, flat-fronted, perpendicular brick building. At each corner stood a tower, soaring as high as the famous lantern or 'little turret' atop Durham House, from which Raleigh had once beheld a 'prospect which is pleasant perhaps as any in the World'.[1] This building was the London residence of Robert Cecil, the Earl of Salisbury, the man who had contrived the downfall of these political giants, shaped the emergence of a new regime under a new king, and cemented his astonishing rise by commissioning the turreted edifice soaring overhead.

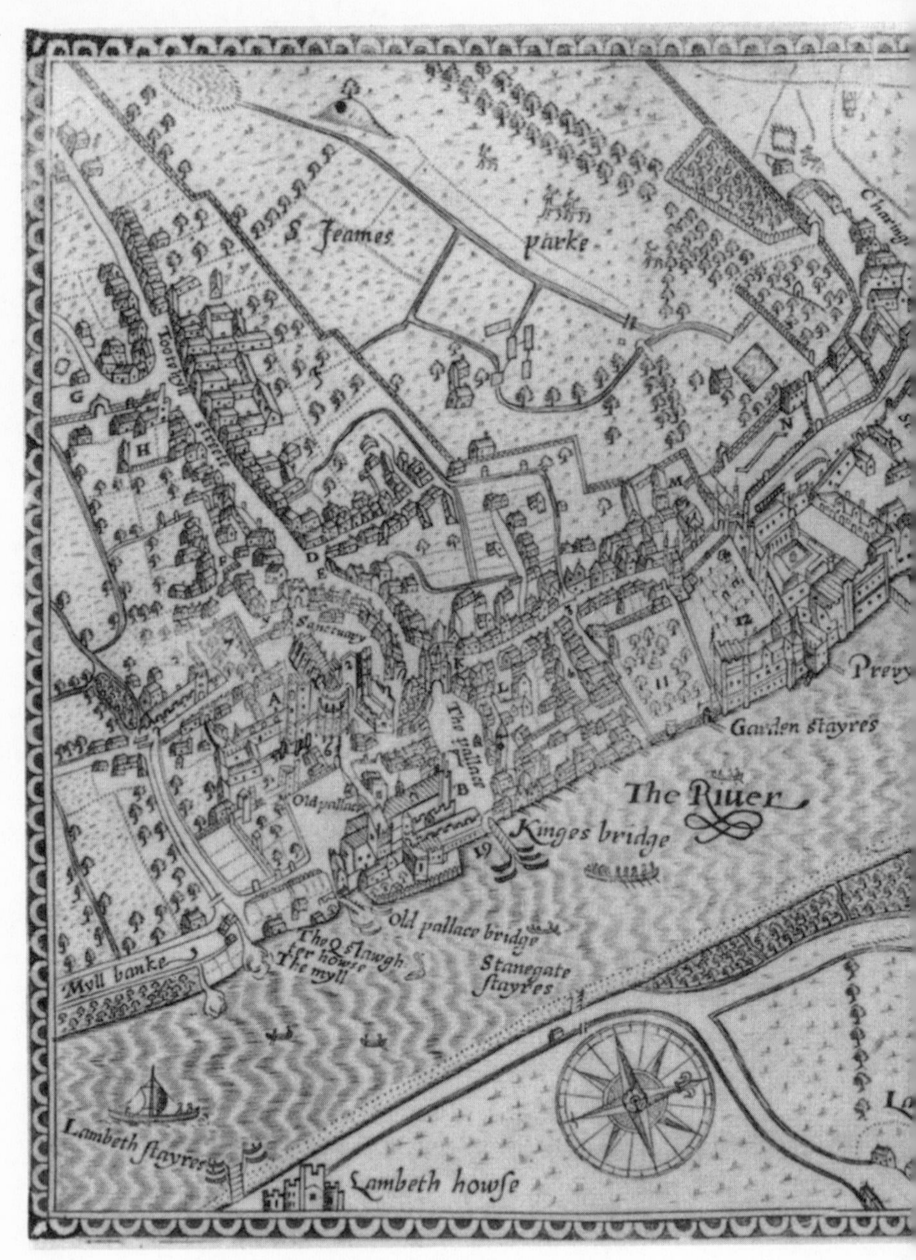

Westminster (1593), showing the Strand to the east, running parallel to the Thames.

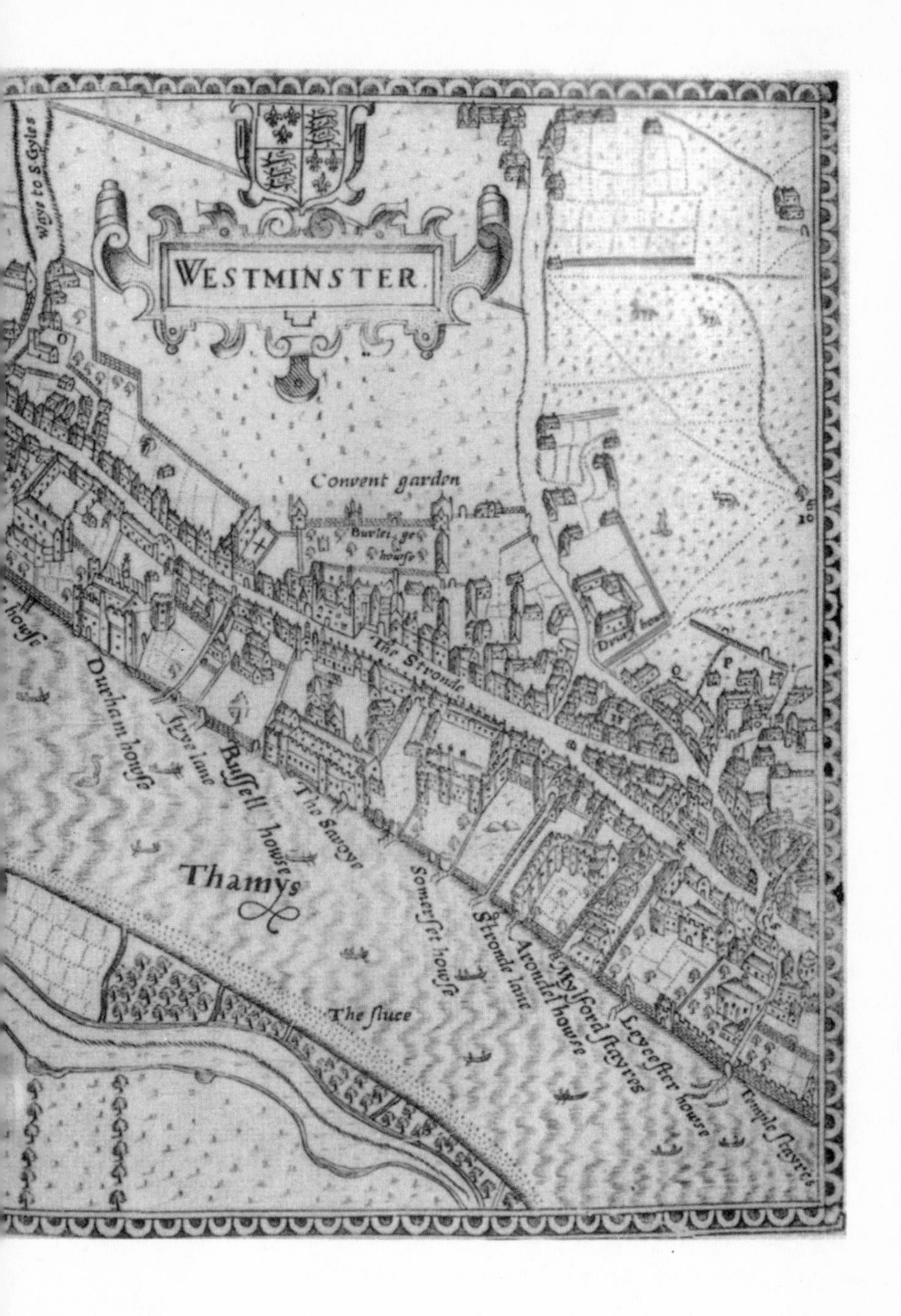

Wayes to S. Gyles
WESTMINSTER
Convent garden
howse
The Stronde
howse
Durham howse
Ivye lane
Russell howse
The Savoye
Thamys
Somersett howse
Stronde lane
Arondell howse
Mylford stayres
Leycester howse
Temple stayres
The sluce

The messenger knocked on the door, and was admitted to a small antechamber. Being Secretary of State to King James, and the monarch's most trusted servant, the Earl was daily assailed by crowds of applicants, supplicants, petitioners and messengers competing for his much divided attention, and the antechamber had been especially set aside to receive them.

A clerk entered the room. He approached the messenger, who handed over the parcel. The clerk broke the wax seal and cursorily glanced at a covering letter, addressed to 'the Right honourable the Earle of Salisbury of his Majestes privie Councell'. It was from Thomas and Edward Hayes, known to Cecil's staff as 'projectors', agents who tipped off the Secretary of State to speculative schemes that might attract a profit or some political advantage. Attached to the letter was a long, formal document. Its title concerned a controversial subject, and marked it for the Secretary of State's personal attention. It read: 'Reasons to move the High Court of Parliament to raise a stock for the maintaining of a colony [in] Virginia'.[2]

The word 'Machiavellian' came into currency in England in the early 1600s, and it was Robert Cecil who in the minds of many personified its meaning. One of the Earl of Essex's servants, defending his master before a commission of inquiry called following the Earl's fall from grace in 1599, described the Secretary of State as 'an atheist, a Machiavel' who literally embodied the warped morality of political opportunism. 'It was an unwholesome thing to meet a man in the morning which hath a wry neck, a crooked back or a splay foot,' said the servant, referring to various deformities with which Cecil had suffered from birth.[3]

Cecil's cousin and long-time ally Francis Bacon argued that it was these deformities that conferred upon the Secretary of State the callousness and determination a great political operator needs, making him 'void of natural affection', and mindful 'to watch and observe the weakness of others'.[4] These were certainly the qualities he had displayed in his dealings with Sir Walter Raleigh.

Raleigh had been one of Queen Elizabeth's most cherished, if exasperating, favourites, and in 1584, to demonstrate her affections, she had granted him an exclusive licence to colonize North America. His efforts had resulted in establishing a small settlement on the island of Roanoke, on the Carolina Banks. On this basis he claimed the entire region for the English Crown, naming it 'Virginia' in honour of the Virgin Queen. But hostilities with Spain had prevented him from sending supply ships to service the fledgling colony, and in 1590, it was found abandoned, the only clue to the inhabitants' whereabouts being a word carved in one of the fort's wooden posts: 'Croatoan', the name of a local tribe. Attempts to find the missing settlers came to nothing, and the supply ship had returned to England.[5]

Even after 1591, when his licence officially expired, Raleigh had continued to claim Virginia as his, on the basis that his Roanoke settlers may have survived, and established a permanent base elsewhere in the region. But several follow-up missions sent to find his 'lost colonists' had proved fruitless. In 1602, a rival mission was dispatched to 'Norumbega', in the area of modern-day New England, to find alternative locations for an English colony. Raleigh knew nothing of it until it returned with a lucrative cargo of cedar wood and sassafras (an aromatic bark used to fumigate bedlinen and treat syphilis). His fury at a flagrant attempt to challenge his monopoly over North America resulted in an attempt to confiscate the goods, and an appeal for help and support from his 'friend' Cecil. 'I shall yet live to see [Virginia] an English nation,' he had promised. But his letter coincided with the final days of Queen Elizabeth's reign, and Raleigh was emerging as an obstacle to Cecil's complex manoeuvres to ensure the succession of James VI of Scotland to the English throne.

Elizabeth had died on 24 March 1603, and within days the Scots King took her place as James I of England. Soon after, Cecil confronted Raleigh with accusations that he had attempted to plot against the succession, and had him arrested. Sir Walter was tried for treason, sentenced to death, and thrown in the Tower to await his fate, leaving Virginia conveniently available for the new regime to dispose of as it saw fit.

And now Cecil had in his hands this new proposal for 'the maintaining of a colony in Virginia'. Thomas and Edward Hayes began in a suitably humble tone: 'Pardon us (right Honourable), that we presume to move this project presented herewith unto you, so remote from the course of your great affairs as America is from England.' These 'great affairs' were the knock-on effects of the Gunpowder Plot, an attempt by a group of disaffected Catholic gentlemen to blow up the House of Commons on 5 November 1605, the day the King and his entire Privy Council, together with most of the English nobility, had assembled at Westminster for the start of a new session of Parliament. Cecil was leading the official investigation into the plot, which was conveniently sweeping up a great many opponents of the new regime.

In their proposal, the Hayeses argued that 'so great a business' as colonizing America 'can never ever be duly effected by private means', as previous experience had shown. So they and certain associates had 'devised another way, where by the cause may be completely set forward': a great public scheme performed under the auspices of Parliament.[6]

Attached to the letter was the motion they intended to be set before the Houses of Parliament. It proposed setting up a large fleet of modern, well-armed and well-equipped ships manned by 'able Mariners, and worthy chieftains', which would be sent across the Atlantic to conquer North America in the King's name. It would be publicly financed, as 'private purses are cold comfort to adventurers and have ever been found fatal to all enterprises hereunto undertaken by the English, by reason of delays, jealousies and unwillingness to back that project which succeeded not in the first attempt'.

A model for such an undertaking could be found, the Hayeses believed, in the Low Countries. This collection of mostly Protestant city states, dominions of the Spanish Crown fighting for their independence, had created Europe's most advanced trading region.

They had 'effected marvellous matters in traffic [trade] and navigation in few years' by financing their expeditions through the issue of 'main' or public stock. In contrast, the English strategy of giving a grandee such as Raleigh a monopoly of trade with a particular region had resulted in failure. 'It is honourable for a state rather to back an exploit by public consent than by private monopoly,' the Hayeses concluded.

They therefore proposed that the King should give his assent to an Act of Parliament to authorize the setting up of a public company to colonize America. This was a matter of urgency, as 'the want of our fresh and present supply of our discoveries [in Virginia] hath in manner taken away the title which the law of nations giveth us' – in other words, use it or lose it.

Cecil was aware of the urgency, as well as the economic advantages, of a colonial venture. Spain's enormous wealth, after all, rested upon her speculative ventures in South America, which were now profiting the new Spanish King Philip III to the tune of two million ducats a year – more than twice James's entire annual income.[7] Cecil was also aware of the growing success of London's trading companies – indeed, James to a degree was coming to depend upon it, as the profits generated by these companies were helping to bankroll his overstretched royal exchequer. The newest, the East India Company, had been founded in 1600 by Sir Thomas Smythe, a close friend and generous supporter of the Crown. It had already sent several ships loaded with bullion to the Far East to buy spices and other exotic goods direct from suppliers. The sale of these items back in England was not only proving highly profitable, but suggested a lucrative new stream of customs revenue for the Crown.

However, the Hayeses had badly miscalculated. There was no way Cecil would countenance a public colonization venture, whatever the trade benefits. The very name 'Virginia' was a reminder why he considered this impossible. During the 1570s and '80s, he and his father Lord Burghley had worked strenuously to marry Elizabeth off to one of Europe's Catholic princes, so she could produce an heir. They had been fought all the way by a group of courtiers who opposed any such match on the grounds that she and therefore her kingdom would come under the sway of Catholic regimes they saw as tyrannical. It was this clique, egged on by Huguenot exiles wanting to avenge the massacre in Florida, that had promoted the cult of Elizabeth's virginity. It became an emblem of Protestant purity and independence, which Raleigh had reinforced with his choice of name for English America.

Cecil had since managed to repair some of the diplomatic damage caused by these struggles, his efforts culminating with the Somerset House Peace Treaty with Spain, which was signed in 1604, a year after James ascended to the throne. But this had done nothing to diminish Virginia's political potency. Just a few months before Cecil received the Hayeses' proposal, a play called *Eastward Hoe* had been staged at the popular Blackfriars Theatre. It featured a dissolute 'thirty-pound knight' (a reference to King James selling off titles to raise ready cash) who had attempted to mount an expedition to the colony. 'Virginia longs till we share the rest of her maidenhead,' the expedition captain had boasted, in an effort to lure footloose and unemployed mariners into joining the mission. 'You shall live freely there,' he added, 'without sergeants, or courtiers, or lawyers, or intelligencers', the apparatus of oppressive government.

The play's authors were arrested for such incendiary lines, and even now, one of them, Ben Jonson, was sending Cecil plaintive letters, begging for his freedom. In such circumstances, the idea of a royally endorsed public venture could not possibly be countenanced.[8]

A private venture, on the other hand – that was a different matter, and a proposal for one just happened to be at hand. Cecil's factotum, Sir Walter Cope, had long held an interest in foreign exploration, and around the same time as the Hayeses, had sent Cecil a private note suggesting a very different venture. Judge Sir John Popham, another prominent member of James's government, 'foreseeing in the experience of his place' as King James's Chief Justice 'the infinite numbers of cashiered captains and soldiers, or poor artisans that would & cannot work, and of idle vagrants that may & will not work, is affectionately bent to the plantation of Virginia'. Cope proposed that a select group of trusted merchants and nobles be approached to fund a small, secret exploratory mission to America to trade with the Indians and search for valuable commodities. All that was required from Cecil was 'two lines from your Lordship in particular, or from the Lords' of the Privy Council in support of the venture, and an expedition would be under way in no time.[9]

This was far more to Cecil's liking. For all its negative political associations, there was something about Virginia that not even Cecil could completely resist. Raleigh's geographer, the cleric and colonial propagandist Richard Hakluyt, had since 1599 adopted Cecil as a patron, and on his new master's behalf uncovered several convincing reports referring to the existence of gold deposits in the 'Appalatcy' or Appalachian Mountains to the west of Roanoke. There were also references to a great salt-water lake lying somewhere to the north-west, which could be reached by river, suggesting there might be a route through the middle of North America to the Pacific – the long-cherished dream of a navigable westerly passage to Cathay and the riches of the Orient. Also, intelligence had revealed that, despite these reports, the Spanish had continued to ignore the areas of America north of Florida.[10]

Cecil was in desperate need of new sources of revenue, not least to cover the cost of glorifying his London residence, which, he admitted to a friend, had 'almost undone me'. Furthermore, to secure his throne, James had been forced to spend heavily on securing the necessary political support among England's anti-Scottish elite, and the royal exchequer was already stretched to breaking point. Any venture that had the potential for private profit and royal revenue could not be ignored.[11]

Nothing more was heard of the Hayeses' scheme. But within a few weeks of Cope's letter arriving at Cecil House, the King had been presented with a royal charter to launch a new Virginia venture as a private enterprise, enveloped in secrecy.

THREE

The Adventurers

As a founding document, providing the legal basis for what became English America, the Virginia Charter that received the royal seal on 10 April 1606 was no Magna Carta or Declaration of Independence. In return for a 20-per-cent share of any precious metal they discovered, the King merely offered certain 'Knights, Gentlemen, merchants and other Adventurers' – the so-called 'patentees' – permission to 'make habitation, plantation and to deduce a Colony of sundry of our people into that part of America commonly called Virginia'.

Lip service was paid to the mission having a higher religious purpose 'in propagating of Christian Religion to such people as yet live in darkness', meaning the Indians, but it was clear that the true motive was monetary, the charter focusing heavily on the 'commodities and hereditaments' to which the patentees, as well as the Crown, would be entitled.

'For the more speedy accomplishment of their said intended plantation and habitation,' the patentees were split into two groups, each given a different region of North America to colonize. The region between 38 and 45 degrees latitude (from modern Philadelphia to the Canadian border) came under the control of a group of merchants from the West Country ports of Bristol, Exeter and Plymouth, led by Chief Justice Sir John Popham. 'Undertakers [i.e. investors], gentlemen, merchants &c' of London made up the other group, who formed what became known as the Virginia Company. They were to develop the region between 34 and 41 degrees, the 450-mile stretch of coastline running from modern Cape Fear to Philadelphia.[1]

Most of London's trading companies were managed by a governor, appointed by investors. The Virginia Company was to be different. Instead of a governor, Cecil set up a 'Royal Council' of thirteen grandees to supervise the company's affairs, which reported directly to him. Its powers were modelled on those of the royal councils long established to deal with strategically sensitive royal possessions, such as Ireland and the North of England.[2] Sir Thomas Smythe, the City's most powerful merchant and one of the royal exchequer's most generous creditors, was made company treasurer, responsible for day-to-day business.

These arrangements soon caused problems. Within a month, there were complaints that the council's interference had 'exceedingly cooled the heat' of investors.[3] Matters were made worse by Cecil insisting that preparations for the expedition be undertaken in complete secrecy, so as not to alert the Spanish.

Even without such impediments, finding backers was difficult. North America had a reputation as a financial slough. Since the 1570s, many a merchant's fortunes had been sunk in the middle of the Atlantic, or been lost at the hands of the Spanish or Indians on a distant shore. Those missions that had returned a profit had done so by engaging in 'privateering', raiding stray Spanish treasure ships in the Caribbean, a practice that had been officially sanctioned in Elizabeth's time, but banned since the signing of the Somerset House Peace Treaty with Spain.

By the summer of 1606, there were few signs of progress, and gloom about the project's prospects intensified with news that the West Country group had managed to equip and dispatch a ship called the *Richard* in August, to reconnoitre a suitable place to settle in the region of Sagadahoc, Maine.

By the autumn, the pressure was intense. If an expedition was not dispatched soon, it would arrive in Virginia too late to plant crops to feed the settlement the following winter.

On or soon after 20 November, a meeting of the London Virginia Company was called at Smythe's offices in Philpot Lane, an inconspicuous alley just north of Billingsgate, London's fish market. Smythe's Great Hall, decorated with souvenirs from his other trading ventures, including an Indian canoe suspended from the ceiling, was a hub of his various global trading interests, and was now inaugurated as the headquarters of the London company's American venture.

There is no record of that meeting. All the relevant paperwork was later seized by order of the Privy Council, and it has not been seen since.[4] Nevertheless, it is likely that the motley crew of 'principal adventurers' that gathered together for the first time that day would not have inspired confidence, nor the final tally of resources that Smythe had managed to assemble.

Smythe had called the meeting because at last there were signs of progress. Three ships had been hired for the expedition, which were now 'ready Victualled, rigged and furnished' for the forthcoming voyage. The government had also decided upon the leadership of the mission, and issued the 'articles, Instructions and orders' finalizing the way the colony would be run.

The document specifying these crucial matters was not, according to later complaints, 'so much as published' by Smythe, who apparently kept certain key details to himself. However, he must at least have now delineated some of its main points.[5]

The settlement in Virginia, the government had decided, would be run not by a single governor, but, as one settler later put it, 'aristocratically', by a local council. This council was to govern the settlement according to laws set down by the Royal Council for Virginia in England, with powers to adjudicate over 'the offences of tumults, rebellion, Conspiracies, mutiny and seditions in those parts which may be dangerous to the estates there, together with murther, manslaughter, Incest, rapes, and adulteries', defendants being given the right to be tried before a jury of 'twelve honest and indifferent persons'.[6]

The membership of the local council, Smythe revealed, would not be chosen by the adventurers, or the wider membership of the Virginia Company, but by the Royal Council for Virginia in England, from among those he had invited to the meeting. The identity of those nominated, however, would not be revealed until they had reached America.[7]

This news must have prompted some speculative glances around the room, as each adventurer sized up those who might soon determine his fortune and fate. Those destined to go on the first mission were 'strangers to each other's education, qualities, or disposition,' one participant noted.[8] Many were 'cashiered captains' in the afternoon of faltering careers, the average age being forty. Most would have described themselves as gentlemen of the Shires, members of a 'middling' class who, relative to their urban counterparts the merchants and lawyers, had not prospered in recent years.

The most senior was a 56-year-old veteran of the Irish and Low Countries campaigns, Edward Maria Wingfield. Of those about to sail, he was the only patentee, the only one whose name appeared on the Virginia Charter as one of the 'humble suitors' given permission to colonize North America.

To many, Edward Maria's distinguishing feature was his middle name, which prompted the impressive explanation that it had originally been given to his father, in recognition of being godson to Henry VIII's sister, Mary Tudor. Edward was the second of many generations to adopt the affectation, proudly showing off the family's ancient lineage and royal connections. Some of his forebears had been loyal to Queen Mary, Henry VIII's Catholic daughter, and were implicated in her suppression of Protestants in the 1550s. This has led to accusations that the family had Papist leanings. But another relation, Sir Edward Wingfield of Kimbolton, had taken part in the Essex Rebellion of 1599, usually considered a distinctly Protestant affair. There is no evidence that Edward Maria veered strongly in either direction. He seems to have adopted the religion of many conservative members of the gentry: pragmatic Protestantism, emphasizing social conformity over religious piety.

His father had died when he was seven, and aged 12 he became the ward of an uncle. When he was in his twenties, he studied law at

London's Inns of Court, but preferred soldiering, and served in the Low Countries. In 1586, he fought with distinction alongside his brother Thomas Maria at Zutphen, hailed by English Protestants as a landmark battle in the fight against Catholic oppression. Later, he was briefly prisoner of the Spanish along with Sir Ferdinando Gorges, who would later become a leading promoter of American colonization.

In the 1590s, Edward Maria had served in Ireland, where members of his family were leading English attempts to 'colonize' areas under Gaelic control. There he would have met Sir Ralph Lane, who had been governor of Raleigh's ill-fated Roanoke colony, and it was perhaps from Lane that he developed an interest in colonial adventures further afield.[9]

By 1604 he was back in England, and soon after became involved in the revived Virginia project, providing not only desperately needed financial support, but helping to secure the royal patent. He had also recruited some of the key personnel, including the mission chaplain, a 38-year-old Sussex vicar called Richard Hunt, and the 'surgeon general', Thomas Wotton. How he came to acquire such a prominent role so early on in the venture is unclear, but a decisive factor was likely to have been his family connection with the famous explorer and privateer Bartholomew Gosnold.[10]

Gosnold had led the 1602 mission to Norumbega which had first challenged Raleigh's Virginia monopoly. His return to England with profitable supplies of cedar wood and sassafras had also demonstrated the potential of North America as a source of natural commodities.

Since then, he and his shipmate, a silver-tongued lawyer called Gabriel Archer, had toured the taverns and company halls of London agitating tirelessly for a full-scale expedition, brandishing a persuasive tract written by Archer about their experiences of the 'goodliest continent that ever we saw'.[11] Their efforts resulted in Gosnold later being described as the 'first mover' of the Virginia venture,[12] and attracting the valuable support of at least one prominent member of the Fishmongers' Company.[13]

However, Gosnold's name had been conspicuous by its absence

from the Virginia patent, and this seems to have been because it had become politically unacceptable. In 1604, a 'Captain Gosnell', probably Bartholomew, had made some intemperate remarks about King James at a dinner party held on the Isle of Wight. One of Cecil's intelligencers happened to be among the guests, and he reported the remarks to his master, prompting a full-scale investigation by the Privy Council. No record remains of the council's conclusions, but following such an episode, it was wise for someone bearing the Gosnold name to keep a low profile.[14] For this reason, Gosnold might have drafted in Wingfield as his proxy, the two families having connections going back generations.[15]

By the time of Smythe's November gathering, Gosnold was able to adopt a more public role in the venture, and had secured a place for Archer in the forthcoming expedition. But as he and Wingfield were now to discover, neither was to be trusted with the role of mission commander. That role was to go to the formidable 46-year-old, one-armed veteran of the Spanish wars who now joined them: Christopher Newport.

Newport had been hired over the heads of the Virginia Company by the Royal Council. He had no previous connection to the Virginia venture, but was certainly qualified for the job. A war hero who had lost his right arm fighting in the West Indies, his reputation reached as far as Spain, where he was known as '*un caballero muy principal*', a very great knight.[16] In 1592, he had helped in the capture of the Portuguese carrack the *Madre de Dios*, the most magnificent prize of the Spanish war, estimated to be worth £150,000.[17] Since the signing of the Somerset House Peace Treaty, he had continued to tour the Spanish Main, but for peaceful purposes, undertaking trade missions on behalf of a number of London merchants. He had returned from one trip with two baby crocodiles, which he presented to the King.

Unlike the other leading members of the venture, Newport was not expected to make any sort of investment in the company, nor to stay in Virginia. He would be responsible for commanding the expeditionary fleet, and leading the initial reconnaissance of the territory. In return, he was to be given sole ownership of any discoveries he made, including deposits of minerals and precious metals.[18]

Such terms caused widespread resentment, because the other men of rank who had volunteered to go, many at this late stage in the preparations, were far more exposed. They were expected to pay not only their own way, but to recruit from their own estates the servants and labourers who would make up the bulk of the settlement's workforce. They were also expected to pay the costs of sending these workers to America, and maintaining them while they were there.

In return for this investment, they were not even to take personal possession of the land upon which they settled. Instead, they were to receive a share in the Virginia Company's overall profits. Their fortunes therefore rested principally on the speedy discovery of some valuable commodity, such as gold, copper, spices or medical ingredients, rather than the long-term development of the colony.

As most of these gentlemen well knew, the odds were unfavourable. The pages of Richard Hakluyt's *Principall Navigations* were filled with horror stories of foreign ventures ending in slaughter or ruin. Just this month, Hakluyt had received reports of yet another disaster. A base set up in Guyana by the English captain Charles Leigh had been deserted following a series of mutinies, and a supply ship sent to relieve him had been forced to abandon sixty-seven passengers on the island of St Lucia, where they had all died of starvation or at the hands of the natives.[19] Such were the risks faced by these planters. But then, in most cases, they were going not because of how much they had to gain, but how little they had to lose.

Despite his elevated status, George Percy was typical. Born on 4 September 1580, he was the sickly, epileptic runt of a litter of eight children fathered by Henry Percy, the eighth Earl of Northumberland. George's family was renowned for its rebellions, which were being replayed every night on the London stage in Shakespeare's history plays. *Henry IV* featured that 'mad fellow of the north' Henry Percy (the first Earl of Northumberland), and his son Harry Hotspur. 'Zounds! I am afraid of this gunpowder Percy,' says one character – a line written before the Gunpowder Plot, so acquiring uncomfortably prophetic force since the revelation that one of George's kinsmen, Thomas Percy, was a ringleader.[20] George was no Hotspur. In fact, he was a disappointment to his family and peers. Someone whispered into King James's ear just before he succeeded to the English throne that George was hated 'damnably' by his brother the Earl, and one official in the Earl's household was moved to describe George's infirmities as 'grievous and tedious'.[21]

George had received the conventional education for a man of his class: Eton College, Oxford University, and the Middle Temple, for legal training. When he was sixteen, his mother had died, leaving him an annuity of around £60 a year, paid by the Earl's staff. This was enough for a comfortable though not lavish standard of living for those who could keep within their means. However, George insisted on extravagance. He had a compulsion for keeping an impressively aristocratic wardrobe and table, having no title or property of his own to demonstrate his elevated rank.

Then came the discovery of the Gunpowder Plot. George was not implicated, but his elder brother and patron Henry, the current Earl of Northumberland, was found guilty of conspiracy, and committed to the Tower. Fined £30,000, Henry no longer felt in a position to support his aimless younger brother, so decided to send him to Virginia. As well as keeping him a safe distance from the political fray, the expedition also promised some alleviation of George's epilepsy, for, as he would later write, 'my fits here in England are more often, more long and more grievous, than I have felt them in other parts nearer the line [equator]'.

He had to pay a high price for a place in the venture. He was apparently forced to hand over his annuity to his brother, and to borrow £8 16*s* from another adventurer to help cover his costs – a debt which he had still to repay years later. Meanwhile, his expenses were far from modest. He subsequently sent home requests to his brother for 'diverse suits' (£32 14*s* 7*d*), knives (3*s*), books, paper, ink and wax (£1 14*s* 9*d*), biscuits (£3 5*s*), cheese (8*s* 2*d*), butter (£1 17*s*), soap, lights and starch (13*s* 6*d*), storage chests (12*s*), assorted boxes (10*d*) and casks (6*s* 2*d*). He even asked for a feather bed, complete with bolster, blankets and a covering of tapestry.[22]

Percy was a name of French origin going back to the Norman invasion of England, breathed in the rarefied atmosphere of royal courts. In contrast, the name that was to become most closely associated with the Virginia story, and intimately linked to the legend of the Indian princess Pocahontas, carried a whiff of the Anglo-Saxon village forge: John Smith.

A wide, barely navigable ocean divided Smith's world from Percy's. Percy was the product of generations of aristocratic breeding and refinement; the stout, bearded Smith prided himself on being a self-made man. 'Who can desire more content that hath small means, or but only his merit to advance his fortunes, than to tread and plant that ground he hath purchased by the hazard of his life,' Smith wrote.[23] Nevertheless, Smith had one thing in common with Percy: a feeling of social exclusion, of otherness, that made the prospect of starting afresh in the New World irresistible.

According to Smith's vivid, if sometimes incoherent and unreliable autobiography, *True Travels*, he was 'born in Willoughby in Lincolnshire'. His father was 'anciently descended from the ancient Smiths of Crudley in Lancashire; his mother from the Rickards at Great Heck in Yorkshire'. He was christened with that most ordinary of names at Willoughby by Alford on 9 January 1580. The ceremony took place in the local parish church, which, like many others in a region of reclaimed marshland that could be treacherous to travel, was dedicated to St Helen, patron saint of travellers.

He described these as 'poor beginnings', so poor as to earn the scorn of his high-born adversaries. But he was not quite as humble as he sometimes claimed. His father, George, was among the better-off farmers in the region, owning the freehold of several acres of pasture in Great Carlton and property in the market town of Louth. He also leased fields off the local lord, Peregrine Bertie, Baron Willoughby of Eresby. Young John was brought up in a substantial farmhouse comprising a hall, three chambers, a 'milkhouse' and a 'beasthouse', and with several servants.[24]

The next stage in John's life, according to his autobiography, was 'his parents dying when he was about thirteen years of age'. They left him with 'a competent means, which he not being capable to manage, little regarded'.[25] This is a curious passage, as both parents were very much alive when he was 'about thirteen': his father died when Smith was sixteen, and his mother many years later, having remarried.

His mind being 'even then set upon brave adventures', he was apprenticed to a merchant in King's Lynn, Norfolk. Bondage to a master did not suit his restless spirit, and soon afterwards he tore up his seven-year indenture and headed off to find new adventures. This probably happened around the time his father died, and produced a rift with his mother, which would explain his decision to write both of them prematurely out of his life.

The elective orphan, free of family ties and the responsibilities of having to manage his father's farm, went to the Low Countries. Unfortunately for Smith, his arrival coincided with a lull in hostilities, a side effect of Cecil's peace treaty with Spain. This forced an early return to England. Smith then embarked on a tour of France in the company of Peregrine Bertie, the son of his father's patron. Returning again to the Low Countries, and finding the opportunities for military glory still limited, he ended up in Eastern Europe, where war between the Holy Roman and Ottoman empires raged more reliably. Though the Holy Roman Emperor Rudolf II was Catholic, this was of no consequence to an English Protestant with military ambitions.

He enlisted with the battalion of a Slovenian warlord, and marched to Transylvania, on the front line of the war. There, during a siege, he claimed to have beheaded three 'Turks' single-handedly before the massed ranks of the opposing Christian and Muslim armies. For this act of bravery, the King of Poland granted him a coat of arms, the title of Captain and the status of 'an English gentleman'.

He was soon after wounded during a skirmish with Tartars and taken prisoner. He and his fellow captives were sent to Axiopolis (modern Cernavodă), a market town on the banks of the Danube, and 'sold for slaves, like beasts in a market-place; where every merchant, viewing their limbs and wounds, caused other slaves to struggle with them, to try their strength'.

A dealer bought the young soldier for a client in Constantinople, who turned out to be the beautiful daughter of a Greek noblewoman whom Smith called Charatza Trabigzanda (probably mistaking the Greek description of her as a girl from Trebizond).[26] She soon 'took (as it seemed) much compassion on him', but not yet being of age, and fearful that he would be sold on, sent him to her brother, a military official working near the Black Sea, 'till time made her Master of her self'. But 'within an hour after his arrival', the brother commanded his servant to strip Smith naked, 'and shave his head and beard so bare as his hand' and place 'a great ring of iron, with a long stalk bowed like a sickle, riveted about his neck'. After enduring several months of this treatment, Smith 'beat out [his master's] brains with his threshing bat', stole his victim's clothes, and made his escape along an ancient caravan route to Astrakhan. He vividly recalled a journey along this intersection of Asiatic trade, each crossroads being marked with a signpost showing the way to the Crimea with a crescent moon, to Moscow with a cross, and to China with a sun. Ending up in Prague, the capital of the Holy Roman empire, he embarked on another epic trek through Germany, France and Spain to the Barbary coast of North Africa, where he hitched a lift with French pirates. Narrowly escaping Spanish capture and being blown up by an on-board explosion, he returned to England.[27]

It is not clear how Smith was introduced to the Virginia venture. At the time Gosnold was promoting the idea in London, he appears to have been staying with or near Robert Bertie at the Willoughby London residence in the Barbican. Robert's father had shown an interest in the Roanoke venture, and he had family connections to both the Wingfields and Gosnolds, so Robert may have effected an introduction.[28]

Whatever Smith's credentials, he, like Newport and Gosnold, had at least made a name for himself in the world of military and maritime affairs. The same could not be said of two mysterious figures mingling among the assembled adventurers and planters on that November day. Smythe introduced them as John Ratcliffe and George Kendall. They were to take a prominent though as yet unspecified role in the forthcoming expedition, the assembly was informed. Nothing further was revealed about these men, other than perhaps the merest hint that their participation was non-negotiable, as they had been appointed at the personal behest of Robert Cecil himself.

FOUR

Departure

BY LATE NOVEMBER 1606, preparations for the Virginia Company's first expedition were well advanced. Edward Maria Wingfield had packed a trunk with reading material, together with 'diverse fruit, conserves and preserves', and dispatched it to Richard Crofts, probably a relative of the Herefordshire landowner and MP Sir Herbert Crofts.

Crofts lived at Ratcliffe, a hamlet on the north bank of the Thames. The stretch of river overlooked by his house was used to moor ships, and on 23 November the *Susan Constant*, the 170-ton flagship for the Virginia fleet, arrived, heavily laden with supplies for her forthcoming voyage. She was tied up alongside the *Philip and Frances*, and Crofts dutifully ensured that Wingfield's precious trunk was safely stowed in one of the cabins. A consignment of clucking hens and a cockerel was also delivered, from which Wingfield hoped to breed a flock to provide himself, and possibly his associates, with fresh eggs and an occasional chicken for the pot once in America.

That night, the *Susan Constant* began to shift with the ebb tide. Being so heavily laden, she was difficult to control, and crashed against the neighbouring ship, damaging the *Philip and Frances*'s bowsprit, sheet anchor and beak-head (defined by Captain John Smith as the part of the ship 'before the forecastle, and of great use, as well for the grace and countenance of the ship, as a place for men to ease themselves in').[1] When the master of the *Philip and Frances* boarded the *Susan Constant* to remonstrate with the crew, he claimed to find them 'tippling and drinking'.[2]

The *Susan Constant* suffered minor damage to two of her portholes, but she was soon patched up, and a few days later moved further downstream to Blackwall, where she was to rendezvous with her two companion vessels: the 40-ton *Godspeed* and the 20-ton *Discovery*. Of the three ships, only the *Discovery*, a small bark or 'pinnace', was actually owned by the Virginia Company. She was to remain in Virginia for use by the settlers, while the other two hired vessels would return to England, laden, it was hoped, with valuable commodities.

On 10 December, ten days before the ships were due to sail, the mission's leaders were once again summoned to Philpot Lane, this time to receive their final 'Orders and Directions' from the Royal Council.

Newport was confirmed as admiral, having 'sole charge' of the venture while he was in Virginia. He was also given the box containing the list of names for the local council, which he was under orders to ensure remained sealed until the fleet arrived in America. Meanwhile, he was left to appoint the 'captains, soldiers and mariners' for the voyage. Reflecting his experience and prominent role in getting the venture off the ground, Gosnold was put in charge of the *Godspeed* while, to everyone's surprise, the mysterious Ratcliffe was given command of the *Discovery*.

The expedition leaders were then handed a set of 'Instructions given by way of Advice', drawn up by Richard Hakluyt. These distilled the collective wisdom of the adventurers' forerunners, and showed that a great deal had been learned from their abundant mistakes.

Hakluyt insisted that, on arrival in America, their first job was to anchor the fleet in a 'safe port' at the mouth of a navigable river. The river was to be the one that 'runneth furthest into the Land' and 'bendeth most towards the Northwest', in the hope that it might be that mentioned by the Indians at Roanoke, leading to the Appalachian Mountains and even the Pacific.[3]

As for deciding the location of the settlement, Hakluyt recalled the experience of the Huguenots at Fort Caroline. The settlers must find 'the Strongest most Fertile and wholesome place' far enough upriver 'to the end that you be not surprised as the French were in Florida'. They should also 'in no Case Suffer any of the natural people of the Country to inhabit between You and the Sea Coast', which might cut off their means of escape in the event of hostilities.

The settlement should be located away from heavily wooded areas, because the trees would provide cover for enemies, and they did not have the resources to clear large swathes of land ('You shall not be able to Cleanse twenty acres in a Year'). They should also avoid a 'low and moist place because it will prove unhealthful'. It was suggested that the best way of finding a suitably wholesome site was to look at the people who lived nearby. If they were 'blear Eyed and with Swollen bellies and Legs', then it was best to steer clear; if 'Strong and Clean' it would be a 'true sign of a wholesome Soil'.

Once a base had been found, they were to erect a 'little sconce' or lookout post at the mouth of the river. The one hundred and twenty settlers were then to be split up into four groups: forty to build a fortified settlement, thirty to prepare the ground for growing 'corn and roots', ten to man the sconce. The remaining forty were to make up exploration parties. Gosnold was to take half of them into the interior, equipped with a compass and 'half a Dozen pickaxes to try if they can find any mineral'. The others were to explore the river to its source, scoring the bark of trees on the river side as they went, to help search parties retrace their route should they go missing. The instructions did not specify who was to lead this mission.

Those left to construct the settlement were advised to lay out streets of 'good breadth' which converged on a central square or marketplace. A cannon was then to be placed in the centre, which could 'command' any street in the event of attack. Before setting up housing, the carpenters and 'other suchlike workmen' were to work on the public amenities, such as a secure storehouse and assembly room.

As for the Indians, they were recognized to be vital to the settlement's success. 'You must have great care not to offend the naturals if you can eschew it and employ some few of your company to trade with them for corn and all other lasting victuals.' 'Above all things, do not advertise the killing of any of your men', and avoid revealing signs of sickness, in case the 'country people' realize the settlers are 'but Common men'.

Trade was both crucial to survival and fraught with difficulties. The settlers should first ensure that the crews of the ships that brought them (which would soon be sailing back to England) should be prevented from having contact with local tribes, 'for, those that mind not to inhabit [the colony], for a Little Gain will Debase the Estimation of Exchange and hinder the trade'. Before the Indians realize that the settlers mean to stay, special representatives should be appointed to barter with them for sufficient 'Corn and all Other lasting Victuals' to last through the first year, the settlers' own crop to be put in store 'to avoid the Danger of famine'.

'The way to prosper and to Obtain Good Success', Hakluyt concluded, 'is to make yourselves all of one mind for the Good of your Country & your own'. Every one of them must 'Serve & fear God the Giver of all Goodness'. They must shun corrupt or anti-social behaviour, as 'every Plantation which our heavenly father hath not planted shall be rooted out'. Finally, they were ordered to keep all matters relating to Virginia secret, and prevent the publication of any material which did not have the Royal Council's prior approval.[4]

Having given their solemn oaths to abide by these orders and instructions, the three leaders of the expedition went to Blackwall to inspect the company that was to be carried across the Atlantic, and settled in the New World.

Instead of the one hundred and twenty envisaged by Hakluyt, they found around one hundred men and boys, plus a few dogs, brought for hunting and as pets.[5]

Thirty-six of the company were identified as 'gentlemen', many of them the footloose younger sons of distinguished families. Anthony Gosnold was Bartholomew's younger brother. Thomas Sandys was the younger brother of the prominent MP Edwin Sandys.[6] Thomas Studley, the man selected to act as 'cape merchant' (in charge of supplies) and an enthusiastic chronicler of the coming adventure, may have come from a line of prolific writers.[7] Kellam or Kenelm Throgmorton was probably related to Bess, the wife of Sir Walter Raleigh.

Other gentleman members of the expedition were from more obscure and modest backgrounds. Nothing is known about the origins of Robert Tyndall, the expedition navigator, beyond the fact that he was the gunner of Prince Henry, James I's son and heir. He may have been the son of John Tyndall, who wrote in 1602 to his 'kinsman' Michael Hicks, Cecil's close friend, 'recommending him to his favour,' but this cannot be verified.[8]

The remainder of the company was made up of an assortment of tradesmen, labourers and young boys. While most of these 'common sort' were being shipped out by their masters, those with a trade or skill were contracted directly by the Virginia Company, to work for a specified period without charge, in return for their transport, tools and maintenance. The surly blacksmith James Read, and a professional mariner, Jonas Profit, probably signed up on such terms, as did Thomas Couper, a barber, Edward Brinto the stonemason, William Love the tailor, and Nicholas Skot, a drummer.

All that is known about the boys is their names: Samuel Collier, Nathaniel Pecock, James Brumfield and Richard Mutton.[9] Some of the labourers and boys were likely to have been 'pressed' into service, an order from the Royal Council providing Newport with the authority to round up suitable candidates from taverns and playhouses, or buy them off gang-masters.

The fleet set sail on Saturday 20 December 1606.[10] It was by design a low-key event. None of the government figures concerned with the venture was apparently in attendance, and whereas Elizabeth had waved off previous missions with a salute of cannon, James did not lift a hand for this departure, nor even seek to be informed.

A few days later, the ships reached the 'Downs', a well-known anchorage off the forelands north-east of Dover, where they awaited a favourable wind.

For weeks the fleet bobbed on the waves, while a relentless westerly whipped up the Channel, spitting sea spray and rain into the faces of the impatient captains. Every so often, a violent winter storm would throw up waves that threatened to overwhelm the ships, or drag their anchors, but, as George Percy, who was aboard the *Susan Constant*, loyally noted, 'by the skilfulness of the captain [Newport] we suffered no great loss or danger'.[11]

The ships might have been unscathed, but passengers, strangers to each other's company and many to the sea, were proving to be less resilient. Boredom and frustration soon crept into the makeshift cabins and cramped quarters. The *Susan Constant*, under 100 foot long and 20 foot wide, was built as a merchantman to carry cargo, not people, so her fifty or so passengers were forced either to bide their time on the freezing, gale-swept decks, or endure the claustrophobia of the airless holds below. The *Godspeed*, about 70 foot long and 16 foot wide, and the *Discovery*, 50 or so foot long and 11 foot wide, were marginally more accommodating, as both had fewer crew and proportionately more space for the passengers, but being smaller, their hulls were jolted even more violently by the churning seas.[12]

For a month the fleet was held in this state of agitated immobility, and boredom began to breed division and distrust among the restless passengers. In particular, Edward Maria Wingfield and George Percy, who found themselves much in each other's company, began to detect in those around them the suggestion of a religious plot, to which Robert Hunt, the mission's chaplain, was somehow connected.

When recruiting for the voyage, Wingfield had visited the Archbishop of Canterbury on behalf of Hunt, to vouch that the cleric suited the role of vicar of Virginia as he was neither 'touched with the rebellious humours of a popish spirit' – a Catholic, in other words, who might be acting for the Spanish – 'nor blemished with the least suspicion of a factious schismatic', a religious independent, who might be tempted so far from his homeland to flout the authority of the Church of England, the bishops and the King.[13] Now, Wingfield began to doubt his own words. Something someone had said, some slight or chance remark, suggested to him and Percy that the vicar might be blemished with a suspicion of factious schismatism after all, which threatened to defile the entire venture.

Meanwhile, another of the passengers, Captain John Smith, had formed an attachment to Hunt, nursing him through a bout of seasickness so severe that, according to Smith, 'few expected his recovery'. During this time, the captain and the cleric fell into conversation about religious matters, and Smith found himself in close accord with Hunt's evangelical leanings, judging him to be 'an honest, religious, and courageous divine'.[14]

It was at this point that Wingfield and Percy decided to confront Hunt with their suspicions. Smith, whose choleric temperament made him as quick to argue as to judge, sprang violently to his new friend's defence, accusing these two 'Tuftaffaty humourists' of trying to hide their own irreligiousness.[15]

During the ensuing row, certain rumours about Hunt's past began to surface, like corpses from the deep. Hunt was by no means the Puritan in his own behaviour as he was now suspected of being in his religious beliefs. Three years before, he had been brought before the court of the archdeaconry of Lewes, the regional administrative body for Heathfield, to answer charges of 'immorality' with his servant, Thomasina Plumber. He was at the same time proceeded against for absenteeism, and there were accusations that he had neglected his congregation, leaving his friend Noah Taylor, '*aquaebajulus*' (water bailiff or customs collector), to perform his duties.[16]

Smith refused to believe such allegations. How could these men, 'of the greatest rank amongst us', circulate such 'scandalous imputations,' he wanted to know. They were 'little better than Atheists'.[17]

Other passengers joined the fray, with at least three gentlemen lining up behind Smith.[18] The argument escalated between these nascent factions until, on 12 February, it was interrupted by a change in the weather. That night, several passengers, Percy among them, clambered up to the deck to gaze into a sky that for the first time in weeks was cloudless. Percy spotted in the glistening firmament a 'blazing star' or comet.

The appearance of such a spectacle over the ship's swinging mast, above a fleet trapped between deliverance into a new world and damnation in the old, was auspicious. The wind turned, and in a flurry of activity, anchors were raised, tillers spun and sails unfurled to catch it. Released from their cyclonic trap, the ships raced across the freezing waters towards the Atlantic. Within a day or so they had reached the Bay of Biscay, and soon after closed on the Canary Islands, off the coast of Morocco.

The six-week delay in the Downs had meant their provisions for the voyage had already been used up, forcing them to break into supplies set aside for their first months in America. So they decided to stop at the Canaries and spend what money they could to make up the deficit.

As soon as the ships dropped anchor, the rows resumed. According to one report, some of the passengers, including one Stephen Calthorp, a gentleman from a prominent Norfolk family, now joined Smith in threatening mutiny.[19] At this point, Newport lost patience, and ordered the ringleader Smith to be 'committed a prisoner' in the belly of the *Susan Constant*, Smith's furious indignation muffled by the heaps of bulging sacks that furnished his cell, and the layers of sturdy oak decking that would wall him in for the coming weeks.[20]

Picking up the trade winds, the fleet headed off into the Atlantic, covering over 3,000 miles in six weeks. The freezing temperatures of a European winter melted into the sultry warmth of the tropics, providing Percy with some hope of respite from his attacks of epilepsy.

On 23 March, they had their first sight of the West Indies, passing Martinique before reaching Dominica, 'a very fair island, the trees full of sweet and good smells'. The island was inhabited by 'many savage Indians', who at first kept their distance. As soon as they realized that the European visitors were not Spanish, the mood changed, and 'there came many to our ships with their canoes, bringing us many kinds of sundry fruits, as pines [pineapples], potatoes, plantains, tobacco, and other fruits'. After weeks of dried biscuits and salted meat, the men consumed the gifts greedily. They also happily accepted an 'abundance' of fine French linen, which the Indians had salvaged from Spanish ships that had been wrecked on the island.

The English gave the Indians knives and hatchets, 'which they much esteemed', and beads, copper and jewels, 'which they hang through their nostrils, ears, and lips – very strange to behold'. During the transactions, the English learned that the natives had suffered a 'great overthrow' at the hands of the Spanish, which explained their initial wariness and subsequent generosity.[21]

The encounter provided a useful introduction to developing relations with locals, and confirmed English assumptions of Spanish barbarism – the infamous 'Black Legend' or *leyenda negra*. The legend had its origins in *A Short Account of the Destruction of the Indies*, a furious indictment of Spanish imperialism written by a Spanish bishop, Bartolomé de las Casas. It had been translated into English in 1583 'to serve as a precedent and warning to the twelve provinces of the Low Countries'.[22] Casas wrote of conquistadors eviscerating seventy or eighty women and young girls, of little native boys being fed to hunting dogs, even of Indian coolies being decapitated after they had performed their duties, to save the bother of having to unlock the clasps around their necks. Such reports had been enthusiastically picked up by Protestant propagandists in France, Holland and England as evidence of the Catholic brutality that had produced the Florida Massacre and now threatened to overwhelm the Low Countries. Hakluyt referred to them on several occasions, quoting Casas's estimate that Spanish actions in the Americas had 'rooted out above fifteen million of reasonable creatures'.[23]

Reassured by the rewards of exercising their higher moral

standards, the English continued their tour of the Caribbean. As they pulled away from Dominica, Percy was transfixed by the sight of a whale being chased by a swordfish and a thresher, a kind of shark identifiable by the enormously extended upper lobe of its tail, which it uses to thrash its prey. 'We might see the thresher with his flail lay on the monstrous blows, which was strange to behold. In the end these two fishes brought the whale to her end,' he observed.[24]

On the morning of 27 March, they arrived at Guadeloupe. A landing party explored up to the foot of the 5,000-foot-high active volcano La Grande Soufrière (the big, sulphurous one), and found a pool of scalding hot water. Newport used it to boil up a joint of salted pork, which was ready to eat after half an hour. They returned to their ships and sailed on for a further 90 miles, in the afternoon reaching the island of Nevis. Here, Newport decided to allow the entire company ashore to make a concerted effort to gather supplies, as the ships' stores were still disturbingly low.

A company of men armed with muskets marched into the densely wooded interior, catching glimpses as they went of the cloud-capped central peak, over 3,000 foot high. Not far inland they found another hot spring, much cooler than the one on Guadeloupe. For the first time in nearly two months, they enjoyed a relaxing soak, as the sun set in a calm Caribbean sea.

'Finding this place to be so convenient for our men to avoid diseases, which will breed in so long a voyage, we encamped ourselves on this isle six days, and spent none of our ship's victual,' wrote Percy. Instead, they lived off rabbits, birds, fish and fruit plucked from the trees, their peace barely disturbed by an occasional glimpse of the locals, who as soon as they were spotted 'ran swiftly through the woods to the mountain tops'. They lost themselves in the forests, slashing through the undergrowth with hatchets and swords, until they came among 'the goodliest tall trees growing so thick about the garden as though they had been set by art, which made us marvel very much to see it'. They saw shrubs with huge tufts of cotton wool bursting from their seed pods, gum trees, and a sort of wild fig, the sap of which made the men 'near mad with pain', forcing them to rush back to the hot spring for relief. They also found the source of a stream at the foot of the mountain. After drinking its sweet, clear water, 'distilling from many rocks', the men 'were well cured in two or three days' of all their ship-borne ailments.

However, even this idyll could not cure every sickness. There was another outbreak of faction fighting, perhaps prompted by Newport's decision to release Smith from the ship's brig and allow him to fraternize with the other men. It seems that he now fell out with some of his former associates, who reported him to Newport. The upshot was, according to Smith, that Newport, fearing a loss of authority, ordered the construction of a 'pair of gallows' on the beach. But 'Captain Smith, for whom they were intended, could not be persuaded to use them', so he was returned to the ship. No other reference to this curious incident survives, and whatever the details, the result was a hasty departure. The ships cast off on 3 April, with water and food supplies still depleted, despite the plenitude that the island had offered.[25]

The fleet sailed past the neighbouring islands of St Kitts, St Eustatius and Saba before anchoring among the Virgin Isles, 'in an excellent bay able to harbour a hundred ships'. A landing party managed to catch enough fish and turtles to feed the fleet for a further three days, but there was no fresh water to be found anywhere on the island.

Passing Puerto Rico, they reached the tiny island of Mona on 7 April. By now, the drinking water in the ships' tanks 'did smell so vilely that none of our men was able to endure it'. A group of sailors managed to find a fresh water supply on the island, and set about filling up barrels to transport back to the ships. Meanwhile, a landing party marched for 6 miles in search of food. They managed to kill two wild boar and an iguana, 'in fashion of a serpent and speckled like a toad under the belly', but the path proved 'so troublesome and vile, going upon the sharp rocks', and the tropical heat so intense, that several men fainted. According to Percy, the adipose fat of Edward

Brookes 'melted within him by the great heat and drought of the country. We were not able to relieve him nor ourselves, so he died in that great extremity', the first casualty of the expedition.

The fleet remained at anchor for two days, while a group took a launch to a nearby rocky islet called Monito, some 3 leagues (9 or so miles) away. They had difficulty finding a landing point along the island's cliff-lined coast, and even more trouble climbing up the 'terrible sharp stones' to open land. However, they were rewarded with the discovery of a fertile plain, 'full of goodly grass and abundance of fowls of all kinds'. White seabirds dived overhead 'as drops of hail' and made such a noise 'we were not able to hear one another speak'. 'Furthermore, we were not able to set our feet on the ground but either on fowls or eggs, which lay so thick in the grass,' and within three hours they had filled their boat, 'to our great refreshing'.

With new supplies of water and food safely loaded, the fleet set off, and on 10 April left the West Indies, heading north for Florida. Four days later, they crossed the Tropic of Cancer, the northerly limit of the tropics.

The following morning, Newport started to take soundings, in the hope of finding the North American continental shelf.

The use of soundings was the old-fashioned method of navigation. A lead weight smeared in tallow and attached to a line knotted at intervals of a fathom was dropped overboard to measure the depth of the sea. When it was hauled up, particles embedded in the tallow were used to tell what sort of seabed lay beneath.

In familiar waters, such as the English Channel, soundings were effective, as a combination of depth measurement and seabed material ('small shingles', 'white stones like broken awls', 'big stones rugged and black') helped to build up a profile of the sea floor that could locate the ship to within a few nautical miles of its position, even when the shore was over the horizon. However, the ocean beneath the fleet's current position was too deep to sound, leaving Newport with no option but to keep sailing.

They continued north-west for ten days, carried over 1,000 miles by the Gulf Stream. Further soundings were taken, but to no avail. By 21 April, Newport had to accept that he was lost. This was probably the moment that Robert Tyndall suggested he try the mathematically based 'new navigation' techniques to plot their position. English mariners, apparently including the crew of the *Susan Constant*, were suspicious of such methods, considering them hocus-pocus. The prevailing attitude was summed up in *Eastward Hoe*, when Sir Petronel Flash uses 'the elevation of the pole' and 'the altitude and latitude of the climate' (garbled descriptions of the relevant techniques) to mistake the Isle of Dogs on the Thames for France.

Nevertheless, traditional methods had failed, so it was time for Tyndall to bring out his cross-staff or astrolabe, and plot a position. Measuring the angle between the horizon and the midday sun, he announced that they had reached 37 degrees north of the Equator, believed to be the latitude of the entrance to Chesapeake Bay. All they now had to do was to use the ship's compass to head due west, and they would eventually reach their destination. The guffaws of sceptical deck-hands probably filled the ships' sails.

That evening, the fleet was hit by a 'vehement tempest, which lasted all the night with winds, rain, and thunders in a terrible manner'. Concerned that the coast was nearby, and the ships might be driven on to the shore, Newport ordered the passengers into the hulls, where they were told they would be safest if the ships collided with rocks or the seabed. They emerged the following morning into the calm, and gazed upon an unbroken horizon. A lead was dropped, to see whether they had yet reached the coastal shallows, but the ocean floor was still beyond the line's 100-fathom reach. Food and water supplies were once again running low. The unpredictable weather threatened another battering. The need to find a safe harbour intensified.

For three days, they aimlessly sounded the seas, doubtful of Tyndall's assurances that their destination lay just beyond the western horizon. Unease developed into panic, and on 25 April, John Ratcliffe, captain of the pinnace, proposed that the fleet head back to England, in the hope that the Westerlies would get them there before supplies gave out.

Then, at four in the morning of 26 April 1607, as the faintest gleam of dawn crept across the placid ocean, the night watch of the *Susan Constant* picked out a disturbance on the western horizon. As the sun lifted behind them, crew and passengers began to gather on deck and squint over the ship's bowsprit. Gradually, the low contours of a coastal plain became distinct, a dark line of trees sitting on the horizon like the pile of a carpet. A few hours later, Tyndall's navigational methods were vindicated. Not only had they reached America, but they were facing 'the very mouth of the Bay of Chesapeake'.[26]

In London, reports reached Robert Cecil that the secret of the Virginia venture was out. The *Richard*, the ship sent by the West Country group to reconnoitre northern Virginia, had been taken off the coast of Florida by a Spanish fleet. A storm had forced one of the Spanish ships to put in at Bordeaux, where its English captives were released on the orders of the French authorities. It was one of these men who had managed to make his way to London and break the bad news. Other members of the expedition, including the mission's pilot John Stoneman, were less fortunate. They had been taken to Spain, where 'rough' interrogation awaited them.[27]

The Spanish at this time had only a hazy understanding of English plans. Around the time the Royal Charter had been issued, King Philip III's ambassador to London, Don Pedro de Zuñiga, had heard of a plan to send '500 or 600 men, private individuals of this kingdom to people Virginia in the Indies, close to Florida'. He had also discovered that ten Indians were being kept in London, who were 'teaching and training' prospective settlers of 'how good that country is for people to go there and inhabit it'.[28]

By 24 January, 1607, Zuñiga was still unaware of the Newport expedition, but had received garbled information that some sort of venture was under way. He wrote an urgent dispatch to Philip III reporting that the English 'have made an agreement, in great secrecy, for two ships to go [to Virginia] every month until they land two thousand men'. He also noted that Dutch rebels were to be sent. There followed a brief but mostly accurate summary of James's charter of the previous April, including a list of those appointed to the Royal Council. The charter itself was a public document, but the order appointing the Royal Council was not, indicating that Zuñiga had found a source close to the Privy Council.[29]

On 26 February, Zuñiga received a response to his previous dispatch. 'You will report to me what the English are doing in the matter of Virginia – and if the plan progresses which they contemplated, of sending men there and ships,' the King wrote, 'and thereupon, it will be taken into consideration here, what steps had best be taken to prevent it.' Over the following weeks, the traffic of intelligence intensified. In April, Zuñiga finally learned that the English had already sent three ships, but he believed the *Richard* to be one of them.

On 7 May, a council of war assembled in Madrid to discuss the implications of the news. The danger, it was decided, was the proximity of the settlement to Spanish interests, since it lay, according to Zuñiga, 'in 35 degrees above La Florida on the Coast'. Though this region of America 'has not been discovered until now, nor is it known', nevertheless it was 'contained within the limits of the Crown of Castille', in other words, Spanish territory. It was therefore concluded that 'with all necessary force this plan of the English should be prevented'.[30]

While these discussions were under way in Madrid, news of the discovery of the *Richard* spread panic. If the Spanish found out what was going on, reprisals might ensue and all the hard-won benefits of peace would be lost. In such a fast-developing situation, it was decided that the Royal Council for Virginia was too cumbersome or too prone to infiltration. Its members were 'dispersed by reason of their several habitations far remote the one from the other, and many of them in like manner far remote from Our City of London'. In response, a new order was published, creating two seperate councils, one for the 'first' or southern colony, the other for the 'second' or northern colony.[31]

Meanwhile, Cecil considered the fate of the Virginia venture in the light of the *Richard*'s capture. Having discussed the matter with the King, he consulted the journal of the Somerset House Treaty negotiations, to see if it might cast any light on the diplomatic ramifications. His conclusion was that, although Virginia was 'a place formerly discovered by us, and never possessed by Spain', the Spanish commissioners had denied that this gave England the right to 'trade' there. With respect to the captured crew of the *Richard*, he advised the King that 'it might be better to leave these prisoners to their inconveniences', though steps should be taken to recover their ship, as it had been captured in international waters. As for those currently on their way 'to a discovery of Virginia', Cecil suggested that they 'should be left unto the peril which they incur thereby'.[32]

Part Two

FIVE

Tsenacomoco

> Even among the most barbarous and simple Indians, where no writing is, yet have they their Poets, who make and sing songs, both of their ancestors' deeds and praises of their gods.
>
> PHILIP SIDNEY, *Defence of Poesy*[1]

THE AREA OF NORTH AMERICA known to the English as Virginia already had a name: Tsenacomoco. The people who lived there left no written record of their culture or history, which even in Thomas Jefferson's time appeared, at least to Anglo-Americans, to be on the point of extinction. 'Very little can now be discovered of the subsequent history of these tribes,' Jefferson wrote in his *Notes on the State of Virginia*. 'They have lost their language,' and several tribes had been forced to merge, reducing themselves 'by voluntary sales, to about fifty acres of land'. The remnants 'have more negro than Indian blood in them,' he noted, anticipating a later practice of merging the two races, in an effort to extinguish any lingering traces of cultural identity.[2]

All that remains which can be traced directly back to the time of the English incursion is what the English themselves wrote about Tsenacomoco. Several of the colonizers took extensive notes, some even tried to learn and analyse the language. What they found was by its nature transient and mutable. The very act of removing it from the realm of voices, songs, dances and dress and committing it to the permanence of paper must have meant that some of its dynamic qualities were lost, and are unrecoverable. But sufficient remains in the historical record to give a hint of what the Tsenacomoco world was like, at least as seen from the perspective of an English Otasantasuwak or 'wearer of leg-coverings' about to step in and destroy it.

A hut stood between the flat sea and the high mountains. It belonged to a god of many shapes and many names. He was most often seen as a mighty Great Hare, and most usually called Ahone.[3]

One day, Ahone stood at the door of his hut, and beheld the emptiness around. So he made a world according to his imaginings, without a fixed form: a world of water, shifting sand, soft mud, trackless forests and tangled vines. The earth contained no metal or rock, nor any hard thing.

He populated the world with creatures. He made fish, which swam in the streams, and a great deer, which grazed in the woods and galloped across the meadows. But one creature he left tied up in a sack, which lay upon the floor of his hut.

Four gods from surrounding worlds peered over the rim of Ahone's, and gazed upon his creation. They were jealous of what he had done, and came to his hut armed with spears to destroy his work. They saw the sack, and they opened it. Men and women sprang out and scurried across the floor, and the four gods tried to catch and eat them. But Ahone returned to his hut and drove those cannibal spirits away.

Hungry and vengeful, the four gods went into the forests of Ahone's creation, and stalked the great deer. They found him grazing quietly in a grove. As they lay in the undergrowth, one of them dressed his arm in the fashion of the neck and head of another deer, and held it above the foliage to catch the great deer's attention. Seeing a companion, the deer did not run.

In this way the four gods caught the great deer and slaughtered him. They butchered and ate the meat, sinews, offal and bones, devouring everything except the hide, which they left at the door of Ahone's hut.

Ahone did not grieve. He picked up the pelt, scraped the skin and sprinkled the hairs across the world, and where each hair fell, a deer sprang up. Then he returned to his hut to fetch his sack, and emptied it over the world, one man and one woman for each country. And so the world took its first beginning of mankind.[4]

The people of each country enjoyed Ahone's bounties, and they multiplied. They called their world Tsenacomoco, because they lived together, so many of them in the land between the sea and the mountains.[5]

The sun rose white as pearl from the sea each morning, and fell copper red into the mountains at night. The plants yielded berries for the summertime, nuts in the autumn, and roots for the cold winters. The streams and rivers gave up fish, and the shores of the great bay they called Chesapeake, the shellfish water, because of the abundant crabs and oysters. The woodlands stretched from the feet of the mountains where rivers tumbled their waters into Tsenacomoco to the shores of the bay, and were full of deer and turkeys.[6] And so every person lived each with the other, in all the corners of the land and creeks of the river, in every grove and every mere, which Ahone had given to each and all of them.

Then, when the mountains had grown, the world cooled and became hard.[7] The Great Hare Ahone left, and Okeus, a scowling warrior, came. The right side of his head was shaven, and from the left side grew a long knot of hair, which draped over his shoulder.[8] Around his neck dangled magic tokens, white pearls of the sea and red copper of the mountains. The love and devotion of Ahone was chased from the hearts of the people of Tsenacomoco and replaced by fear and awe. They stopped dawdling in the forests and treading the waters, and huddled at their hearths. Their world that once stretched from the sea to the mountains was confined to the glow of their fires.

Now, what Ahone had once freely given they had to make for themselves. To replace the forest canopy they had slept beneath, they wove sapling branches, thatched with leaves, or hides when the hunting was good. For woodland groves and clearings, they burnt undergrowth and slashed trees. For berries and seeds once harvested wild, they planted beans and corn in the winter to fetch from their gardens in summer. For fish once speared in the running waters, they built ingenious weirs to catch them.

The face of Okeus stared at them from the darkness of the forest, goading them to come to him, threatening them if they did. He made them fearful and brave, adoring the things that hurt them beyond their prevention, such as the fire, flood, lightning and thunder.[9]

To make each of their spirits strong, the people of the Tsenacomoco mingled them together with hectic dances around the fire, upon which they cooked great feasts. And this excited Okeus.

And he divided the men from the women. The men he made to face him. To mask their fear, they dressed themselves as he did, and wore their hair as he wore his, with a long black lock on the left side hanging down. They no longer ran with the single deer, but like Okeus himself, they used fires and noise to make whole herds flee to their archers, who would stand in a half-circle to receive them, like a great maw with arrows for teeth. The women watched the home, tending the gardens and the children, keeping the fire alight for the return of the men with their meat.

Then one spring day, as the mountains opened their bowels and emptied them into the rivers, a hunter found pieces of crystal scattered in the tinctured waters.[10] In the subtle world of Tsenacomoco, nothing so hard or sharp had ever been touched. The hunter took one of his arrows, and where a turkey spur or sliver of bone had once been its tip, used a fleck of the crystal, which he fixed with antler glue and bound with the sinew of a deer. When he fired the arrow at a tree, it pierced the wood like muscle. Like everlasting teeth, it never lost its edge.

Men now yearned for these hard, shiny things, which gave those who possessed them the power called *manitu*. *Manitu* was not like the deer in the forest or the fish of the sea; it did not die or decay, nor was it replenished with the flying of the geese or the return of the leaf. *Manitu* did not grow old. It was not washed away by the water or worn away by use. It could be given, taken, hoarded, seized, stolen, and those who had more were lords over those who had less. And so in time he who had most was made chief, and called *weroance*, to whom the rest must give any precious things they had, and tribute of venison, corn or counsel, so they might live under his protection and benefaction.[11]

Some *weroances* rich in *manitu* built circles of wood around their villages, so the people were protected. But they could not move freely any more. Other *weroances* built long huts in the forest, for their secret receptions with Okeus. The special men, the *quiyoughcosucks*, who painted their bodies black and red in the colours of Okeus, would talk to the god in a dark vault at one end of the hut, where only the *quiyoughcosucks* and not even the *weroances* could go.

Okeus grinned at the attention given to him, and he craved more. He taunted the *weroances* at their desire for *manitu*. He goaded each to prove that he was the great one. But to become the great one, each needed more of the hard things foreign to Tsenacomoco that bestowed *manitu*. So one had to fight the other, to take the things in their possession, or to protect their own. The time of war followed.

The noise of battle brought people from other worlds. The Monacans came from the mountains and stood where the rivers fall. They did not speak with the same tongue nor did they know Okeus. They lived as the people of Ahone once lived, moving freely through the woods and the seasons, having no abode but the forest. For each Monacan man had all the glittering stone from the rivers and red copper from the mountains he needed, and so his own *manitu*.[12] They came to the place called Powhatan, the place where the river falls. There, the *weroance* bought with corn and hides all the blue crystal and red copper the Monacans had.[13]

Then, within the memory of men, the Otasantasuwak, the ones who wear trousers, came across the sea upon great swans.[14] They were ghosts, very pale, weak and bony. Many had beards and whiskers, and they spoke a strange language, though some old men thought they had heard these voices before.

The Otasantasuwak carried sticks which could spit fire, but not with the accuracy or speed of an arrow. The fire sticks were fed with a seed that would not grow in the ground. They could not in their heavy clothes run through the forests, or move stealthily, or hunt down the stag. They had no means of sustaining themselves, but believed they lived by the bread given to them by their *quiyoughcosucks*, who chanted to their god using a leather pouch stuffed with leaves of white and black markings. They had crystal and copper, some of better quality than that of the Monacans, and piles of rock which lay in the gizzards of their great swans. They gave places their names, and their names places, and once these were chosen, the names stayed the same, even as everything changed.

In these dangerous times, a villager of Powhatan, Wahunsunacock, grew great in stature from these wars and encounters, and came to be chief of *weroances*, the *mamanatowick*, with power to save Tsenacomoco in these troubled times.[15]

Wahunsunacock understood the people of other worlds, who called him Powhatan after the place of his origin that lay at the heart of his power. From them, he got more precious things than were had by any other *weroance*. The touch of the Otasantasuwak became deadly to any who approached them, unless they were by him permitted, be they people in his power, or other people jealous of him.

Under this rule, he fetched copper from the Monacans.[16] From beyond the swamp, in the land of a wise chief impotent in his limbs, he fetched pearls of the purest white, unlike the grains gathered from the Tsenocomoco shores, which were the colour and hardness of rotten teeth.[17]

It is said Wahunsunacock had received secret knowledge from the Otasantasuwak of more precious and strange things than any that had ever been known, for which he had prepared the great temple in Tsenocomoco, at the branch of a river atop a red sandy hill among the woods, in a place called Uttamussack.

Wahunsunacock had placed there seven grave men who were the *quiyoughcosucks*. Their chief was dressed in regalia of a stuffed snake and weasel skins, a cloak and a crown of feathers. Wahunsunacock set these priests over the things he had won, and the bones of his forefathers, and a great rock, which they planted in the ground.[18]

At Uttamussack, Okeus was called and in a mighty rage made it known to Wahunsunacock's priests that from the great bay a people would arise who would overthrow the *mamanatowick*. So Wahunsunacock declared that the people of the Chesapeake, who lived on an opposite shore of the bay that took their name, were his mortal enemies, for they had been touched by the Otasantasuwak, and so he made them extinct.

But Okeus spoke again. He said Wahunsunacock might overthrow and dishearten attempters and such strangers as should invade his territories or labour to settle amongst his people. He would do this twice, and their tribute of copper, crystal and other precious things that were due to him would make his power even greater. But of a third attempting, these people would defeat him, and he would fall unto their subjection and under their conquest.[19] Then would white clouds of those ghosts blow across the great sea and blank out the sun, and the soft mud would dry, and the trees and plants turn to stone, and all that was supple would become hard, and all the sounds and songs of the Tsenacomoco would become trapped in the black marks upon white leaves.

SIX

Soundings

THE FRAGILE FLEET of three wooden 'eggshells' sailed past a cape on the southerly lip of the Chesapeake Bay, and dropped anchor on the south shore. To the north, a vast body of water stretched to the horizon. To the west, low-lying land receded as far as the eye could see. The English had found Virginia.[1]

'There we landed and discovered a little way,' wrote Percy, who was a member of the first thirty-strong landing party to go ashore that warm spring day in late April 1607. 'But we could find nothing worth the speaking of but fair meadows and goodly tall trees, with such fresh waters running through the woods as I was almost ravished at the first sight thereof.'

They stayed all afternoon, using the remains of a long day to take in their new surroundings. As darkness fell, they made their way back to the beach. Then, 'there came the savages creeping upon all fours from the hills like bears, with their bows in their mouths, [who] charged us very desperately'. The English let off volleys of musket fire, which to their surprise the Indians 'little respected', not withdrawing until they had used up all their arrows. Gabriel Archer was injured in both his hands, and Matthew Morton, a sailor, was shot 'in two places of the body very dangerous'. The casualties were carried back on to the boats, and the English withdrew to their ships.[2]

Later that evening, there was a solemn gathering of all the leading members of the expedition aboard the *Susan Constant*. According to their orders from the Royal Council, within twenty-four hours of their arrival at Virginia, they were to 'open and unseal' the secret list nominating the settlement's ruling council, 'and Declare and publish unto all the Company the names therein Set down'.[3]

Newport announced the names chosen by the Royal Council in England. His own was listed first, followed by Wingfield's and Gosnold's. Another nomination was John Martin, the sickly son of Sir Richard, a prominent goldsmith and the Master of the Mint. Martin's election was probably a foregone conclusion, given the wealth of his family and their generosity as patrons of this and other ventures.[4]

More surprising was the appearance of the mysterious George Kendall and John Ratcliffe on the list, together with Captain John Smith. The latter's inclusion might have suggested his immediate release from the brig, but Newport decided to keep him there for the time being.

George Percy and Gabriel Archer were not nominated. Percy may have been excluded because of worries about the loyalty of his brother, the Earl of Northumberland. The reasons for excluding Archer, who had worked so hard with Gosnold to promote the venture in its early years, were more opaque, and he took the news badly.

The council's first job was to nominate a president, who would supervise the taking of the oath of office. But rumbles of recrimination, puzzlement about the roles of Ratcliffe and Kendall, and the difficulty of deciding what to do about Smith discouraged such finalities.

The following day the mariners brought up from the hold the expedition's collapsible three-ton barge or 'shallop', and started to assemble it. Meanwhile, a landing party continued reconnoitring the surrounding territory. Several miles inland, they spotted smoke. They walked towards it, and came upon a campfire, still alight with oysters roasting on a barbeque. There was no sign of the picnickers, so the English polished off the meal, observing as they licked their fingers that the oysters were 'very large and delicate in taste'.[5]

The next day, 28 April, the shallop was launched, and Newport, together with a group selected from the ranks of the 'gentlemen', set off in search of a river or harbour suitable for settlement. The only map they had was one based on explorations made during the Roanoke expedition of 1585. This showed the south shore of the bay as interrupted by a series of inlets, one leading to the village of Chesapeake, the other to 'Skicoac'. The shoreline ended at what appeared to be the mouth of a river, which flowed west. The river's northern bank was a peninsula, which was marked with a dot, indicating an Indian settlement. To the north of the peninsula was another river, flowing north-west. Along the top of the map was what appeared to be the northern shore of the bay, populated by two villages, 'Mashawatec' and 'Combec'.

What they might have imagined to be the inlet leading to Skicoac proved to be nothing but shoal water, barely suitable for their shallop, let alone a ship. As they coasted deeper into the bay, they apparently passed the first river mouth, and came to the peninsula. They disembarked on one of the beaches, and explored its perimeter. They found a huge canoe, 45 foot long, made from the hollowed-out trunk of a large tree. They also found beds of mussels and oysters 'which lay on the ground as thick as stones', some containing rough pearls.[6]

Exploring inland, they found the land became more fertile, 'full of flowers of divers kinds and colours,' remarked Percy, 'and as goodly trees as I have seen, as cedar, cypress, and other kinds'. They also found a plot of ground 'full of fine and beautiful strawberries four times bigger and better than ours in England'. Great columns of smoke could be seen rising from the interior, and they wondered anxiously whether the Indians were using fires to clear land for planting, or 'to give signs to bring their forces together, and so to give us battle'.

Returning to the shallop, they sounded the surrounding waterways, but could find none deep enough for shipping. They returned both buoyed up and dragged down by the day's discoveries. The land was so inviting, but worthless to them if they could not find a suitable river or safe harbour.

Later that evening, a group rowed out to examine further the river mouth they had passed earlier that day. They started to zigzag across the water, taking soundings as they went. Slowly, painstakingly, their measurements built up a profile of the river bed beneath, revealing a channel 6 to 12 fathoms deep, enough for heavy shipping. So great was their relief, that Archer named the neighbouring point of land 'Cape Comfort'.

With this discovery, the decision was taken to commit the settlement's fortunes to the Chesapeake. To mark the decision, a group rowed to the southern side of the mouth of the bay, and on the promontory they had passed when they arrived erected a large cross, facing out to the ocean. They named the land upon which they stood Cape Henry, in honour of Tyndall's patron, James's 13-year-old heir.

On 30 April 1607, the fleet nosed past Cape (later known as Point) Comfort and into the broad river that lay beyond. Five Indians appeared on the shore, running along the beach to keep up with the ships. Newport called to them from the deck of the *Susan Constant*. At first they did not respond. Newport then laid his hand upon his heart as a gesture of friendship. They laid down their bows and arrows, and waved to him to follow them. Newport, together with Percy and a few others, clambered into a boat, and rowed towards the shore. The Indians dived into a tributary and swam across with their bows and arrows in their mouths. The English followed, until they found themselves floating towards a group of warriors waiting for them on the bank. From there, they were escorted to a town, which the English understood to be called 'Kecoughtan'.

Kecoughtan comprised a cluster of twenty or so dwellings built 'like garden arbours', interspersed among the trees. The walls of each house were made of saplings, the roof by the branches bent over to create a vault. The entire construction was covered with reed mats and in some cases with bark, a free-hanging mat acting as the door. The English were intrigued by the elegant simplicity of the buildings, and the lack of permanent structures or even of locked doors.[7]

The Indians who emerged from the houses presented a sight not altogether unfamiliar to some of the English, as they wore clothes and followed customs similar to those at Roanoke, whose appearance

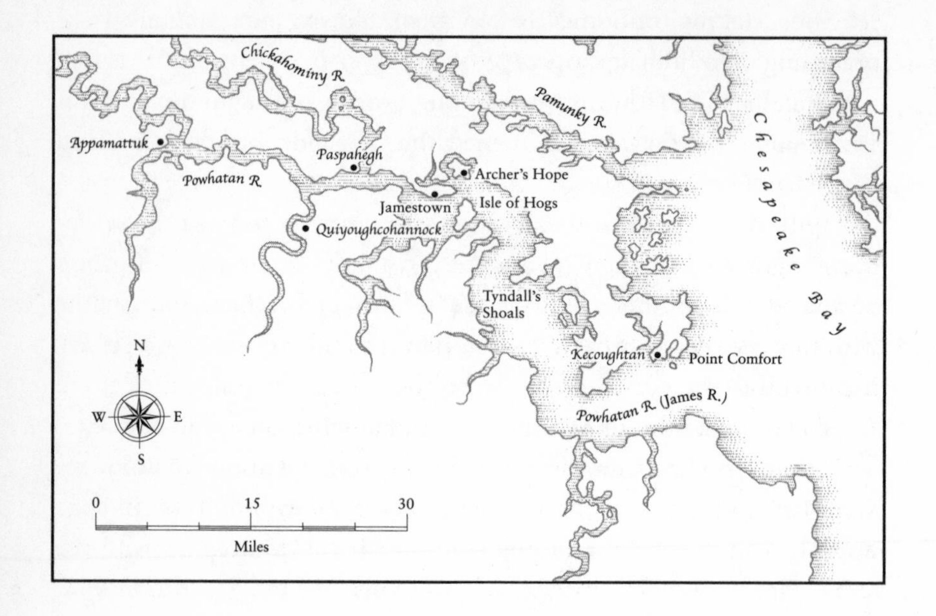

Powhatan River: initial reconnaissance, 8–12 May 1607.

had been carefully recorded by the painter John White. The most distinctive feature was the hair. The men shaved the right side of their heads, and let the left side grow to the length of an 'ell' (3 foot 9 inches), which they tied in an 'artificial knot' and decorated with feathers. In the intensifying heat of the Virginian late spring, they dressed sparingly, covering their 'privities' with an animal hide decorated with teeth and small bones, but were otherwise naked. To welcome the English, some had painted their bodies black, others red, 'very beautiful and pleasing to the eye', and wore turkey claws as earrings.[8] Gabriel Archer, standing among the exhausted, louse-ridden, poorly-nourished English could only admire the strength and agility of the 'lusty, straight men' whom they now encountered.[9]

As soon as the English entered the village, the men greeted them with a 'doleful noise', and approached 'laying their faces to the ground, scratching the earth with their nails.' Percy, a man of insecure religious convictions, was alarmed, fearing that the Indians were practising their 'idolatry' upon him.

Once the welcome was over, the Indians brought mats from their houses and lay them on the ground. The elders sat in a line, and were served with corn bread, which they invited the English to share, but only if they sat down. The English crouched awkwardly on the mats 'right against them', and accepted the offer. The elders then produced a large clay pipe, with a bowl made of copper, and filled it with tobacco. It was lit, and they offered it to their guests, who puffed it appreciatively.

To complete the ceremonies, the Indians put on a frantic dance, 'shouting, howling, and stamping against the ground, with many antic tricks and faces, making noise like so many wolves or devils'. The display lasted half an hour, during which Percy, drawing on a knowledge of courtly dance acquired as a child, noticed that they all kept to a common tempo with their feet, but moved to an individual rhythm with the rest of their bodies.

When the dance ended, Newport presented the elders with beads and 'other trifling jewels'. The English then returned to the fleet, content that, as instructed by the Royal Council, they had not only taken 'Great Care not to Offend the naturals', but laid the basis of a fruitful trading relationship.

Over the following few days the fleet carefully felt its way up the broad river, which even 20 or so miles upstream was wider than any in England, the channel deep enough for large ships. Tyndall carefully mapped the route of the channel, naming a large and treacherous sandbank about 30 miles upstream 'Tyndall's Shoals'. Just beyond, the river turned sharply, forming a loop similar to that of the Thames around the Isle of Dogs, which was later named the 'Isle of Hogs' (now home to a wildlife reserve, and a nuclear power plant). Opposite it, Gabriel Archer spotted an island which looked ideal as a location for their settlement. He had a fondness for puns, so he dubbed it 'Archer's Hope', a 'hope' also being a stretch of land cut off from its surroundings by marsh or fenland.

Continuing slowly upstream a further 20 miles, they reached a point where the river forked. They anchored the fleet in the broad waters of the confluence, and a landing party was sent to reconnoitre. They came to a town called Paspahegh, on the northern shore of the wider tributary, where they were entertained by an 'old savage' who made 'a long oration, making a foul noise, uttering his speech with a vehement action, but we knew little what they meant'. While witnessing this unimpressive spectacle, they were approached by the chief or *weroance* of the tribe on the southern bank of the river, who had paddled over in a canoe to remonstrate with them for favouring the Paspahegh over his own people. Gratified by this competition for their attention, and dismissive of the Paspahegh's efforts, the English said they would visit him the next day.

The following dawn was heralded by the arrival of a canoe alongside the *Susan Constant*, paddled by a messenger, who signified that he had come from the chiefdom on the opposite bank of the river. He beckoned the English to follow him. They duly manned their shallop, and pursued the canoe to the southern shore, which some had begun to call the 'Salisbury Side', in honour of Cecil, the Earl of Salisbury. As the English drew up by the river bank, Percy beheld the chief and his warrior escort waiting for them, 'as goodly men as any I have seen of savages or Christians'.

The chief, called Chaopock, presented a particularly impressive spectacle, his body and visage a map of promising commodities. His torso was painted crimson, which was perhaps how he got his name, the local word *chapacor* being the name of a root used to produce red dye. His face was painted blue, 'besprinkled with silver ore as we thought', probably a paste made of antimony, which was mined further north. He wore 'a crown of deer's hair coloured red in fashion of a rose fastened about his knot of hair', and on the shaven side of his head, a 'great plate of copper'. He played a reed flute as the English clambered ashore, and then invited them to sit down with him upon a mat, which was spread out for them on the bank. There, sitting 'with a great majesty', he offered them tobacco, his company standing around him as they watched the English puff on the pipe. He then invited them to come to his town, which the English understood to be called Rappahannock (but which they later learned was called Quiyoughcohannock). He led them 'through the woods in fine paths, having most pleasant springs', past 'the goodliest cornfields that ever was seen in any country', up a steep hill to his 'palace', where they were entertained 'in good humanity'.[10]

The next day, Newport left the fleet riding at anchor before Paspahegh and continued in the shallop up the wider branch of the river, stopping at a point where it once again divided. There he encountered another tribe, the 'most warlike' Appamattuck, who came to the banks of the river and confronted the English, their leader crouched before them 'cross-legged, with his arrow ready in his bow in one hand and taking a pipe of tobacco in the other' uttering a 'bold speech'. After an exchange of peaceful gestures, the English were allowed ashore, giving them an opportunity to admire the swords the Appamattuck carried on their backs, a unique weapon with a blade of wood edged with sharp stones and pieces of iron, sharp enough 'to cleave a man in sunder'.

Having established that there was no suitable location for the settlement upstream, the shallop returned to the fleet. On 12 May, Wingfield and other members of the council took the pinnace back downriver to Archer's Hope, to assess its viability as the best site for their settlement. A landing party found that 'the soil was good and fruitful with excellent good timber'. They found vines as thick as a man's thigh running up to the tops of the trees, turkey nests full of eggs, hares and squirrels in the undergrowth, birds of every hue – crimson, pale blue, yellow, green and mulberry purple – fluttering through the forest canopy. Returning to the ship, the council assembled to consider whether this should be the site for their settlement. Gosnold and Archer, perhaps with Percy's support, argued strongly for its merits: the abundance of natural resources around, its defensible location, 'which was sufficient with a little labour to defend ourselves against any enemy'. But Wingfield was worried about the sandbanks, which prevented shipping coming close enough to the shoreline to allow cargo and people to transfer directly on to land, a handicap that many considered the reason the Roanoke settlement had proved unsustainable. The other worry was the position of the Kecoughtans. Hakluyt had advised that the settlers should 'in no case suffer any of the natural people of the Country to inhabit between you and the sea'.

After a heated debate, they sailed around the Isle of Hogs to look at an alternative site. It had been brought to the settlers' attention by the Paspahegh, who claimed it to be part of their territory. It was a promontory sticking into the river, creating a constriction that forced the deep-water channel right up to the shore. In a daring test of its suitability, one of the ships was sailed up to the river bank, close enough to be tethered to the trees.

The settlers disembarked and investigated the area. They were standing on a peninsula which dangled like a piece of fruit from the mainland, its stalk a thin causeway across an otherwise impassable swamp. They also noted that the shoreline used to moor the pinnace was shielded by woods from enemy vessels coming upstream. It had once been inhabited by Indians, and some evidence may have survived of their presence, but they were long gone.

The site presented difficulties. The river and surrounding creeks and streams were brackish, so a well would have to be dug, or fresh water supplies brought in by ship. The site was also covered with large trees and thick vines which would take considerable effort to clear. There was also the issue of the Kecoughtans controlling land access to the bay.

After further arguments, during which Gosnold repeated some of his objections, the fateful decision was taken to make this the site of their settlement. As tools and supplies were unloaded from the ships, the council now gathered to constitute itself. More arguments erupted over whether Smith and even Archer should be co-opted, but in the end it was decided they should be excluded. Each council member then took the oath of allegiance, and elected Wingfield their president.

'Now falleth every man to work,' as Smith later put it, labourers to clearing the 'thick grove of trees' covering the western end of the site, soldiers to scouting the mainland, farmworkers to tilling soil and planting seeds, and President Wingfield to tending the breeding flock of thirty-seven chickens he had brought from London.

Conditions were tough. Most of the settlers slept under trees. Those who could afford to bring one enjoyed the comforts of a tent. The elderly Wingfield benefited from this and other indulgences, including the 'divers fruit, conserves, and preserves' he had packed into his trunk, though he later claimed that some had been pilfered.[11]

For spiritual sustenance, the settlers erected a makeshift chapel out of an 'awning (which is an old sail)' hung between the trunks of 'three or four trees to shadow us from the Sun'. 'Our walls were rails of wood, our seats unhewed trees, till we cut planks, our Pulpit a bar of wood nailed to two neighbouring trees,' Smith recalled, and 'in foul weather we shifted into an old rotten tent, for we had few better.'[12] In honour of their King, the council dutifully called this ramshackle collection of tents pitched on a sweaty island in the middle of a teeming forest 'Jamestown'.

The activity on the island attracted the attention of the Paspahegh people, who visited the site on several occasions to see what was going on. To prevent causing offence, Wingfield banned the building of any permanent defensive structures and the performance of military exercises. This was in line with Hakluyt's instructions, which urged the settlers to develop trading relations with the Indians 'before they perceive you mean to plant among them'. Smith, however, had different ideas.

It is unclear whether Smith had yet been released from his shackles, but he was still excluded from the council's deliberations, and he railed at Wingfield's decision, claiming it was motivated by 'jealousy', his possessiveness of power. According to the Charter, the

King had accepted the patentees' request to settle Virginia principally so that they 'may in time bring the infidels and savages living in those parts to humane civility and to a settled and quiet government'. This for Smith was royal absolution for a great civilizing mission, which should begin as it was supposed to proceed.[13]

For the time being, Wingfield's low-key policy prevailed, though Kendall, who it now transpired had some experience of military engineering in the Low Countries, managed to erect one bulwark, ingeniously constructed out of 'the boughs of trees cast together', possibly towards the isthmus joining the island to the mainland.

On 18 May, the Paspahegh chief himself paid a regal visit to the island with a hundred of his warriors, who 'guarded him in a very warlike manner with bows and arrows'. The chief asked the English to lay down their arms, which they refused to do. 'He, seeing he could not have convenient time to work his will, at length made signs that he would give us as much land as we would desire to take,' Percy claimed. Whether or not it had been properly understood, the chief's gesture relieved the tension, and the English put away their weapons. Then a fracas erupted between one of the English soldiers and an Indian, apparently over a stolen hatchet. One struck the other, and soon bystanders were joining in, provoking the nervous English to 'take to our arms'. In response, the chief left 'in great anger', followed by his retinue.

Two days later, forty of the Paspahegh arrived with a deer carcass, apparently a peace offering. At the invitation of one of the English soldiers, they also put on an impressive demonstration of their shooting skills. The soldier propped a 'target' or shield 'which he trusted in' up against a tree, and gestured to the Indian to take a shot at it with his bow and arrow. To the soldiers' astonishment, the arrow penetrated the shield 'a foot through or better, which was strange, being that a pistol could not pierce it'.

By this stage, life for the settlers was settling into a routine, and some were beginning to feel a little at home. A group surveying the surrounding land found 'the ground all flowing over with fair flowers of sundry colours and kinds as though it had been in any garden or orchard in England'.[14]

Many of the settlers were by now anxious that Newport should head back home at the earliest opportunity to fetch more supplies. The mission plan had been drawn up on the basis of a return by May of that year 1607, and provisioned accordingly. But on Thursday 21 May, Newport declared that, rather than embark for England, he would lead the 'discovery' of the river commissioned by the Royal Council. From his perspective, his own future, as well as that of the mission, rested upon the discovery of precious metals, or navigating a new route to the South Seas. He could not afford to go home empty-handed.

He chose to accompany him George Percy, Robert Tyndall, Gabriel Archer – who was to keep a journal – and Thomas Wotton, the surgeon. He also decided to release Smith from custody and take him too, perhaps to keep him out of Wingfield's way, or to offer him an opportunity to redeem himself. The rest of the company was made up mostly of the crew of his ship, the *Susan Constant*. The Royal Council had also called for Gosnold to 'cross over the lands' with twenty men, 'carrying half a dozen pickaxes to try if they can find any mineral'. No such expedition was mounted.

As the time approached for Newport's departure, he called his men before him, and pledged that none would return until they had found 'the head of this river, the lake mentioned by others heretofore, the sea again, the mountains Apalatsi, or some issue', by which he meant the legendary saltwater lake somewhere in the American interior, which might provide navigable access to the 'South Sea' or Pacific, and the Appalachian mountain range, from which was said to run the 'stream of gold or copper'.[15]

At noon, they climbed aboard the shallop, hoisted its sails, and, beneath a hot, early summer sun, began their slow progress upstream.

By the first night they had managed 18 miles, reaching a 'low meadow point' on the south side of the river. There they met the Weyanock people, who claimed to be hostile towards the Paspahegh.

The following morning, they set off early, managing 16 miles before breakfast. Stopping at an islet created by a large loop in the river, they found 'many turkeys and great store of young birds like blackbirds, whereof we took divers which we break our fast withal'.

A canoe carrying eight Indians appeared on the river, and the English hailed them with one of the first words they had learned: '*wingapoh*', which they understood to mean 'good fellow'. The Indians approached, and 'in conference by signs', the English asked them for guidance on the river's course. One of the Indians, apparently the leader of the group, stepped forward and offered to help. Archer was unable to discover his name, so dubbed him the 'Kind Consort'.

Using his toe, the Kind Consort started to draw a map in the sand of the river bank. Archer stopped him, offering him a pen and paper, and showing how he could use it. Immediately understanding what he had to do, the Kind Consort began to draw, laying out for the delighted English 'the whole river from the [Chesapeake] Bay to the end of it, so far as passage was for boats'. He indicated that, upstream of 'Turkey Isle', as Archer had dubbed their current location, lay another islet, and beyond that, a series of waterfalls, marking the end of the navigable river. A day's march beyond the falls, the river divided in two, both branches coming from the mountains. This was the land of two other 'kingdoms'. 'Then, a great distance off,' stood the mountains of 'Quirank', which, he whispered, had rocks containing veins of *caquassan*, understood to be the Indian word for red earth, which might signify the presence of copper or even gold. Furthermore, the Kind Consort confirmed that just beyond these mountains lay 'that which we expected', as Archer coyly put it in his journal, referring to the saltwater lake.

Anxious to proceed, Newport declined offers of hospitality from the Kind Consort, and sailed on. However, the Indian 'with two women and another fellow of his own consort' was anxious to track the English, and continued to follow them in the canoe, proffering dried oysters from a basket as they went. Eventually, the English

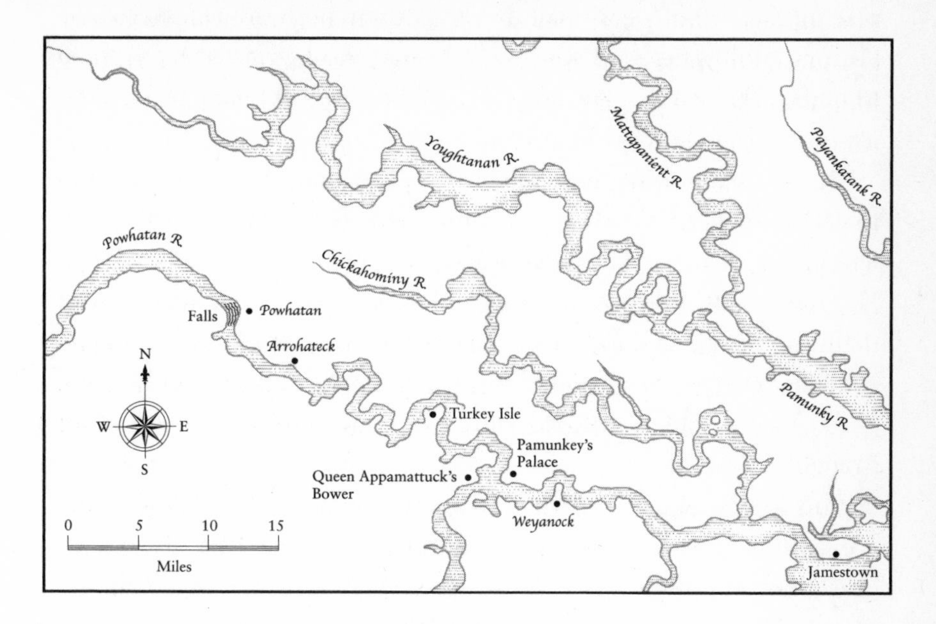

Powhatan River, renamed 'James' by the English,
explored as far as the Falls, 21–26 May 1607.

found they could no longer resist, and, rendezvousing at a 'point' on the river bank, 'bartered with them for most of their victuals'.

Two miles further up lay the first signs of the rockier world ahead, the shore being 'full of great cobblestones and higher land'. Once again, the Kind Consort appeared, this time offering 'sweet nuts like acorns (a very good fruit), wheat, beans, and mulberries sod [soaked] together'. Newport bought what he could, after which the Indian disappeared.

The following day, a further 5 miles on, the English landed, and 'found our kind comrades [the Indians] again', who escorted them to a town on the north bank of the river called Arrohateck. There they received their most opulent welcome yet. A feeling spread among the English that, as they proceeded west, they were closing in on the centre of Indian power.

'King Arrohateck', as Archer named the town's *weroance*, honoured Newport by laying a reed mat across his shoulders and placing a crown of 'deer's hair dyed red' upon his head. He offered the visitors 'mulberries, sod wheat and beans, and he caused his women to make cakes for us'. He also volunteered some of his men as guides for the journey into the interior.

It was from King Arrohateck that the English first discovered that there was a 'great king', a supreme ruler or *mamanatowick* to whom all the chiefs around the river paid homage and tribute. His name was Powhatan. 'Now as we sat merry banqueting with them, seeing their dances and taking tobacco,' the chief's warriors suddenly got to their feet, and formed a guard of honour for another guest, whom they welcomed with a 'long shout'. This, the English thought, must be the *mamanatowick* Powhatan himself. It was in fact his son Parahunt, known as 'Tanxpowhatan' or 'Little Powhatan', the confusion arising because the town he ruled was also called Powhatan, perhaps because it was the *mamanatowick*'s birthplace.[16]

Noting that King Arrohateck had remained seated, the English guests did likewise, 'our captain in the middest'. However, acknowledging the status of the new arrival, Newport offered 'gifts of divers sorts, as penny knives, shears, bells, beads, glass toys, etc., more amply than before'. In gratitude for the generosity, 'King Powhatan' offered provisions and guides to escort the English to his town, further upriver. 'Thus parting from "Arrohateck's joy",' wrote Archer, brimming with optimism, 'we found the people on either side the river stand in clusters all along, still proffering us victuals.'

A further 10 miles upstream, the river narrowed. On the northern bank rose the most striking landmark they had yet seen: a tall mound in the midst of fields full of wheat, beans, tobacco and other crops, intermingled. On top of the mound there was a collection of houses, which for the English had the appearance of the Indian capital. Archer called it 'Powhatan's Tower'.

They landed, and walked up to the town, where they found the Powhatan and Arrohateck chiefs waiting for them. The English noticed that they now sat apart from their own people. The only other person with them was a man who appeared to be a counsellor, who sat beside Parahunt. Once again, food was offered, 'but our best entertainment was [the] friendly welcome'. There followed a discussion by 'words and signs' during which King Powhatan explained that all the 'kingdoms' on the north bank of the rivers were *cheisc*, which the English understood to mean 'all one with him' or 'under him'. But the Chesapeake people, who lived on the southerly shore of the bay, were the enemy 'generally to all these kingdoms'. Archer showed the chief the scars on his hands, barely healed, which had been inflicted by these Indians when the English first landed, and 'for which we vowed revenge, after their manner pointing to the sun'.

Parahunt now placed his own gown upon Newport's shoulders, which the English understood to mean he was offering a 'league of friendship'. Putting his hand on his heart, he said to Newport, '*Wingapoh chemuze*,' Archer taking this to be the kindest of all salutations in the Indian language.

It was now late, and the English said they needed to return to their ship. They were sent on their way with six of Parahunt's men, and the English left behind one of their own as a gesture of trust. Rather than return directly to the shallop, they rowed 3 miles up the river, where they found what the Indians had warned them of: a great cataract that was clearly impassable. Even this far inland, the river was still tidal, rising 4 foot between high and low tide, and suitable for vessels with a draft less than 6 foot. But beyond, the only way forward was on foot.

'Having viewed this place between content and grief,' Archer concluded, 'we left it for this night, determining the next day to fit ourselves for a march by land.'

The guides who were with them were sent home, except one, called Navirans, who asked if he could sleep on board the shallop with the English. Newport agreed, a gesture of faith rewarded with the safe return of the Englishman left at Powhatan's Tower, 'who coming told us of his entertainment, how they had prepared mats for him to lie on, gave him store of victuals, and made as much [of] him as could be'. A close relationship developed between Navirans and the English, particularly with Archer. The Indian 'had learned me so much of the language,' Archer wrote, 'and was so excellently ingenious in signing out his meaning, that I could make him understand me, and perceive him also well-nigh in anything'.

The following day, Whitsunday, 24 May, Newport decided to return some of the hospitality the English had received. His men built a fire on the shore, and they boiled two pieces of pork and some peas, the best that could be offered from their dwindling supplies. Newport invited the two chiefs to join him. Parahunt accepted, but Arrohateck excused himself on the grounds that he needed to return to his village.

As Arrohateck was about to leave, the convivial mood abruptly changed. An English mariner reported that two 'bullet bags' containing 'shot and divers trucking toys' had gone missing.

The chiefs acted quickly and decisively, ordering the immediate return of all stolen property. The speed with which the items reappeared was impressive evidence of the chiefs' authority. Everything that had gone was now laid at Newport's feet, including a knife the English had not even realized was missing. 'So Captain Newport gave thanks to the kings and rewarded the thieves with the same toys they had stolen, but kept the bullets.' Newport also warned that the custom in England was to punish theft with death.

Good relations apparently restored, the Powhatan *weroance* sat down to the feast, 'and we fed familiarly', Archer reported, 'without sitting in his state as before'. The relaxed atmosphere was helped by quantities of beer, aqua vitae (spirits) and sack (Spanish white wine). Alcoholic drinks were not part of the local diet, and this first exposure to some potent European brews had an unusually strong effect on Newport's guest. This might explain why the chief fell into such an uninhibited mood, talking about the copper, iron and other rich and rare commodities to be found in the mountains beyond the waterfalls.

As the merrymaking was drawing to a close, Newport said he

wanted to embark on a three-day expedition further inland to see if he could find these commodities. The chief, perhaps prompted by a sobering word whispered in his ear, suddenly fell silent. He got up to leave, promising only that he would rendezvous with the English later that day at the foot of the falls.

In the afternoon, the English rowed upriver. They found the Powhatan chief sitting on a bank next to the lower reaches of the cascading water.

At this point, the nameless Kind Consort who had appeared to the English at Weyanock approached in a canoe, continuing his mysterious knack of reappearing at significant moments of the English exploration. He told Newport's men to 'make a shout'. They were unsure why, but they did as they were asked and cried out. They assumed it was to welcome King Powhatan, though it may have been to acknowledge some other power that inhabited the falls, or *paquacowng*, as the Indians apparently called them.[17]

Newport led a group across the rocks to talk to Parahunt. The expansiveness had evaporated. The chief 'sought by all means to dissuade our captain from going any further'. It would be tedious travel, he claimed. Ahead lay the Monacan people, who were enemies, and liable to attack Powhatan guides if not the whole party, and even if they got past them, the Quirank mountains that lay beyond were difficult and dangerous, devoid of the food supplies they would need for a proper exploration. The Monacans 'came down at the fall of the leaf', he told Newport, and attacked his people's villages. Newport offered five hundred English troops to fight alongside the Powhatan people upon the Monacans' return, 'which pleased the king much'.

To Archer's surprise, Newport agreed not to proceed any further, 'holding it much better to please the king, with whom and all of his command he had made so fair way, than to prosecute his own fancy or satisfy our requests'.

The *weroance* now departed, followed by all his men except Navirans, who accompanied the English to an 'islet' in the middle river, which stood before the falls. There, Newport announced that the river would henceforth be known as the James, and ordered the soldiers to erect a large cross, as they had done at Cape Henry. It bore the Latin inscription 'Jacobus Rex. 1607': King James 1607. As the cross rose into the sky, Navirans gave out a great cry. Newport was reassuring, explaining 'that the two arms of the cross signified King Powhatan and himself, the fastening of it in the middest was their united league'. This explanation apparently 'cheered Navirans not a little'.

The English prepared for their journey back down the river they had renamed the James, and Newport sent Navirans to invite 'King Powhatan' for a farewell meeting. The chief duly appeared with Navirans and his retinue on the river bank, and Newport rowed alone from the shallop to the shore, to present a gown and a hatchet as farewell gifts.

The mood had changed. The chief seemed angered by the appearance of the cross, casting its long evening shadow across the river's sacred waters. Percy noted that the 'savages' now 'murmured at our planting in the country'. Newport prompted Navirans to pass on the explanation that the cross symbolized peace. Parahunt appeared to be reassured, and, according to Navirans's translation, rebuked his people: 'Why should you be offended with them as long as they hurt you not, nor take anything away by force? They take but a little waste ground which doth you nor any of us any good.'[18]

With feelings of reassurance mixed with uncertainty, the English left and headed back for Jamestown. As night fell, they stopped at Arrohateck. The chief was ill, complaining that the 'hot drinks' the English had given him had made him sick. Newport confidently predicted that he would feel better in the morning, which was duly the case, and to celebrate his recovery, the chief ordered venison to be roasted for the visitors.

While they were there, some of the Arrohateck people offered to show the guests their homes and gardens. Passing among the houses, scattered around groves of tall trees, the English entered a world hidden from them by the diplomatic formalities experienced so far. This was the domain of women. While the men 'fish, hunt, fowl, go to the wars', the women kept the home and hearth, and tended the fields. They also raised and educated the children, shaved the men, foraged for fuel, chopped wood, spun flax, ground corn, baked bread, butchered meat, gathered medicinal herbs, healed the sick and mourned the dead, 'which', Smith observed, 'is the cause that the women be very painful [i.e. burdened] and the men often idle'.

Being allowed to mingle with them provided the curious Englishmen with their first proper exposure to female company since leaving London. The effect was powerful. The women appeared natural and unaffected. They had 'handsome limbs, slender arms, and pretty hands; and when they sing they have a delightful and pleasant tang in their voices'. They wore make-up to enhance their features rather than disguise their blemishes, and shared cosmetic tips and recipes freely, unlike the 'great ladies' of English society, who kept secret from one another 'their oil of talcum or other painting white and red'. The Indian women wore clothes not to hide their age, but, as Captain Smith put it, to be 'agreeable to their years'. Their bodies were not trussed, bustled and costumed, but flaunted. Their arms, thighs and breasts could be openly admired, being elaborately advertised with 'cunningly embroidered' tattoos. They were 'voluptuous', fully developed sexual beings, scantily dressed and approachable, yet retaining that feminine virtue most vaunted by European men, 'modesty'. It was a combination that aroused the Englishmen's imaginations and starved libidos.

Mothers breast-fed their babies, a practice that English women of all but the poorest classes avoided, favouring the use of wet-nurses. They also loved their children 'very dearly', but were tough as well as tender, making them 'hardy' by washing them in the river even on the coldest mornings, and 'tanning' their skins with ointment until 'no weather will hurt them'. Their houses, Smith observed, were as 'warm as stoves, but very smoky', due to the fire in the centre of the floor, which vented through a simple hole in the roof. It being summertime, the mats covering the walls may have been rolled up to let in the air, but the women continued to tend the fire, as, 'if at any time it goes out, they take it for an evil sign'.

A set of simple bedsteads was the only recognizable domestic furniture to be found inside an Indian house. They were made of short posts stuck into the ground, with 'hurdles' or frames made with sticks and reeds placed on top to act as the mattress. There was a bed for each member of the family, upon which they would sleep 'heads and points', head to feet, in a circle around the fire. Mats acted as bedlinen, and, while they slept, the perpetual smoke, which darkened their skins but did not sting their eyes, kept away mosquitoes and fumigated clothes.[19]

While he was being shown around one of the huts, an Indian woman took Archer's hand and pressed the leaves of a herb into his wounded palm. The plant was *wisakon*, he was told (the word was in fact a general term for medicinal herb). It looked to him like liverwort or bloodwort, two well-known medicinal herbs used in England. He was also shown a root which contained the poison that had laced the arrowheads.

The visitors watched the women bake rolls and cakes, and a demonstration of 'the growing of their corn and the manner of setting it'. The fields, just a few hundred foot square, were more like gardens, cleared by burning sections of the surrounding woodland. The soil was neither cultivated (the Indians had no draft animals or tilling equipment) nor manured, producing, in the opinion of Smith the farmer's son, 'so small a benefit of their land'. He was sure the most basic English agricultural techniques would multiply the yields.[20]

Finally, the chief took the English into what Archer dubbed the 'Mulberry Shade', a hunting lodge set apart from the village, where King Arrohateck laid on a meal of 'land turtle' while his men went into the surrounding woods to see if they could catch a deer. The chief also asked for a demonstration of English firearms. Newport duly ordered 'a gentleman discharge his piece soldier-like before

[King Arrohateck], at which noise he started, stop'd his ears, and express'd much fear, so likewise all about him'.

There was also an incident, confusingly documented, which enabled the two nations to compare their methods of summary discipline. Navirans drew attention to one of Arrohateck's people who, as Archer put it, 'press'd into our boat too violently upon a man of ours'. Newport, 'misconstruing the matter, sent for his own man, bound him to a tree before King Arrohateck, and with a cudgel soundly beat him'. The chief intervened, saying one of his men was responsible for the 'injury'. He went over to the culprit, who tried to run away. The chief set off in pursuit, running 'so swiftly as I assure myself he might give any of our company 6 score in 12' (i.e. beat them ten times over). The offender was brought back, and the rest of the king's retinue brandished cudgels and sticks 'as if they had beaten him extremely'. Archer does not mention if the punishment was actually executed, or only threatened.

These violent proceedings did not dampen the convivial mood, but seemed to draw the two leaders, Newport and King Arrohateck, closer together. As the day drew to an end, Newport presented the chief with a red waistcoat as a farewell gift, 'which highly pleased him'. The English boarded their shallop and cast off, the Arrohateck men saluting them with two hearty shouts as they pulled away from the shore.

That night, they anchored near Appamattuck, home of those 'most warlike' people Newport had visited while reconnoitring the site of the English settlement. The following day, they went ashore and were led by Navirans through fields newly planted with corn to a 'bower' of mulberry trees. They sat down to await the Appamattuck chief, but were instead surprised by the regal approach of a 'fat, lusty, manly woman' clothed in deerskin, covered in copper jewellery, including a crown, and attended by a retinue of women 'adorned much like herself, save they wanted the copper'. This, the English decided, must be a queen, as she was treated with the same reverence as the Powhatan and Arrohateck chiefs, 'yea, rather with more majesty'. Her name, though they did not yet know it, was Opussoquonuske.[21]

The English, struggling to readjust their assumption that Indian royalty must be exclusively male, were anxious to discover her role. She explained that she was a chief under the authority of Powhatan, 'as the rest are', but the visitors noted that 'within herself' she was 'as great authority as any of her neighbour *weroances*', if not greater. For two hours the English gazed upon this Indian Elizabeth, while feasting on the 'accustomed cakes, tobacco, and welcome'. They offered to demonstrate their weapons, and Archer noted that, when a musket was fired, 'she showed not near the like fear as Arrohateck'. Newport then decided they should leave for the final leg of their expedition.

Navirans led them 5 miles downstream, and persuaded them to put in for one final meeting. The location, he said, was 'one of King Pamunkey's houses', a structure that may have been a hunting lodge, or even specially constructed for the occasion, as the Pamunkey homeland lay 20 miles away, along the banks of a river neighbouring the James. The English seemed unaware that such encounters were now being carefully orchestrated, nor did they realize the importance of the man they were about to meet. He was Opechancanough, whose name meant 'man of a white (immaculate) soul'. Later described as possessing a 'large stature, noble presence, and extraordinary parts', he was said by some to be Powhatan's brother, but by others to have come from 'a great way from the south-west . . . from the Spanish Indians, somewhere near Mexico'. In his forties at the time of this first encounter, he acted as the military chief of Tsenacomoco, the Indian name for the Powhatan empire. He had come to meet the English to assess their intentions and strength.[22]

After the magnificent Queen Appamattuck, Archer found 'King Pamunkey' a ridiculous figure, 'so set [upon] striving to be stately as to our seeming he became a fool'. He claimed to come from a 'rich land of copper and pearl', and showed off a pearl necklace and a sample of copper 'the thickness of a shilling' which Archer managed to bend round his finger 'as if it had been lead'. Affecting nonchalance at this information, the English asked what other commodities his land offered. The king obligingly boasted it was also 'full of deer' though added that 'so also is most of all the kingdoms'.

Archer called the venue for the encounter with Opechancanough 'Pamunkey's Palace', mocking the king's extravagant claims, and ignoring Navirans's hints that the name was inappropriate.

Continuing their journey home, they spent the night at the 'low meadow point' where they had anchored the first night of the expedition, 18 miles away from the settlement. The following morning, they went ashore with Navirans. They encountered a hunting party of ten or twelve Indians who were camping on the shore, and Navirans arranged for them to go fishing for the English. 'They brought us in a short space a good store,' Archer noted, who accounted them 'good friends'.

Then, without warning or explanation, Navirans 'took some conceit' of the English, and refused to go any further with them. 'This grieved our captain very deeply,' Archer observed, 'for the loving kindness of this fellow was such as he trusted himself with us out of his own country.' By now, Newport imagined that he had managed to establish a rapport with the Indians, that his diplomacy had been embraced, his honourable intentions accepted, the superiority of his weaponry acknowledged and admired. Navirans's sudden change of heart punctured this presumption. Newport ordered the shallop's immediate return to Jamestown, 'fearing some disastrous hap at our fort'.

They arrived back on 27 May, to find the settlement in chaos. It transpired that the day before, two hundred Indian warriors had mounted a sustained attack. Following his meeting with Newport, Opechancanough had evidently decided the English presence must be eliminated, before it became permanent.

At the time of the attack, the settlers had been planting corn in the newly cleared fields. Most of their weapons were still packed in 'dryfats', waterproof storage casks, so they only had a few pistols and swords to defend themselves. As the ranks of Indian warriors descended upon them, they had been forced to run for cover, to few finding shelter behind the island's single defensive bulwark. Led by President Wingfield, all five council members apparently put up a fight with hand weapons, but were forced to retreat. In the ensuing skirmish, which 'endured hot about an hour', one boy was slain and as many as seventeen labourers wounded. Every single member of the council sustained injuries, except Wingfield, who had a miraculous escape from an arrow which passed through his beard. According to later reports, the entire company would have been slain, had not the sailors loaded one of the ships' cannons with a 'crossbar' (round shot with a spike embedded in it), and fired it towards the Indian position. The projectile had hit a tree, bringing down one of its branches, which apparently fell among the attacking Indians and 'caused them to retire'.[23]

'Hereupon the president was contented the fort should be palisaded,' Smith noted dryly.

George Kendall was put in charge of designing the defences, Archer of laying out the street plan for the town. The best-known work in English on military architecture at the time was a translation of a French work by William de Bellay called *The Practice of Fortification*. It was sufficiently influential for Christopher Marlowe to lift a section verbatim for his popular first play, *Tamburlaine the Great*, and for Percy's brother the Earl of Northumberland to hold a copy in his library. Rule one for Bellay was: 'the figure triangular is not to be used at all' in the 'delineation' of a fort, because it resulted in long, penetrable walls and vulnerable bulwarks or ramparts at each corner. Similar advice had been offered to the Roanoke settlers, who had been told to build their fort in the shape of a pentangle. Despite these warnings, Kendall chose a triangular shape. This was possibly because the river acted as a natural defence, and de Bellay had also advised that 'if any part [of the fort's proposed location] may be better assured of the situation than the rest', that was the side to have any sharper angles or longer walls. Kendall proposed putting the fort's longest side, which measured 140 yards, along the waterfront, with two shorter sides, 100 yards each, jutting into the island, enclosing an area totalling just over an acre.[24]

Meanwhile, Archer had been working in the pinnace, drawing up plans for the town that would lie within the defensive walls. Wingfield had gone to inspect them. He rejected Archer's work, and a confrontation ensued, defeating the president's efforts to give Gosnold's restless, rebellious friend a useful role. Alternative plans were drawn up, which, from archaeological remains, appear to have allowed for a row of barracks next to the southern palisade, officers' dwellings on the western side, a storehouse to the east, and a church in the middle.

The labourers worked around the clock, with the reluctant help of the ships' crews, and within days Kendall's design was taking shape. The curtain wall was made of rows of split logs, sunk into a ditch that was backfilled to keep them upright. At each corner, large crescent-shaped bulwarks were built using the same method, upon which the company's carpenters constructed stout platforms to carry lookouts and artillery. Winches and cranes were erected to lift some of the 'four and twenty pieces of ordnance' brought from England, and soon the bulwarks bristled with culverins, enormous cannons with barrels 11 foot long, capable of shooting 18-pound shot over a distance of nearly 2,000 foot.[25]

While excavating around the neck of land connecting the island to the mainland, workers found a stream flowing down a small bank. In the glint of the trickling water they saw what looked like yellow crystals. Captain John Martin was immediately summoned to inspect what they had found. Samples were taken and, using apparatus brought from London, he performed an assay or test to see if any metal could be drawn off. The news that he had managed to extract a small amount of what appeared to be gold 'stirred up in them an unseasonable and inordinate desire after riches'. A barrel was filled with soil taken from the surrounding area, for testing back in England.[26]

Meanwhile, the Indians kept up their attacks. As the fortifications rose, their tactics changed from full-frontal assaults to harassment.

On Friday 29 May, they managed to shoot forty or so arrows 'into and about the fort' from the cover of the surrounding woodland, before being repulsed by a volley of musket fire. 'They hurt not any of us,' Archer wrote, 'but finding one of our dogs, they killed him.' The following Sunday, while feverish construction continued, 'they came lurking in the thickets and long grass'. Eustace Clovill, a hapless offspring of Norfolk gentry, was found 'straggling without the fort', and was shot six times. His brief moment in history came when he staggered back to the fort with the arrows still stuck in his body, and shouted, 'Arm! Arm!' before collapsing. He died eight days later.

On Thursday 4 June, as dawn was breaking, one of the settlers left the fort 'to do natural necessity' in the latrine just beyond the palisade. As he squatted on the ground, he was shot with three arrows by Indians who had 'most adventurously' crept under the bulwark blocking access to the island.

The relentless assaults soon took their toll on English morale, and the acrimony that had been stewing since the ships were stuck on the Downs erupted with new vigour. The council began to disintegrate. 'The cause of our factions was bred in England,' Smith later observed, but 'grew to that maturity among themselves that spoiled all'.

Religion remained a point of contention. On arrival, the settlement chaplain Robert Hunt had implemented a puritanical regime of daily common prayer, and lengthy sermons on Sunday mornings. Wingfield made his dislike of such pious earnestness known, and on several occasions found opportunities to miss, and even to cancel, the sermons.

In fact, according to their accusers, Wingfield and Newport had no interest in the venture's spiritual mission, even though it featured prominently in the Royal Charter. They were merely 'making Religion their colour, when all their aim was nothing but present profit'.

The most forceful critic was Captain Smith. Having kept uncharacteristically quiet during the James river expedition, he was driven by an overwhelming sense of vindication to confront Wingfield and the council. He demanded that the president take manful possession of this virgin land. The Indians had not cultivated it, so it no more belonged to them than to the wildlife in the undergrowth or the birds in the trees. In any case, these vast expanses contained 'more land than all the people in Christendom can manure, and yet more to spare than all the natives of those Countries can use'.[27]

This land was waiting to be fashioned 'by labour' into a new 'commonwealth', a new England. What could command more honour for a man, Smith later wrote, 'with only his own merit to advance his fortunes ... than planting and building a foundation for his posterity, got from the rude earth by God's blessing and his own industry without prejudice to any?' He continued:

> What can he do less hurtful to any, or more agreeable to God, than to seek to convert those poor Savages to know Christ and humanity, whose labours with discretion will triple requite thy charge and pain? What so truly suits with honour and honesty, as the discovering things unknown, erecting Towns, peopling Countries, informing the ignorant, reforming things unjust, teaching virtue and gain to our native mother country?[28]

Several of the gentlemen excluded from the council could identify with these sentiments, and began to agitate for a more active, aggressive policy. Archer started 'spewing out ... venomous libels and infamous chronicles' about Wingfield's government. Wingfield had 'affected a kingdom,' he claimed. Archer was trying to summon a parliament, Wingfield retorted. On 6 June, Archer led a group, including several soldiers who had left their positions, who 'put up a petition to the council for reformation'.[29]

The content of the petition was not recorded, but its consequence was that four days later, Smith was finally absolved of all charges and made a member of the council. According to Smith's later testimony, Wingfield was also required to pay him compensation to the value of £200, an enormous sum, which Smith claimed he magnanimously donated 'to the store for the general use of the colony'.[30] The following day, 'articles and orders for gentlemen and soldiers were upon the court of guard and content was in the quarter', meaning that peace had been restored among the unsettled ranks.

On Saturday 13 June, to reinforce a new but still fragile mood of accord, Newport presented the soldiers and labourers with a sturgeon 7 foot long, which had been caught by the crew of his ship. The next day, the settlement was approached by two unarmed envoys. Archer recognized one of them to be the 'Kind Consort' who had followed them in the early days of the James river expedition. The reappearance of this mysterious individual comforted the settlers, and he was invited into the fort, which was in its final stages of construction. The envoy endeavoured to explain what had been going on, reassuring the embattled, bewildered Englishmen around him that the chiefs of the Pamunkey and Arrohateck were still their friends, as were those of two tribes the English had not yet encountered, the Mattapanient and Youghtanan, who lived on western tributaries of the Pamunkey, the river north of the James. However, several other tribes were their 'contracted enemies', including the Quiyoughcohannock, the Weyanock and the Paspahegh, upon whose land the English had now settled. 'He counselled us to cut down the long weeds round about our fort and to proceed in our sawing,' Archer reported. 'Thus making signs to be with us shortly again, they parted.'

The day after this encounter, the fort was pronounced complete. It was crudely made and far from impregnable, but probably sufficient to frustrate a full-scale attack. Inside, however, conditions for the settlers were not much better than when they had arrived, possibly worse. Having failed to extract any trade or tribute from the Indians, their food was little more than bran. 'Our drink was water, our lodgings castles in the air,' as Smith put it. There were no permanent buildings, except perhaps for a store to keep provisions and armaments.[31]

Newport decided the time had come for him to leave. The ships were now fully laden with clapboard timber, the barrel of soil to test for metals, and upwards of two tons of sassafras, the highly profitable commodity brought back by Gosnold on his 1602 voyage.

The council assembled to draft a progress report for the Royal Council in London. It was brief and began on an optimistic note:

> Within less than seven weeks, we are fortified well against the Indians; we have sown good store of wheat; we have sent you a taste of clapboard; we have built some houses; we have spared some hands to a discovery; and still as God shall enable us with strength we will better and better our proceedings.

There followed a complaint about the sailors, 'waged men' who fed off their supplies, and gathered valuable commodities such as sassafras for their own private gain, losing or damaging many tools in the process. However, mindful that these same sailors might be recruited for future relief missions, the council asked that they be 'reasonably dealt withal, so as all the loss neither fall on us nor them'.

'The land would flow with milk and honey if so seconded by your careful wisdoms and bountiful hands,' the report continued. 'We do not persuade to shoot one arrow to seek another, but to find them both. And we doubt not but to send them home with golden heads. At least our desires, labours, and lives shall to that engage themselves.'

They then listed the bountiful glories of their new home, the mountains, the rivers, the fish, the fruit, the miraculous medicinal herbs, before ending on a note of desperation:

> We entreat your succours for our seconds with all expedition lest that all-devouring Spaniard lay his ravenous hands upon these goldshowing mountains, which, if it be so enabled, he shall never dare to think on.
>
> This note doth make known where our necessities do most strike us. We beseech your present relief accordingly. Otherwise, to our greatest and last griefs, we shall against our wills not will that which we most willingly would.

The report was dated 'James town in Virginia, this 22th of June' and signed 'your poor friends' followed by a list of council names.

Smith's name appeared after President Wingfield's.

As the council handed their report to Newport, another was pushed into his hand by one of the settlers, William Brewster. Brewster may have been related to the Pilgrim Father of the same name who set sail for America aboard the *Mayflower* 13 years later. He had been secretly commissioned by Cecil to write a private report on the expedition and in particular the conduct of the council.[32]

Like the council, Brewster began with an optimistic assessment of the settlement's prospects, and added an equally desperate plea for a supply mission to ensure their fulfilment:

> Now is the King's Majesty offered the most stately, rich kingdom in the world, never possess'd by any Christian prince. Be you one means among many to further our seconding to conquer this land as well as you were a means to further the discovery of it.

This was just the garnish for what followed: a report on the conduct of the council, describing the faction fighting that had nearly destroyed it. Unfortunately, although the beginning of the letter is still to be found among Cecil's papers, the confidential part was torn off, and has never been recovered.

On Sunday 22 June, Hunt led the company in Holy Communion, the first to be held since their arrival. In the evening, Newport invited some of the gentlemen aboard his ship for a last supper. Final preparations were made on the *Susan Constant* and *Discovery*, while the *Godspeed*, which was to remain in Virginia, was decommissioned, her sails removed to the fort to prevent her being taken by renegades or attackers.

Newport set sail the following morning, promising to return within twenty weeks.[33] The settlers lined the shore and watched the fleet cast off and make its way down the majestic James. The ships disappeared within minutes behind the thick canopy of trees covering the eastern end of the island.

Anxiety, if not dread, gripped those left behind, as they walked back into their makeshift accommodation. Living conditions were still rough. Their 'houses' were for the most part fragile tents, devoid of home comforts. The weather, Indians and supplies were erratic, preventing all attempts to find a settled or familiar routine. Worse yet, the councillors in whose hands their fate now rested seemed to be infected with the scheming and plots, petty rules and brutal punishments they had hoped to leave behind in England.

'You shall live freely there [in Virginia], without sergeants, or courtiers, or lawyers, or intelligencers,' a character in *Eastward Hoe* had promised. The hollow laughter aroused by those scurrilous words at the Blackfriars Theatre each night echoed all the way across the Atlantic.

SEVEN

The Spanish Ambassador

SOMETIME IN JULY, 1607, Robert Cecil was strolling under a warm summer sun in the gardens of Whitehall Palace, the royal residence in Westminster. One Lanier, either John, son of a Huguenot refugee and a royal musician, or his son Nicholas, also a musician, approached him, accompanied by a soldier called Captain Hazell. A discussion took place between this Hazell and Cecil about which nothing is known, except that it concerned Hazell embarking on a secret mission to Spain accompanied by someone 'best experienced' in the coasts of Virginia.

Hazell apparently recommended for the job one George Weymouth, 'a special favourite of Sir Walter Cope's', and an experienced mariner who as recently as 1605 had led a reconnaissance mission to Cape Cod. Weymouth was duly hired, and a few days later, he and Hazell slipped out of London, and made their way towards Deal in Kent, where they were to pick up a ship bound for Spain.[1]

Meanwhile, on 29 July, the *Susan Constant* slipped into Plymouth harbour. Newport scribbled a letter to Cecil, announcing his arrival, and the discovery of 'a river navigable for great ships one hundred and fifty miles'. 'The country is excellent and very rich in gold and copper,' he reported, adding that 'of the gold we have brought a Say' or sample, which he hoped to present to the King and the Privy Council. 'I will not deliver the expectance and assurance we have of great wealth, but will leave it to Your Lordship's censure when you see the probabilities.'

The remainder of the letter contained excuses for not leaving his ship and making the journey overland to London. His 'inability of the body' detained him, he wrote, possibly referring to the difficulty of making a long journey on horseback with a missing arm. So he would sail his ship round to London as soon as 'wind and weather be favourable'.[2]

His desire to spend some time at Plymouth might not have been motivated by medical needs alone. He had probably decided to allow the crew time to sell some of the sassafras they had brought from Virginia. They would have found on the quayside at Plymouth any number of merchants happy to buy the precious commodity without declaring it to the port authorities. The lack of any subsequent references to it in the official correspondence would suggest this is what happened, depriving the settlers and investors of some of the cash needed to pay for the promised supply mission.

The *Susan Constant* took two weeks to make the final leg of the journey to London. Sir Walter Cope was waiting impatiently for her, eager to get his hands on the 'sperm', as he called the ore sample, which he hoped would fertilize England's new American possession.

In a letter of 12 August, reporting the ship's arrival to Cecil, Cope's excitement was uncontainable: 'If we may believe either in words or Letters, we are fallen upon a land that promises more than the Land of promise: instead of milk we find pearl, & gold instead of honey.' He acknowledged that Cecil might treat such claims with 'slow belief'. There was, after all, 'but a barrel full of the earth' to show for this first mission. But he hoped that tests to be conducted that very day would reveal 'a kingdom full of the ore'.[3]

'I could wish your Lordship at the trial,' Cope continued. Cecil's 'word and presence may comfort the poor citizen of London'. Many had evidently refused requests to continue backing the venture, despite an attempt the previous March to enlarge the Royal Council and thereby increase the investors' representation.[4] In Cope's opinion, they might be persuaded to 'adventure much more in this most hopeful discovery', but would need 'a little help' from Cecil, probably meaning further reform to the council and the company's unusual structure. There was no shortage of money, he noted. Sir Thomas Smythe (who, Cope added in a marginal note, would benefit from 'a word of thanks' from Cecil 'for his care and diligence'), had recently persuaded 'fifty citizens' to offer £500 apiece for a share in an East India Company expedition to the Far East.[5]

By the following day, Cope's mood was very different. Four trials of Newport's soil samples, conducted in various laboratories around the city, had produced not so much as a grain of gold. 'In the end, all turned to vapour.' There had been suspicions the previous day that John Martin, who had tested the soil in Jamestown, had not done so properly. Now, Martin was accused of having 'cousined' (fooled) not only Newport, but the King, the State and his own father – the latter, Cope suggested, in a desperate attempt to persuade the mean old plutocrat to send some private supplies, 'which otherwise he doubted never to procure'.[6] Newport at this stage appears to have given up hope, and announced that he did not intend to return to Virginia.

By mid-August, things were looking a little more hopeful. On 17 August, Sir Thomas Smythe, braiding every strand of influence in the hope of keeping the project together, informed Cecil that Newport was back on board. The captain now claimed he must have brought back the wrong sample. He pledged to lead the supply mission back across the Atlantic, 'never to see your Lordship before he bring that with him which he confidently believed he had brought before'.[7]

Others remained sceptical. A week later, the diplomat Dudley Carleton, one of Cecil's protégés, sent his regular correspondent John Chamberlain a downbeat assessment of the settlement's prospects. 'They write much commendations of the air and the soil and the commodities of it, but silver and gold have they none,' he wrote, 'and they cannot yet be at peace with the inhabitants of the country.'

He seemed well-informed, having somehow managed to see a copy of a letter from George Percy smuggled to his brother the Earl of Northumberland, who was still languishing in the Tower for his involvement in the Gunpowder Plot. 'They have fortified themselves and built a small town which they call Jamestown,' Carleton added, 'and so they date their letters.' He then proceeded to produce a series of convoluted puns ridiculing the settlers' efforts. Never mind a supply from England, they could expect a 'double supply' from the Spanish – a mission sent to wipe them out. 'The town methinks hath no graceful name,' he mused, pointing out that it was not only dangerously close to Spanish Florida, but came 'too near Villiaco', meaning villainy. George Percy, he noted, called the town 'James-fort, which we like best of all the rest, because it comes near to Chemesford', Chelmsford, a town in Essex which one Puritan inhabitant described as a 'dunghill of abomination'.

Carleton then added a postscript about Captain George Weymouth. Weymouth had apparently been 'taken' the week before, 'shipping himself for Spain, with intent as is thought to have betrayed his friends and showed the Spaniards a means how to defeat this Virginian attempt'.[8]

Carleton, who was usually well informed on political matters, was obviously unaware that Weymouth had been 'shipping himself for Spain' on Cecil's orders. However, he was correct in guessing that the mission somehow threatened 'to defeat this Virginian attempt', for as soon as Newport arrived back in England, the government made strenuous efforts to prevent Weymouth from leaving the country.[9] For the sake of his cordial relations with Spain, Cecil had evidently been poised to abandon England's claim to North America, and only the belated appearance of Newport's ship had won a stay of execution.

Don Pedro de Zuñiga, the Spanish ambassador, had remained poorly informed about the Virginia mission since the issuing of the patent. He was aware that some sort of venture was under way, but felt it was too insignificant to bother his King. News of Newport's return seemed to reinforce his complacency. 'They do not come too contented,' he told the King in a monthly report, 'for in that place there is nothing other than good wood for masts, pine-tree pitch and resin, and some earth from which they think they can extract bronze'.[10]

A month later, however, Zuñiga's tone was transformed. 'They are mad about the location,' he reported to the Spanish King, 'and frightened to death that Your Majesty will throw them out.' Preparations were in hand for a supply mission, and 'many here and in other parts of the kingdom . . . are already arranging to send people there'. 'It would be very advisable for Your Majesty to root out this noxious plant while it is so easy,' Zuñiga concluded.

Zuñiga had undergone such a radical change of mind since recruiting one of the members of the Royal Council for Virginia as an informer. The spy's identity is unknown, but a candidate is Sir Herbert Crofts, who had joined the council in March 1607. Crofts, probably a kinsman of the planter Richard Crofts, is known to have become disenchanted with James's regime, and a decade later fled to the Spanish Netherlands, where he 'turned popish'.[11]

In the following month's dispatch, Zuñiga reported that ships were ready to take a new supply to the colony, together with one hundred and twenty fresh settlers. He once again urged Philip to preempt the enterprise by launching an attack. 'It is thoroughly evident that it is not their desire to people [the land], but rather to practise piracy, for they take no women – only men.' He also reported that James was 'urging the Scots to go there'.[12]

Philip commanded Zuñiga to speak to King James about the matter, 'expressing regret on my part, that he should permit any of his subjects to try and disturb the seas, coasts, and lands of the Indies'. Zuñiga duly asked Cecil for a royal audience, but for several months, his requests were ignored.[13]

Then, on the night of Saturday 26 September, an invitation from the Royal Chamberlain unexpectedly arrived at the ambassador's residence in Highgate. Having been indisposed for a week, following the death of his infant daughter Mary, His Royal Highness now felt able to grant Zuñiga an audience at 2 p.m. the following day. The venue was to be Hampton Court, the magnificent royal retreat on the bank of the Thames west of London.

Zuñiga arrived at the palace at the appointed time, and was shown into the State Apartments, where James was waiting for him. He was welcomed 'as usual, very courteously'. The ambassador passed on condolences on the death of the King's daughter, for which James thanked him 'very much'.

The formalities dispensed with, Zuñiga got straight to the point. King Philip, he said, considered it 'against good friendship and brotherliness' for English subjects 'to dare to want to settle Virginia, since it is a part of the Indies belonging to Castile, and that this boldness could have inconvenient results'. 'Inconvenient' was the diplomatic word for deadly. Spain, he was suggesting, was prepared to go to war over the issue.

James affected insouciance, claiming 'he was not informed as to the details of what was going on, so far as the voyages to Virginia were concerned'. He had 'never known' Philip 'had a right to it [Virginia], for it was a region very far from where the Spaniards had settled'. Nevertheless, he did not want the matter to become a point of contention between him and Spain. The settlers, he said, 'went at their own risk, and if they were caught there, there could be no complaint if they were punished'. In other words, if Spain chose to attack the colony, there would be no English reprisals.

Zuñiga was not satisfied; at least, not according to the version of the conversation he reported to Philip. The whole enterprise was, the ambassador told James, a 'shabby deceit, for the land is very sterile, and consequently there can be no other object in that place than that it seems good for piracy'. James was inclined to agree. He too had heard 'that the land was unproductive, and that those who thought to find great riches there were deceived'. He also thought that the sort of people who had gone there were 'terrible'.

There followed a discussion on other matters, including Irish affairs and the indiscipline of James's English Parliament, a body which they agreed had become a '*poblacho*' (dump). Zuñiga then prepared to take his leave. Thanking the King for his time, he urged him again to find a remedy 'for the Virginia affair' as soon as possible. James said he would 'look into the matter', and promised that the Privy Council would give Zuñiga 'satisfaction'.[14]

Over the following few days, Zuñiga sent several messages to the Privy Council at 'Moptoncurt' (his Spaniolated rendering of Hampton Court), urging decisive action. Cecil eventually agreed on a meeting to finalize matters. Having studied the relevant treaties, Cecil (according to Zuñiga) agreed that the English 'cannot go to Virginia'. Furthermore, he accepted that 'if something bad happens to them, let it be their fault'.

With typical inscrutability, Cecil simultaneously told Smythe he wanted 'a speedy dispatch of the ship intended to be sent to Virginia'.[15] This instruction, together with Newport's decision to rejoin the expedition, energized frantic efforts to equip and man two ships hired for the purpose, the *John & Francis*, and the *Phoenix*, both 250 tons.

As always, the expedition was desperately underfunded. To provide the settlers with cloth to keep them warm through the winter, Sir Thomas Smythe was reduced to buying moth-eaten samples left over from a previous East India Company voyage.[16]

By the beginning of October 1607, Newport was ready to sail. Smythe had managed to gather more than one hundred new settlers, including six tailors or cloth workers to look for opportunities for textile manufacturing, two apothecaries to find medicinal ingredients such as herbs, rare earths and animal parts, two goldsmiths and two refiners, to ensure more reliable tests of ore samples. A physician, a surgeon, a tobacco-pipe maker, a perfumer and a white greyhound made up the complement.[17]

The two ships sailed from the port of Gravesend, east of London, on 8 October. They stopped off at Plymouth to await favourable winds, and two weeks later set out into the Atlantic.[18]

EIGHT

Bloody Flux

IN THE LIST OF SETTLERS who went to Virginia on the first voyage, William White is described as a 'labourer'. This did not necessarily mean he was humbly born, as the term was used in passenger lists to identify those brought at another's expense, such as sons brought by their fathers, or servants brought by their masters.[1]

However, nearly all the labourers have one feature in common: very little is known about them. The only trace of biographical information relating to William White is a marginal note by George Percy describing him as a 'made man', a phrase used to refer to someone of low social standing who has unexpectedly come into a fortune, an *arriviste* or *parvenu*.[2]

Percy's snooty assessment of this social inferior probably arose from White's decision, soon after Newport's fleet had first sailed up the James, to jump ship and join the Indians. He was one of several 'renegades' who understandably found the prospect of life with the Indians preferable to the dangers and depredations of the English camp.[3]

The renegades were rarely mentioned in official records or letters home. Their very existence was not officially recognized until 1612, when a regulation was introduced making it a capital offence to live without permission in the town of 'any savage *weroance*'.[4] The conspiracy of silence arose out of a combination of envy and resentment. When they slipped from the settlement, the renegades entered into the feverish imaginations of those left behind. They became voices from beyond the veil of trees and vines, taunting the settlers for putting up with such hardships, ridiculing English assumptions of cultural and technological superiority, beckoning their former comrades to drop their tools, and wander through the woods into the alluring Indian embrace.

White had somehow fallen into that embrace. He ended up ten miles upstream of Jamestown, at the town of Quiyoughcohannock, visited by the English during their first reconnaissance of the river. It sat on a high bluff overlooking the James (modern Claremont, Virginia). Its name meant 'priest of the river', and it served as the religious centre for those living along the banks. This may have been why its people were so curious about the English and their religion, and why they welcomed the nervous, bewildered interloper who had come among them.[5]

White would have been placed in the care of one of the senior women of the village, probably Oholasc, the stately queen of the Quiyoughcohannock, the most 'handsome a savage woman as I have seen', an English admirer noted. She was raising a son named Tatacoope, whose father was believed to be the *mamanatowick* Powhatan. Tatacoope was heir to the village chief, Pepiscuminah, or 'Pipisco', as the English called him. But Pipisco was in disgrace, having 'stolen away' one of Opechancanough's wives. For this impertinence, he had been exiled to a nearby settlement, 'with some few people about him', including the offending woman. As a result, Chaopock, who was Pipisco's brother from the neighbouring village of Chawopo, had been made acting *weroance* until Tatacoope came of age.[6]

Life for White in the court of this great queen would have been comfortable, at least compared to the conditions being endured at Jamestown. Just being part of a rhythm of daily life was a relief, particularly among such positive people. White noted that before dawn, the men and women, together with children older than ten years, would leave their beds and run down to the river. There they would bathe and wash, and await the sun's arrival sitting on the bank, within a circle of dried tobacco leaves, such was their delight in the coming of a new day.[7]

After these daily ablutions, mothers would set out the breakfast, and summon their youngest boys to come to them with their toy bows and arrows. Only when they had managed to shoot a piece of moss tossed into the air could they tuck into a meal typically comprising boiled beans, fresh fish and corn bread, garnished with venison if the hunting was good. Older boys then went off with their fathers, to learn how to weave weirs and nets to catch fish, or to stalk game and butcher carcasses.[8]

And so one day passed into another, until White began to notice this idyll's fragilities. In recent years, hunting had been bad. For reasons that mystified the Indians, the deer population in the vast woodlands surrounding the river was dwindling. As supply continued to shrink, demand was becoming desperate, if only to fulfil an annual tribute demanded by the *mamanatowick* Powhatan. Adding to the pressure was concern about a continuing lack of rain. The village store of corn and beans was running low, but the plants in the fields were not yet a foot high, suffering from the second year of a drought more extreme than even the elders could recall.[9]

The Quiyoughcohannock *weroance* Chaopock was also concerned about the amount of loot rival chiefs across the river were extracting from the starving English. The blue beads, copper and glittering minerals being offered in an ever greater multitude were valuable as symbols of chiefly power, and good substitutes for venison as tributes to offer Powhatan. The circumstances that had led to Chaopock's promotion to chief made him sensitive about matters of status, and going to the English with his begging basket would be demeaning. But, with so many of his fellow *weroances* proving so greedy for the interlopers' wares, Chaopock felt he had no choice but to go downriver to see what he could get. He went sometime in the summer, and returned a few days later, showing off a bright red waistcoat presented to him by the English *weroance*, Wingfield.[10]

Then the English were forgotten, as the village became preoccupied with internal matters. White had no idea what was going on, but frantic preparations signalled an important and rare event. The first sign that it was taking place was the appearance of community leaders in heavy make-up. 'The people were so painted that a painter with his pencil could not have done better,' White noted. 'Some of them were black like devils, with horns and loose hair, some of divers colours.'

One day, White awoke to find the town deserted. Probing the surrounding woodland, he discovered the entire community congregating in a clearing, preparing for some grand event. There followed two days of frantic dancing, intensifying the mood of anticipation. The horned satyrs, carrying tree branches in their hands, danced in a quarter-of-a-mile circle around the village fire, one group moving in the opposite direction to the other, both emitting a 'hellish noise' when they met. The branches were then thrown to the ground, and the satyrs 'ran clapping their hands into a tree', from which they would tear another branch. Anyone who lagged behind was beaten by Chaopock's personal guard with a 'bastinado' or cudgel made of tightly packed reeds. 'Thus they made themselves scarce able to go or stand.'

On the third day, fourteen strong boys aged ten to fifteen years, painted white from head to foot, were led into the village's central arena. For the rest of the morning, the adults danced around them, shaking rattles. In the afternoon, women arrived with dry wood, mats, skins and moss. White, with a growing sense of foreboding, noted that they had also brought funerary goods used in the preparation of corpses. The women then began a terrible wailing.

The boys were led to the foot of a tree, where they were made to sit, watched by warriors brandishing bastinados. Presently, a guard formed up into two lines facing each other, creating a path leading from where the boys were sitting to another tree. Five young men dressed as priests were allowed to fetch the boys one by one, and lead them along the path. As they went, the guards subjected them to a hail of blows, forcing the priests to shield the boys with their naked bodies, 'to their great smart'. Once all fourteen boys had been transferred from the foot of one tree to another, this ritualized abduction was repeated, then repeated a third time. Finally, the pummelling ceased, and the guards tore down the last tree under which the boys had been seated, and decorated themselves with its dismembered branches and twigs.

White could not make out what had happened to the boys in the midst of the mêlée, and it was at this point in the proceedings that he was asked to leave. Later, however, he glimpsed them again, somewhere near the torn tree. They were 'cast on a heap in a valley as dead'. His shock was compounded by Chief Chaopock, standing 'in the midst' of them, summoning his warriors to bring wood to build a great pyre, 'set like a steeple'. White was convinced that these were preliminaries to sacrificing the children 'to the devil, whom they call Kewase [Okeus], who, as they report, sucks their blood'.[11]

Later that day, everyone returned to their houses, as if nothing had happened. Okeus's work with the Quiyoughcohannock was evidently done, and the frenzied, gruesome mood that had gripped the town was allowed to subside. The smoke of the smouldering town fire carried the wanton spirit away from the houses, through the trees and out on to the river, where it turned towards the bay, following the smell of fresh blood.

On 6 August 1609, a Jamestown settler named John Asbie succumbed to 'the bloody flux'. His death marked the start of a month of mortality in the English fort.

Sickness and desertion were already rife. Supplies were all but spent. There was nothing left to drink, and in the absence of a working well, the men were forced to use the river, 'which was at a flood very salt, at a low tide full of slime and filth, which was the destruction of many of our men'.[12]

Rations from the 'common kettle' were 'half a pint of wheat and as much barley boiled with water for a man a day, and this having fried some 26 weeks in the ship's hold'. This put pressure on other sources of sustenance, including Wingfield's flock of chickens. Only three survived to peck at the hard ground between the frail tents, when they were not being chased by ravenous labourers.[13]

Responsibility for this sorry state of affairs has traditionally been laid at the feet of the English class system. Unlike the hard-working, clean-living Pilgrim Fathers who arrived in 1620, the 'gentlemen' who made up such a large proportion of Jamestown's population have been portrayed as work-shy fops and dandies who 'were used to social strata but not to discipline', and who preferred the 'narcissistic contemplation of heredity' to getting their hands dirty.[14]

In fact, most of the gentlemen of Jamestown had military backgrounds, and, while some hated the idea of heavy manual labour, they by no means saw themselves as due a life of ease.

Captain John Smith was himself a gentleman, and proud to be called so, advertising in his autobiography that the title had been endorsed by no less a figure than the King of Poland.[15] Most of the others similarly classified had less heredity to contemplate than even Smith. He, at least, had a 'competent means' from his father and the patronage of a prominent member of the English nobility. Many had lived hard lives on the battlefields of the Low Countries, France and Ireland, or on the edge of destitution. Some were quite poor, the family of one military captain, for example, boasting an estate worth just twenty shillings.[16] Perhaps one or two were feckless opportunists who preferred to rely on their charm and wits rather than honest toil to make a living. But their behaviour was a consequence of the relative lack of social privilege, not an excess of it. Similarly, they had not crossed the Atlantic and put themselves in the predicament in which they now found themselves because life at home had been so easy, but because it had been so hard.

Smith's peers, particularly the squabbling council members, may have contributed to the settlement's first crisis, but even he acknowledged that the damage they caused was collateral. The origin of Jamestown's woes, at least in these early stages of settlement, lay in the simple lack of food. Calculations for provisioning the expedition had inevitably erred on the side of stinginess, with enough to cover a two-month crossing, and six more months in America while a planting of crops ripened, the fort was built, and trade relations with the Indians were established. According to this timetable, a harvest would be ready by the time the supplies started to run out, which could supplement, and ultimately replace, food bought from the Indians.[17] Unfortunately, the crossing had taken not two months but five. This triggered a cascade of problems: by the time they arrived, the planting season had been all but missed, so there was no hope of a proper harvest in August; the crews of the *Susan Constant* and the *Godspeed*, who were due to return to England, had run out of their own supplies, so had to be fed out of the settlement's store while Newport reconnoitreed the river. This left the settlers fewer than twelve weeks to acclimatize to their new surroundings and become self-sufficient.[18]

Both at the time and since, critics have wondered why they could not live off Virginia's natural resources, given the abundance apparently at their disposal. Smith, Archer, Percy and countless successors boasted about the fat vines and fruitful trees, the nuts and berries, the deer running through the forest and the rodents scampering in the undergrowth, the sturgeon in the river and the oysters on the shore. Could that not feed a hundred or so men, in the middle of summer? The Indians' own experience demonstrates otherwise. Hunting, fishing and scavenging, which they performed with astonishing efficiency, provided only a third of their daily needs, averaged out over the year; the other two thirds coming from their staple crop of cultivated corn and beans. The English, being in an alien environment, were bound to rely on a staple crop to provide a much higher proportion of their diet – four fifths at least, which meant finding between one and two tons of corn every week, just to survive.[19]

Back in June, there had been hopes that the Indians would make up the deficit. A messenger had arrived at the fort, claiming to be an emissary of the 'Great Powhatan'. The *mamanatowick* wanted peace with the English, and 'desired greatly' their friendship. Furthermore, he had commanded all attacks on the fort to cease, so that the English 'should sow and reap in peace'. If any Indians ignored this command, he would, he pledged, 'make wars upon them with us'. 'This message fell out true,' Wingfield later observed, 'for both those *weroances* have ever since remained in peace and trade with us.' It was soon after this that Wingfield had been approached by the Quiyoughcohannock chief Chaopock, promising peace and offering food supplies in the autumn, for which Wingfield offered him his red waistcoat, knowing English clothes to be among the possessions most prized by the Indians.

However, peace with the Indians did not yield the hoped-for influx of food. Skirmishes continued with the Paspahegh, and a high level of mutual suspicion and misunderstanding made trading difficult with more distant towns.

By August, each man was receiving fewer than five hundred calories in his daily ration, and the result was a lethal combination of rampant disease and seething discontent. Rumours began to spread that some of the gentlemen, in particular the suspiciously healthy Wingfield, were hoarding private supplies. There was talk of the council hiding a stash of oil and alcoholic drink, and of the councillors' 'privates' or favourites getting preferential treatment. One of the gentlemen, Jehu (or John) Robinson, was accused by Wingfield of plotting to seize the shallop and escape with associates to Newfoundland.

As the death toll began to rise, it became clear that social status provided little protection. Percy kept a doleful muster of the dead for those summer months, and it makes solemn reading: John Asbie, the first casualty, was followed two days later by George Flower, gentleman, who died 'of the swelling'. Next was William Brewster, 'of a wound given by the savages', who was buried on 11 August.

August 14 was one of the worst days. Francis Midwinter, gentleman, and Edward Morris, 'corporal', died 'suddenly', and Jerome Alicock, the settlement's standard-bearer, was dispatched by a 'wound'. A skeleton with a bullet-shattered shin was dug up at Jamestown nearly four hundred years later, raising speculation that it may have been Alicock. Another candidate is the following day's casualty: Stephen Calthorp, the man who had been accused of instigating the 'intended and confessed mutiny' involving Captain Smith.[20]

Over the following three days, four more settlers died, including John Martin, son of councilman John Martin senior. Martin senior had been confined to his tent for some time, his various ailments making him too weak to leave his sickbed. But bitterness over the death of his child stirred him into action, and he accused Wingfield of 'defrauding' his son of the rations he needed to survive.

A day later, Drew Pickhouse, an impoverished former Sussex MP, followed the others into a makeshift grave. He had hoped to find good fortune abroad, having lost his estate at home following a legal dispute with a local aristocrat. He had left behind a wife and eleven children, presumably in rented accommodation. They would not learn of his and their fate for another eleven months.[21]

A two-day respite did not prepare the settlers for the most distressing death of all: that of Captain Bartholomew Gosnold on 22 August, after three weeks of illness. Just before Newport had taken his leave of the settlement to return to England, President Wingfield had confided that he feared only two rivals taking over the presidency, that 'ambitious spirit' Archer, who 'would if he could', and Gosnold, who was so 'strong with friends and followers' he 'could if he would'. Now he could not, depriving the settlers of the one man who seemed to command the confidence and respect that might lead them out of their troubles and miseries.

The next day, flocks of birds burst into the air as the boom of Jamestown's huge culverins saluted Gosnold's burial. He was interred with full military honours in a grave next to the river. A week later, Thomas Studley, the 'cape merchant' responsible for trade with the Indians, followed him.

The gloom lifted briefly when a 'boy', one of the 'renegades', was returned by the Paspahegh chief, with 'the first assurance of his peace with us'. But it was not enough to prevent the council's long-expected disintegration.

Under the stress of malnutrition, surviving council members began to argue violently with one another, as the conviction formed that there was a saboteur in their midst. All the suspicions and sectarianism that they had thought they had left behind them, suddenly and violently erupted in the midst of their misery.

Kendall was the first to be accused. It somehow emerged that he was a spy who had consorted with Sir William Stanley, a renegade English captain who had switched to the Spanish side in the Low Countries campaign. Kendall's remonstrations that he had been working undercover for Cecil, and revelations that there were others on the council who were doing the same, were not enough to save him from being arrested for such 'heinous matters'.[22]

Kendall's incarceration failed to produce an improvement in conditions, and the hungry eye of suspicion began to glance around the remaining council members. It came to rest on the plump form of President Wingfield, who seemed to be surviving the privations of recent weeks suspiciously well. His regular refusal to attend Robert Hunt's alfresco services also raised doubts about his religious loyalties. The other three active members of the council, John Ratcliffe, John Smith and John Martin, voted for him to be removed from the council. His replacement as president, Ratcliffe, then arrested him for a list of crimes against the colony carefully tabulated by Gabriel Archer, who was still smarting at his own exclusion from the council. Wingfield was swiftly tried, and confined to the pinnace following the inevitable sentence of guilt.

But with each allegation, the paranoia intensified rather than diminished, and now a wave of rumours about Ratcliffe spread through the camp. Old questions about his unexpected selection as a commander of the fleet, and nomination to the council, began to take on a new significance. Who was this man who had so skilfully manoeuvred himself into such a powerful position? Where did he come from? Was he the Ratcliffe who, like Kendall, had acted as a spy in the Low Countries, penetrating, or perhaps being a member of, a powerful Catholic cell? Or was he the Ratcliffe who had been imprisoned in the Tower alongside Guy Fawkes following the Gunpowder Plot? Or the Ratcliffe who was a close friend of Cecil's cryptographer and secretary Richard Percival?[23]

Kendall, who in the chaos of the camp had manage to get access to other members of the council, had his own scurrilous answers to such questions, and by sharing them, hardened speculation that it was Ratcliffe who, all along, was the source of their sorrows.

A group of men, Percy and Smith among them, began to agitate for Ratcliffe's removal. They proposed that the blacksmith James Read, who had access to the pinnace to maintain the metal fittings, approach Wingfield to see if he would back a plot to restore him to the presidency.

Ratcliffe learned of these intrigues, and gave Read a public thrashing for his 'misdemeanour'. The smith, a qualified craftsman rather than a manual worker, considered someone of his status deserved more respectful treatment, and 'offered' to strike the president with his sledgehammer in return.

Once again, the rotting canvas of the settlement's tents became the walls of a makeshift courtroom, and the stump of a tree the judge's bench, as Read was tried for mutiny. He was found guilty and sentenced to death.

Permanent gallows were among the many fixtures of a settled community that the company still lacked, so a rope was tied to the branch of a tree and a ladder propped against the trunk, to be kicked away once the noose was around the neck of the condemned. When it came to carrying out the sentence, the blacksmith was naturally 'very obstinate', and put up a fight. Finally, he was forced up the ladder, where 'he saw no other way but death with him', and became 'penitent'. He begged for a word with Ratcliffe regarding a private matter.

Ratcliffe granted his request, and the smith revealed to him details of Kendall's involvement in a plot to restore Wingfield. Ratcliffe granted Read a pardon, and, for reasons yet to be disclosed to the rest of the company, ordered Kendall's immediate arrest and confinement aboard the pinnace, alongside Wingfield.

In the midst of this turmoil, Smith replaced the late Thomas Studley as cape merchant. In his view, this put him in charge of dealing with the Indians, and eager to escape the broils tearing the company apart, he set off on a number of expeditions up the James, to trade for fresh supplies. Even Wingfield was impressed with the captain's vigour and dedication to the task, which at this time of direst need 'relieved the colony well'.

Smith faced formidable obstacles. Few men were in a fit state to accompany him, they were all inadequately equipped, none was proficient in the local language, and Smith lacked the skills of a mariner to sail the shallop. But, ever willing to confront overwhelming difficulties with confident resolve, he set off downriver, heading for Kecoughtan, at the mouth of the James.

His reception by the people of Kecoughtan was very different from the one received by Newport when the fleet first arrived. As his boat approached the shore, Smith claimed he was 'scorned' like a 'famished man', the Indians offering him 'in derision . . . a handful of corn, a piece of bread for their swords and muskets, and suchlike proportions also for their apparel'. In typically boisterous style, Smith drove the boat on to the beach, and ordered his men to 'let fly' their muskets, even though he knew this was 'contrary to his commission,' as set out in the Royal Council's instructions. The Kecoughtans melted away into the woods.

Smith headed towards the village, passing what he claimed to be 'great heaps of corn', recently harvested from the surrounding fields. Then he heard a 'most hideous noise'. Sixty or seventy villagers, 'some black, some red, some white, some parti-coloured, came in a square order, singing and dancing out of the woods'. They carried before them what Smith saw as a diabolical doll 'which was an idol made of skins stuffed with moss, all painted and hung with chains and copper'. Smith thought it was Okeus, the most powerful god of the Powhatan pantheon, who another English observer noted could look 'into all men's actions and, examining the same according to the severe scale of justice, punisheth them with sicknesses, beats them, and strikes their ripe corn with blastings, storms, and thunderclaps, stirs up war, and makes their women false unto them'.[24] If it was Okeus, his appearance in such a manner, before the Otasantasuwak, the wearer of leg coverings, was unprecedented. This god, the English were later told, had prophesied their coming to Virginia, and his appearance now must have been designed to stage a momentous confrontation: to frighten the invaders off, perhaps, or possibly the opposite: to lure them in, integrate them into the Powhatan world, to see what havoc they would wreak.

Smith at this moment had little interest in spiritual speculations, and ordered his men to attack the oncoming parade 'with their muskets loaden with pistol shot' until 'down fell their god, and divers lay sprawling on the ground'. Smith snatched the idol, and the Indians disappeared into the woods. Presently, a priest approached offering peace for the return of the *okee*. Smith told them if six of them came unarmed and loaded his boat, he would 'not only be their friend, but restore them their *okee*, and give them beads, copper'. This the Kecoughtans did, according to Smith, loading his boat with venison, turkeys, wildfowl and corn, while 'singing and dancing in sign of friendship'.[25]

Smith set off back for Jamestown, congratulating himself that his more robust approach to Indian relations was already paying dividends. En route, he stopped off at Warraskoyack, a few miles upstream of Kecoughtan, on the opposite bank of the river. There he managed to extract some more corn, a total, he claimed, of thirty bushels, getting on for a ton.

Back at the fort, he distributed the corn, but found that the thanks and appreciation he felt he deserved were muffled by the ravenous stuffing of mouths.

In any case, this was no long-term solution. Thirty bushels of corn would feed forty or so men for four or five days. With barely two weeks' worth of food left of the store brought from England, a more drastic solution was called for.

According to Smith, Ratcliffe suggested that he take the pinnace back to England 'to procure a supply'. Memories were still fresh of Ratcliffe's proposal that the fleet return home even as it was on the threshold of Chesapeake Bay, and to Smith this new idea suggested some darker design, perhaps to deprive the colony of its only means of escape. After the inevitable bout of violent argument, it was agreed that instead the pinnace and shallop should be taken upriver to the falls, in the hope that sufficient supplies could be extracted, violently or otherwise, from the Indian villages sitting among those fertile lands. 'Lots were cast' to decide who would command the expedition, and the lots somehow contrived to make the obvious selection of Smith.

The mariners set about rigging the pinnace, an operation that would take a few days, as the ship's masts and sails had been stowed in the fort to prevent it being taken. Meanwhile, Smith continued his search for food. He set off in the shallop for Quiyoughcohannock. He found the village abandoned, except for 'certain women and children who fled from their houses'. 'Corn they had plenty,' Smith observed, but he had no commission to 'spoil' or loot, so he left the village unmolested. On the return journey to Jamestown, he visited Paspahegh. With the English settlement now so firmly entrenched on their land, relations with these people were still bad. Smith described them as 'churlish and treacherous', and accused them of trying to steal English weapons as he traded for ten bushels of corn.

The pinnace was now ready for the expedition to the falls, and, arranging to rendezvous with the ship by the next tide at Point Weanock, Smith set off in the shallop to explore the Chickahominy, the tributary of the James.

He left Jamestown on the morning of 9 November, and reached Paspahegh, at the confluence of the James and Chickahominy, that afternoon. The tide was low, so the captain and his company of eight

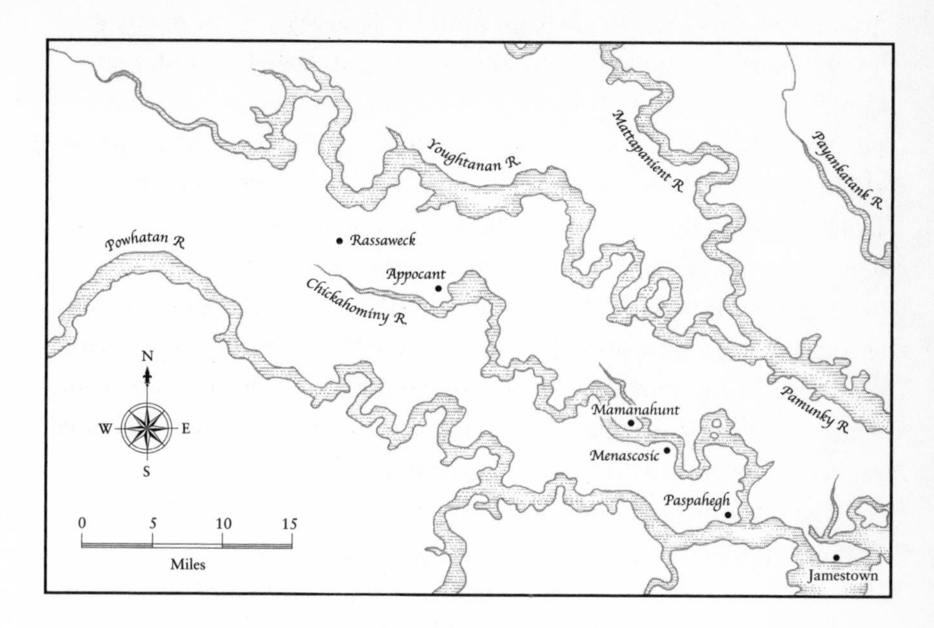

Smith's exploration of the Chickahominy River,
9–10 November 1607.

or so men waited at the Indian village. As evening approached, an Indian from one of the villages along the Chickahominy came to Paspahegh and offered to guide the English up the river. The Paspaheghans 'grudged thereat'. Smith, observing an opportunity to snub his ungrateful hosts, accepted the invitation. By the light of the moon, he took the shallop up the Chickahominy, reaching Menascosic, his guide's village, by midnight. 'The next morning,' Smith records, 'I went up to the town and showed them what copper and hatchets they should have for corn, each family seeking to give me most content.'

According to Smith, the people of Menascosic would have sold him all the corn he wanted, but 'lest they should perceive my too great want', he refused further offers, and continued upriver looking for other people to trade with, passing along the way a grove of plane trees 'watered with many springs', and 'a great marsh of 4 or 5 miles circuit, divided in 2 islands by the parting of the river, abounding with fish and fowl of all sorts'. Further on he discovered a series of villages, 'at each place kindly used, especially at the last', which was Mamanahunt, 'being the heart of the country, where were assembled 200 people with such abundance of corn as having laded our barge as also I might have laded a ship'. Smith triumphantly set off back to Jamestown, his sojourn vindicated by seven hogsheads of food, equivalent to nearly fifty bushels, at least a week's supply.

The shallop arrived at Jamestown in the middle of the night. As it slipped through the water of the river towards the flickering beacon of the fort's night watch, Smith noticed something odd. The pinnace, which should have used the high tide to sail upstream for the intended rendezvous, was marooned on a sandbank near the fort.[26]

The sun rose the following morning upon a settlement once more in the throes of mutiny. Having managed to 'strengthen' himself with the ship's crew, Kendall had hijacked the pinnace, and set sail for Spain, in order to reveal to King Philip 'all about this country and many plans of the English which he knew'. He was somehow prevented, incompetent navigation or the crew's intervention driving the ship on to the mud.

Kendall was removed from the pinnace, tried for mutiny, and sentenced to death by firing squad. In a desperate attempt to escape his punishment, he revealed what only a former Cecil agent would know: that the president, in whose name the sentence was passed, was not called Ratcliffe. He had been operating under an alias all the time. His real name was John Sicklemore.

This revelation added yet a further layer of mystery to this heavily laminated individual. Sicklemore was a name much rarer than Ratcliffe, and he may have been forced to abandon it following some crime or indiscretion. In the State papers of the time, the only Sicklemore of note was a Catholic priest operating under the alias John Ward. As part of the Gunpowder Plot investigation, he was discovered to be conducting secret Masses in a series of households in Northumberland – curiously including a family named Ratcliffe. But that Sicklemore was thought to have escaped to the Continent.[27]

Could a further change of identity somehow have transformed a papist agitator into a colonial adventurer?

Speculation was pointless. The gravity of Kendall's crimes made Ratcliffe's name change seem a mere technicality, and Gabriel Archer, who had now emerged as the president's most loyal lieutenant, had the legal training and natural cunning to circumvent pseudonymity. Under the Royal Council's instructions, Ratcliffe could delegate his judicial powers to fellow councilior John Martin, whose name was his own. So it was Martin who condemned Kendall to death, and a few days later, the prisoner was led out of the fort and shot.[28]

These events once more threw the council into disarray, prompting further efforts to reinstate Wingfield, led by a group of 'best sort of the gentlemen'. Wingfield refused to countenance the idea whilst Ratcliffe and Archer were still at large, and when efforts to arraign them failed, he attempted to commandeer the pinnace so he could sail to England and 'acquaint our [Royal] Council there with our weakness'. Smith claims to have stopped him with rounds of musket and cannon fire, which forced him to 'stay or sink in the river'.[29]

This all happened during November 1607. The onset of winter, which was colder than the English had expected after such a hot summer, brought an unexpected bounty of food. 'The rivers became so covered with swans, geese, ducks, and cranes that we daily feasted with good bread, Virginia peas, pumpions, and putchamins, fish, fowl, and divers sorts of wild beasts as fat as we could eat them,' reported Smith. It was such a cornucopia even the 'Tuftaffaty humourists' Martin, Ratcliffe, Wingfield and Percy, lost interest in returning to England.

To take advantage of the sudden increase, and perhaps because the pinnace was not currently serviceable, Smith decided upon one further expedition up the Chickahominy in the shallop, this time with the intention of reaching its source. Smith still harboured hopes of discovering a navigable way to the South Sea, and wondered whether the Chickahominy might bypass the geological obstacle that blocked the James at the falls.

He left on 10 December, taking with him Thomas Emry, Jehu Robinson, George Casson and three or four others.

Sometime in late November or early December, as the warmth of autumn subsided into a bitterly cold winter, William White found himself in the midst of another outburst of activity, as the Quiyoughcohannock started packing mats, hides, weapons and supplies in preparation for an expedition.

All the men, including White, and several of the women, set off with their luggage to the river, where they loaded up a fleet of long canoes. They were joined on the bank by the deposed Quiyoughcohannock Chief Pipisco, together with his 'best-beloved', the wife he had 'stolen' from Opechancanough. Under the relaxed terms of Pipisco's exile, the couple was allowed to travel 'in hunting time'.[30]

The entire troupe boarded the canoes, and set off down the James. Arriving at Paspahegh, they turned up the Chickahominy tributary, heading north-west, continuing forty or so miles upstream, the waterway becoming almost impassable beneath a canopy of low branches and fallen trees. Eventually, as the river course approached a series of cataracts, the fleet drew up on the northern bank and disembarked. Close by, in a woodland clearing, they set up a large encampment, comprising forty or so tents made of sapling branches covered with mats. They named the camp 'Rassaweck', which was to be their base for a series of hunting expeditions, the deer being more plentiful in this remote area of the forest, at the hem of the foothills leading into the great western mountain range.

They were joined by people from villages near and far, and by Opechancanough. White observed the man dismissed by the English as a pompous fool being received at the camp as a great general, attended by twenty guards in the finest garb, brandishing swords made of bone edged with slivers of precious rock.

Conditions in the camp were challenging. The cold was 'extreme sharp', and a freezing north-west wind whipped through the makeshift dwellings. The Indians seemed unaffected, claiming that their red body paint, made from the root *pocone* mixed with oil, made them impervious (this, not their natural complexion, was the reason they became known as 'red' Indians). To an Englishman, still clad in the summer clothes he had been wearing when he absconded, or wearing the scant Indian garments he had been forced to adopt since, thoughts might have started to stray to the warmth of a winter coat and even a settler's cabin.[31]

A week into December 1607, the mood in the camp suddenly became agitated. After a flurry of activity around Opechancanough's tent, a group of warriors entered the camp, dragging with them a captive Englishman. White recognized him as George Casson, one of three Cassons, probably brothers, who had accompanied him on the journey from England.

The man White now beheld was not the one who just a year ago had stood alongside him on the dockside at Deptford, awaiting departure to the New World. He was undernourished, badly injured and terrified for his life. For some reason, he had aroused fury in his captors. They violently jostled him towards the campfire, where Opechancanough awaited him.

White was an obvious candidate to act as interpreter in the subsequent interrogation. Casson, it transpired, had been caught at Appocant, a village lower down the Chickahominy. The vessel used by the English for exploration had been found anchored there, with Casson left on board to guard it. He had been 'enticed' ashore by some women from the village, and then captured.[32] The reaction of his captors suggested he might have attacked or even raped one of the women.

Opechancanough was already aware that one of the English captains had entered his territory. This Otasantasuwak or one-who-wears-trousers had come into the Powhatan heartlands not with the stealth of an Indian, but like a lumbering pachyderm, crashing through the trees, scattering flocks of birds and herds of woodland creatures before him. The same man had already engaged in several raids up the James and Chickahominy for food, Quiyoughcohannock being among the targets.[33]

Casson had little more to add, other than pathetic appeals for clemency, and the name of the captain: John Smith. This information extracted, he was stripped of his clothes until he stood naked before the gathering assembly of men, women and children, his front frozen by the winter chill, his back heated by the crackling fire. Two wooden stakes were driven into the ground either side of him, to which his ankles and wrists were bound.

Opechancanough continued to interrogate the terrified captive about this captain's intentions. A warrior or priest then approached, brandishing mussel shells and reeds. Using the edges of the shells as blades, and the reeds as cheese-wires, the executioner systematically set about cutting through the flesh and sinews of Casson's joints, stretched out between the staves. As each of his limbs was removed, it was cast upon the fire, until only his head and trunk were left, writhing helplessly on the blood-soaked ground.

Turning the torso over, so Casson faced the ground, the executioner carefully cut a slit around the neck, then slipped a mussel shell beneath the skin. He proceeded to ease off the scalp, and, turning the body back over again, gently unpeeled Casson's face from the skull. He then slit open Casson's abdomen, and pulled out his stomach and bowels, which steamed in the cold winter air. Casson's remains then joined the rest of his body to burn on the fire, until only his dried bones were left, which, according to White, were gathered up and deposited in a 'by-room' in one of the tents.[34]

The punishment of being hanged, drawn and quartered was well-known in Europe, being reserved as retribution for the worst crimes against the monarch. The Indian equivalent was, if anything, more refined in its brutal theatricality. It was not the usual form of capital punishment (murderers were beaten to death with sticks, thieves knocked on the head), and there are only two instances of its use recorded in Virginia. It may have been reserved for foreigners, or for particular crimes, or, in the case of Casson, to be witnessed by an outsider – a lurid warning to share with his countrymen when, in terror, he ran back to them.[35]

White trekked or canoed the difficult twenty-five miles from Rassaweck to Jamestown, and staggered into the fort to find not the refuge he would have hoped for, but a midden of disease and destitution. The council was too weak to discipline the young man. Instead he was debriefed, in the hope that his experience of life with the Indians would reveal their motives and intentions. The terrifying finale of Casson's torture and execution demonstrated that such hopes were in vain.

Soon after, members of Smith's party somehow made their way back to the fort with news that they had lost contact not only with several members of their company, but Smith himself, who had disappeared further upriver in a canoe. A search party was sent soon after to discover Smith's fate, and returned with the corpses of Emry and Robinson, the latter found with as many as thirty arrows in his body. Of Smith, there was no sign. The settlers must have feared – and some of them hoped – for the worst.

At around the same time, Ratcliffe appointed Gabriel Archer to the council. He did this without John Martin's consent, and despite howls of protest coming from Wingfield's cabin in the pinnace.

The winter had by now set in with a vengeance, the onset, research has subsequently revealed, of a 'Little Ice Age' that plunged the northern hemisphere into one of the coldest spells for centuries. In London, one of the Thames's rare 'frost fairs' was in progress, the river freezing over so firmly that innumerable booths were soon to be found 'standing upon the ice, as fruitsellers, victuallers, that sold beer and wine, shoe makers and a barber's tent'.[36] Virginia was on the same latitude as southern Spain, and the English had assumed the climate would be similarly Mediterranean. Now they were learning otherwise, and as they endured the bitter night frosts in their drafty tents, and snow flurries during the day, most assumed Smith must have died of exposure, if he had not been tortured and killed by the Indians.[37]

One particularly cold day, when the ground was encrusted by a thick hoar frost and snow danced in the air, three lightly clad warriors stepped nimbly out of the forest and approached the fort. Guards chased them off, but as night fell, they returned, and were discovered to be carrying a piece of paper torn from a notebook. It bore a message from Smith.

The note revealed that Opechancanough, the chief Archer had dismissed as a fool at 'Pamunkey's Palace', had taken him prisoner. Smith wrote that he had been offered 'life, liberty, land and women' if he provided information on Jamestown's defensive weaknesses, which must mean that a full-scale attack on the fort was being planned. He ordered the soldiers to let off some of the field artillery, to demonstrate to the messengers the strength of English arms.

The message also listed a series of articles which the messengers were to take back with them. They had been told before their departure what had been asked for, so were intrigued to see these very items produced, as though Smith 'could either divine or the paper could speak'.[38]

As Ratcliffe, Martin, Percy and the others watched the messengers run back into the forest, the camp was seized by a sense of common purpose. A thorough review of defences was ordered. A gruelling roster was drawn up for the watch, each man bearing arms to serve once every three days, watching all night 'lying on the bare cold ground, what weather soever came', and all the following day, 'which brought our men to be most feeble wretches'. Smith's expeditions up the Chickahominy, one or two undertaken by Martin, and the abundance of fish and waterfowl had eased the food situation, allowing some further reinforcement of the fort's defences, as well as the construction of a few huts and common buildings.

On 2 January 1608, a company of warriors came out from the woods, two carrying baskets, another a coat and a knapsack, another walking alongside an Englishman. Even in the gloomy dawn of a January morning, guards could tell from the stocky frame and bushy beard that the Englishman was John Smith, and that the men around him were not captors, but an escort.[39]

Smith marched into the fort and immediately ordered the guard to fetch a millstone and two demi-culverins for the Indians to take home. Puzzled at what on the face of it was a violation of the instruction preventing the Indians having access to English arms, they duly produced the items, and Smith was able to amuse himself as the Indians struggled to lift the stone and five tons of cannon. Once they had given up, he ordered that the culverins be charged and loaded with stones. They were fired into a nearby copse of trees 'loaded with icicles', and the 'ice and branches came so tumbling down that the poor savages ran away half dead with fear'.

Smith's dramatic reappearance provoked a mixed response. 'Each man with the truest signs of joy they could express welcomed me,' was Smith's recollection, 'except Master Archer and some 2 or 3 of his [friends].' Archer, exercising his questionable powers as a councillor, called for Smith's arrest 'for the lives of Robinson and Emry', the two men killed by the Indians while accompanying Smith into the upper reaches of the Chickahominy.[40]

Archer insisted that Smith be tried for the charges 'upon a chapter in Leviticus'. This curious choice of legal device hints that Smith, who had returned full of excited talk of pagan rites, comely queens and Indian embraces, was suspected of some sexual impropriety with his captors, which provided a pretext for avenging the death of his two companions.[41]

Whatever the purpose behind the charges, they resulted in Smith being tried the day of his return. He was pronounced guilty, and in sentencing denied even the dignity of a soldier's death before a firing squad. The rope was once more flung over the tree branch, the ladder once more propped up against the trunk. 'But it pleased God,' wrote Wingfield, 'to send Captain Newport unto us the same evening, to our unspeakable comforts; whose arrival saved Master Smith's life.'

NINE

True Relations

NEWPORT'S 'GOOD TALL SHIP' the *John & Francis* arrived at Jamestown on 2 January 1608. The crossing had taken three months, a stop at Dominica providing a chance to supplement the stores brought from England with a variety of more exotic goods, such as potatoes, bananas and pineapple, plus some parrots, to entertain the settlers, if not to feed them. Though the arrival of so many more mouths to feed must have aroused some apprehension, the belated fulfilment of Newport's promise to return was met with relief and delight, spoiled only by the news that the consort vessel the *Phoenix*, under the mastership of Captain Francis Nelson, had gone missing in fog just 30 or so miles from the mouth of the Chesapeake.[1]

As soon as he was ashore, the admiral set about re-establishing his authority. As well as halting Smith's execution, he released Wingfield from the pinnace, and stopped Archer calling a 'parliament', an assembly of all the settlers, which he may have tried to summon in a final attempt to overthrow Smith. Newport confirmed Smith's membership of the council, but stripped him of the office of cape merchant, that job being given to one of the new arrivals, John Taverner. Smith no doubt resented the demotion, not least because it weakened his claim to being in charge of relations with the Indians.

Another newcomer, Matthew Scrivener, 'a very wise, understanding gentleman', was sworn in as a new councillor. John Martin was put in charge of recovering the correct sample of gold-bearing earth to take back to England. Percy, meanwhile, remained on the sidelines. Despite the council's lack of numbers, he was still discounted from office.[2]

Newport also took swift steps to improve the lot of the long-suffering settlers. He commanded his crew to set about building a 'fair storehouse' for the bounties brought from England, a 'stove' or warm room to provide respite from the cold, and a proper church, no doubt to the delight of the neglected chaplain Hunt.

The relief was short-lived. On 7 January, a fire swept through the fort, which destroyed all but three of the existing buildings, together with the palisades, ammunition, clothes, food supplies and Preacher Hunt's library of books. One of the new settlers, Francis Perkins, 'gentleman', who had arrived with his son, also Francis, 'labourer', was distraught to discover that all their possessions had perished, except for a mattress, which had yet to be unloaded from the ship. Meanwhile, President Ratcliffe was badly hurt in a shooting accident, his pistol blowing up as he attempted to fire it. It 'split his hand', leaving him with injuries that would take months to heal.[3]

After such a setback, the settlers needed a distraction from their woes, and Smith readily supplied it, by telling them the extraordinary 'relation' of his capture by the Indians, and his meeting with the great Powhatan. It was a story that only he could tell. But it would revive their 'dead spirits', he promised, as well as capture imaginations for generations to come.

Captain Smith's Relation of his being Taken Prisoner by the Indians, how they Conjured Him, Powhatan entertained Him, would have slain him, and how his daughter saved his life.[4]

HAVING RISKED his life gathering food for the settlement, and saving his countrymen from starvation, Captain Smith had determined to explore the Chickahominy to its source. So off he had set with a small company of men, following the river along its lazy course. For 40 miles they floated up the wide waterway, which for the most part was a quarter of a mile broad, three and a half fathoms deep, exceedingly oozy, and surrounded by great low marshes and high lands – a vast and empty wilderness.

Passing a place called Appocant, the highest inhabited town, he continued a further 10 miles until his way was blocked by a great tree. Here the river suddenly narrowed to just 8 or 10 foot at high water, 6 or 7 at low. The flow was exceedingly swift, the river bed firm, and the land around for the most part a low plain covered in sandy soil. He was obviously nearing the source, but where were the mountains from which such rivers spring? The captain began to wonder whether it might not flow from the great inland lake mentioned by the Indians.

If he had continued, his critics were bound to accuse him of endangering the shallop, so he resolved to return to Appocant. There he left George Casson in charge of the vessel, and hired a canoe and two Indians to row him upriver.

With his two guides, and two of his company, Thomas Emry and Jehu Robinson, he explored 20 miles further into this wilderness, where the river still kept its depth and breadth,

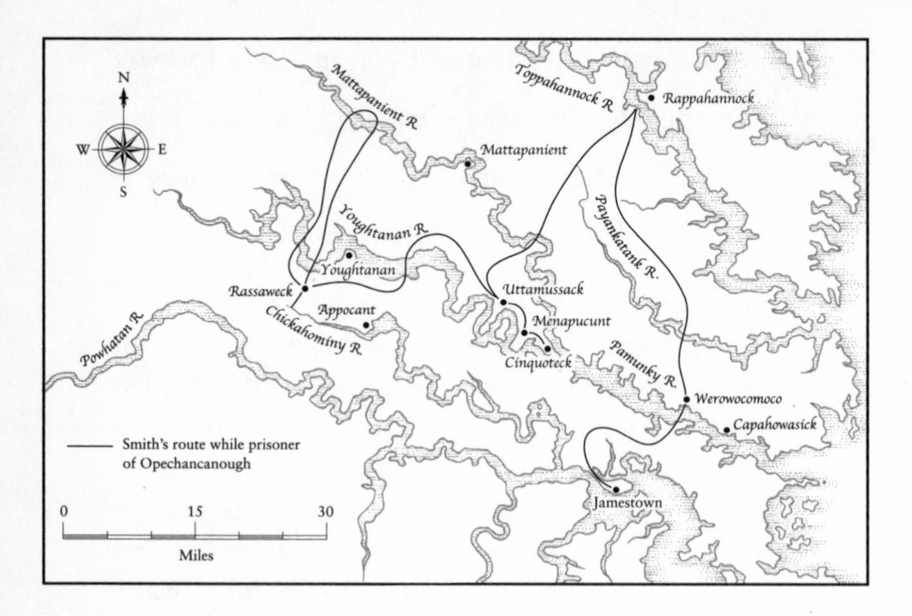

Smith's capture, 10 December 1607–2 January 1608.

but became much more encumbered with trees. Here they all went ashore to refresh themselves. A campfire was lit, and the captain went off to see the nature of the soil, and survey the course of the river. He took with him one Indian. The other he left with Emry and Robinson by the fire, telling them to keep their firearms ready, which they were to discharge at the first sign of any attack.

Within a quarter of an hour, he heard one of the soldiers cry out. Since he had not heard a musket fire, he supposed they had been attacked by their Indian companion. Smith seized his guide, and bound his arm fast to his own. He aimed his fine French pistol at the Indian's head, and demanded to know what was done. The terrified guide told Smith to flee, but as he spoke, an arrow hit the captain in the right thigh. He spied two Indians in the bush drawing their bows, and shot at them with the pistol.

Suddenly, the air was thick with arrows, some passing through Smith's clothes but, miraculously, not his body. Using his guide as a shield, Smith fired his gun three or four more times, while attempting to make his way back to the river. Then he found himself surrounded by two hundred warriors – no, three hundred – with bows drawn and arrows aimed.[5]

Realizing that even he could not overpower such opposition, Smith lowered his weapon. The Indians lowered theirs. Smith told his guide to inform his captors that he was an English captain, and wished to be allowed to return to his boat. His demand was refused. His English companions were already dead, they said, and Smith must surrender. But the captain would not. Keeping the enemy in his sights, and his guide skilfully in his clutches, he tried to edge towards his boat. But he lost his footing and fell into a freezing quagmire.

Smith had no option but to surrender, and threw his gun aside. Still the Indians dared not approach him, but waited until he was near dead of cold. Seizing him, they dragged him from the bog, and escorted him to the campfire, beside which Robinson lay dead, his body bristling with arrows. There was no sign of Emry.

Ever expecting his execution, Smith was nevertheless treated with kindness by the Indians, who allowed him to warm himself before the fire. They even rubbed his limbs to restore feeling to them. Smith demanded to be presented to their leader. He was taken to meet the man he recognized as the King of the Pamunkey, Opechancanough.

Smith realized that, despite his vulnerable position, he must impress his captor with a demonstration of the civilized world from which he came. He drew from his pocket an ivory compass and a notebook. Using signs and a few words, he tried to explain to Opechancanough how the compass worked, but the Indians were mystified by it, marvelling at the elaborate markings and the movement of the needle. They were bewildered that they could see the needle, yet not touch it. Glass was a substance yet unknown to them.

As Opechancanough admired this token of European ingenuity, Smith revealed to him his people's conception of the cosmos. He described how the earth was not flat, but a globe, surrounded by concentric spheres, the innermost carrying the moon, the next the sun, chasing the night around the world continually, the other spheres the planets, which appeared as stars in the sky moving against the fixed backdrop of the zodiac. He tried to explain the vastness of the world, its continents and oceans, how there was a diversity of nations and variety of complexions, and how his homeland, England, lay at the opposite end of the earth. They all stood amazed with admiration at Smith's learned revelations.[6]

Notwithstanding, an hour later, they tied him to a tree. As many warriors as could fit fanned around him, raised their bows, drew back their strings, took their aim, but as they were about to fire, Opechancanough held up the compass, and they laid down their bows.

The warriors then formed up in a line, twenty bowmen at the front, five on each flank and five at the rear, with Smith in the middle, guarded by three great savages on either side, who held him fast by the arms, and carried their bows with arrows ready notched should he attempt to escape.

They walked for nearly 6 miles across the freezing mire until they came to a clearing, where thirty or forty tents were erected, forming a hunting camp the Indians called Rassaweck. Women and children came out from behind the mat coverings and gaped at the prisoner paraded before them.

They brought Smith to the fire at the camp's ceremonial centre, and Opechancanough stood before him. A ring of warriors formed around them, and they started to dance, singing and yelling with hellish notes and screeches, making an exceedingly handsome show. One of the Indian captains then led Smith to one of the principal huts in the camp, which was longer than the others, where thirty or forty tall men stood guard over him. He was brought a haunch of venison and 10 pounds of bread. His stomach was unused to consuming such quantities of rich food, and he could not eat all that he was offered. The remains were placed in baskets, and at midnight, he was offered them again, but he still felt full. The following morning, his guards finished off what he had left, while he was offered three platters of fresh venison and bread, enough to feed ten men – perhaps twenty.

Despite such princely treatment, Smith was desperately cold in his flimsy quarters, and managed to persuade an Indian called Moacassater to fetch him some clothes that had been stored in his canoe, including his winter cape. Opechancanough also restored to him the compass together with his notebook.

And so he continued to feast and to live at ease for several days, the conviction forming that they might be fattening him up to eat him.

Then, a man crept into his quarters with a sword to kill him, and was only prevented by the guards. The man wanted to avenge his son, whom Smith had shot with his pistol during the arrest. Smith offered to examine the victim, who was breathing his last. He had a special water at the fort that might save him, he said, but a request to fetch it was refused, and the boy died.

Smith devised another tactic. He proposed contacting his compatriots to tell them how kindly he was being treated, in case they should launch reprisals. Opechancanough agreed to this, and Smith wrote a message on his notepad, including a list of items the king desired. Three near-naked messengers braved the coldest weather to make the 25-mile journey on foot. They returned three days later with the things Smith had requested, which amazed the king, who did not understand how making marks on the notebook's leaves could carry Smith's commands over such a distance.

A journey now began through the frozen landscape, across two rivers, the Youghtanan, and the Mattapanient, which Smith learned were tributaries of the great river running from the bay north of the James. Smith learned that this river was called the Pamunkey, the same as Opechancanough's nation. They passed two other hunting camps, and houses which Smith was told belonged to the great emperor, Powhatan. Smith said he wanted to meet the emperor, but his requests were ignored. After a four- or five-day march, they presently returned to Rassaweck, where Opechancanough's men were ordered to break camp. They all set off cross-country again, back towards the Youghtanan River, which they crossed in canoes at a place where it was as broad as the Thames at London.

For two days they followed the course of the Youghtanan into the Pamunkey heartlands, until they reached the town of Cinquoteck, Opechancanough's great city. It stood near where the Youghtanan and Mattapanient rivers flowed into the Pamunkey. Smith spent a night in the town, but was woken early the following morning and led into the woods. Presently, he found himself at the foot of a hill of red sand, on top of which stood a long house, a barrel-vaulted structure made in the same manner as the Indians' ordinary dwellings, but nearly 100 foot long. Nearby, perched on lower hills, were two similar but smaller buildings, each about 60 foot long.[7] This was Uttamussack, the holiest place of the Powhatan people.

Smith was led up the hill and taken into the long house. Within, a great fire was burning, and mats were spread on either side. Smith was seated on one of these mats, and his escort left. For some moments he was alone, the fire crackling before him. Then, from out of the darkness sprang a priest, his skin covered in black body paint. Snake skins and weasel pelts stuffed with moss dangled over his face. He let out a hellish shriek, shook a rattle, and danced around the fire, scattering a circle of cornmeal.

Three of this fiend's assistants rushed in carrying rattles. They were painted half red, half black, with white around their eyes and red above their mouths and across their cheeks. As they joined the priest in his frantic dance, another three appeared, painted as the others, but with the colours reversed. Presently they all sat down before Smith in stately array, the priest having three of his imps to one side, three on the other.

After a brief interval, they sprang to their feet and continued their antic dance, shaking their rattles and chanting a song. When the song had ended, the priest laid down five grains of corn. Then he strained his arms and hands with such violence that veins bulged, and he began an oration, of which Smith understood not a word. When it came to an end, his six assistants chorused a short groan, and three more grains were laid down. There followed another song, another oration, and the careful laying down of more grains of corn, until the entire fire was encircled with little heaps. That done, the spirits repeated the sequence, substituting little sticks for the corn.

They continued in this manner without food or drink throughout the day, and as night fell, they feasted. The rite was repeated the following day, and once again the day after that. When Smith asked of its meaning, the Indians told him that the circle of cornmeal signified their world, flat and round, like a platter. They were the fire in the centre. The sticks were the English, entering their world.

When these strange conjurations had come to an end, Smith was taken to the home of Opitchapam, Opechancanough's brother.[8] Where Opechancanough was lusty and warlike, Opitchapam was frail and lame. He invited

Smith into his house, to feast upon platters of bread, fowl and wild beasts. Smith sat down and ate, but those around him refused to join in. Once Smith had finished his meal, the remains were taken away in baskets, and distributed among the king's women and children.[9]

Their ceremonials thus completed, Smith was returned to Menapucunt, where he was welcomed by Kekataugh, another brother, who now became Smith's host. Kekataugh had a high regard for the captain, who had offered kindness when the chief had visited the fort several weeks before. Smith was led along a path to the king's royal residence, atop a high, sandy hill. There he was feasted, and people flocked from all around to meet him, and offer their friendship.

The king took Smith on a tour of the four or five long houses that made up the royal enclosure, and afterwards they walked along a brow of the hill. They looked down upon the hundred houses of Menapucunt, and beyond, the river Youghtanan, snaking through a fertile, flat country stretched as far as the eye could see. A pleasanter place could not be imagined.

Kekataugh had shown the captain his dominion, and now wished for a favour in return. He summoned his guard of forty bowmen, and commanded them to give Smith the pistol taken from him when he was captured. Kekataugh then pointed at a target 120 paces away, and invited Smith to fire at it. Smith tried, but the pistol's cock snapped, arousing much disappointment among his eager audience.

The following day, Smith was marched nearly 30 miles to the north, to the shore of a third river, as wide as the James. On the opposite bank they saw a town, called Rappahannock, or perhaps Toppahannock. This was confusing, as Smith understood Rappahannock to be the name of the village on the southern shore of the James, upriver of the fort, also called Quiyoughcohannock. But he was certain from his compass and the surroundings that this was not the James river.

Kekataugh took Smith across the river to meet the people of this Toppahannock nation. They came out of their houses to look at him. Smith was told that some years past a captain like him had come in a great ship up this river, who had been made welcome. But this captain had slain the Toppahannock king, and the people wanted to know if Smith had performed the deed. They quickly knew it was not him, as the foreigner who had struck down their king had been a large man.

Before departing, Smith learned that the great river upon which the town stood, also called Toppahannock, was home to many more Indian nations, one called Cuttatawomen, which stood where the river joined the bay, and another called Appamattuck. He was also told that the river sprang from many mountains in the far west, in a land that belonged to the Mannahoack people, who were not under Powhatan's dominion.

The following day, Kekataugh led Smith back south, and rested that night in a hunting lodge which Smith learned belonged to Powhatan. The next day, they continued their journey to a destination unknown to Smith, crossing a small river between the Pamunkey and Toppahannock called the Payankatank, passing through many thousands of acres of fertile country, across hills and dales, full of good timber, and in each valley a crystal spring. All these lands were deserted, yet with English ploughs and manuring they would become the richest in the world.

They approached the northern shore of a bay, fed by three creeks. Across a great expanse of ooze stretching a mile and a half, lay the channel of the Pamunkey river. Next to one of the creeks there stood a town, which they called Werowocomoco.[10]

Two hundred warriors were waiting to receive Smith at this town. They seemed shocked by his appearance, and gazed at him, as though he were a monster. Behind them, frantic preparations were under way.

Eventually, a savage dressed in most warlike apparel appeared, and escorted Smith into one of the houses. As Smith's eyes adjusted to the gloom and smoke, he could see a large bedstead, covered with richly decorated mats. Upon it lay a tall, well-proportioned old man, with grey hair and a thinning beard. Pearl necklaces dangled around his neck, and a mantle made of raccoon skins was spread across his legs. On either side of him sat a maid aged sixteen or eighteen, and ranked alongside was a row of men painted red, their heads decorated with the white down of birds, and their necks with chains of white beads. This, Smith learned, was the Great Powhatan.

Until this moment, Smith had believed that the Indian emperor of these rich, fertile domains of Virginia was the chief the English had met at the falls during Newport's reconnaissance of the James. He also learned that the emperor's proper name was not Powhatan, but Wahunsunacock.[11] Smith noticed Opussoquonuske, the comely young queen the English had met at Appamattuck, standing to one side. She approached him, offering a bowl of water to wash his hands, while another woman brought him a bundle of feathers to dry them.

The emperor then spoke some words, which from their manner of expression were welcoming. He gestured servants to bring platters of food of all sorts, and as they were set before Smith, the emperor gave assurances of his friendly intentions, and Smith understood from them that he would be released soon – Powhatan indicated four days.

As they ate, Powhatan asked why the English had come to his country. Smith replied with signs and words that they had been driven to these shores after being defeated in a sea battle against their enemies the Spanish. They had first landed at

Chesapeake where they had been attacked by the local people, and then moved to Kecoughtan, where they had been received in friendlier fashion. They had been forced to sail further upriver because they needed to find somewhere safe to repair their leaking pinnace, and, finding refuge on the island, had built a temporary wooden fort to protect themselves. Soon Smith's father, Captain Newport, would return with supplies, and they would then all return to England.

The emperor demanded to know why, if the stay was temporary, he had been exploring the Chickahominy river in his boat. To discover if there was navigable access to the South Sea, on the other side of the mainland, Smith replied. Also, one of Newport's children had been slain by the Monacans of the mountains, which crime Smith had been sent to avenge.

After some stroking of his sparse and greying whiskers, Powhatan told Smith that he had heard of such a sea deep in the west. He had been told of it by an Indian he had taken prisoner. The salty waters of this sea dashed against the stones and rocks, and, during a storm, spilled over into the river, which became brackish. Some said this sea was five days' march beyond the falls, others that it was eight days. It could only be reached by passing through lands of a mighty people called the Pocoughtronack, a fierce nation of shaven-headed cannibals, who had fought wars against the people of Moyaonce and Potomac, the territories north of Powhatan lands. Another mighty river, which issues into the top of the bay, runs through these lands, fed by the mountains that stand between the two oceans.

Beyond these lands were people dressed in short coats and sleeves to the elbows, Powhatan claimed. They possessed ships like the English, and a land called Anone, abundant with brass, with houses made of stone walls.

As for the slaughter of Newport's son, he knew that the

people who did this were the Atquanahucke, who inhabit the Canada river in the north.[12]

Powhatan delivered this discourse on his great and spacious dominions with such pride, it invited Smith to respond with a description of his world and all its advancements. He told Powhatan of the great King who was his sovereign, and of an innumerable multitude of ships under his royal command. Impersonating trumpet voluntaries and thundering cannons, he demonstrated the military might the English possessed, and its deadly expression in the hands of his one-armed *weroance* Captain Newport, whom others called the King of all the Waters.

No sooner had Smith finished his presentation, than he was seized by as many guards as could hold him, and two great boulders, rarities in these soft and watery lands, were brought into the room. Smith was then dragged to one of the rocks, and the guards held his head over it like an infant's over a font, the sprinkling of holy water to be a deadly rain of blows.

It was in this vulnerable position, expecting death, that he first beheld the peerless beauty of Powhatan's twelve-year-old daughter, Pocahontas. She was a nonpareil, her features and proportions exceeding all the rest of her people.[13] When no entreaties would prevail with the king to save Smith, she ran to the stone upon which Smith's head was about to be dashed. She clasped his head in her arms, and laid her cheek upon his, to save him from death, interposing her countenance between the deadly intentions of her father Powhatan and the welfare of the brave captain.

At this, the emperor suddenly cast aside his savage disposition, and was contented that Smith should live to make hatchets for him, and bells, beads and copper for her, not as a royal slave, nor as tribute of a subordinate to a superior, but as the emperor did himself, as Indian kings, unlike English ones, made their own robes, shoes, bows and arrows.

Two days later, Smith was taken to one of the long houses in the woods, and seated on a mat before a fire, as he had been before. Then, from behind a mat screening off part of the house, came the most doleful cry Smith had ever heard. Powhatan, painted black and dressed more like a devil than a man, appeared, followed by as many as two hundred others. The *mamanatowick* approached Smith, and said that they must be friends. He esteemed him as much as his son, Nantaquoud, he said.

In recognition of the mutual understanding between them, Powhatan then released Smith. The emperor announced that he would make Smith chief of Capahowasick, a territory an hour's march downriver of Werowocomoco. All he asked was that the English surrender Paspahegh, and pay him tribute of copper hatchets and other metal goods. He also asked that when Smith returned to Jamestown, he should send Powhatan two objects: the grindstone the English use to make flour of corn, and two cannons.

The following day Smith set off for Jamestown, accompanied by four, or perhaps twelve men, one carrying his coat and knapsack, two carrying baskets loaded with bread, and another acting as his guard and companion.

Thus Smith not only procured his own liberty and saved the fort from attack, but aroused in the Indians such admiration that they considered him a demi-god.[14]

Reactions to Smith's story were varied. Some were sceptical. Why did both Opechancanough and Powhatan decide to kill the captain one moment, and embrace him the next? Why had Powhatan's daughter so thwarted her father's will to save a stranger's life?

Smith silenced these 'malicious tongues'. The sceptics did not understand, as he now felt he did, the impetuosities of the savage mind, which, as far as he was concerned, was driven more by instinct than reason. In any case, that was not the point. He had uncovered a vast land draped across the banks of three rivers greater than any in England, ruled by an ageing but still powerful emperor, populated by hunters and warriors, comely queens, dancing devils and a beautiful princess. And this, according to Smith's own published account, was what his fellow settlers concluded from the captain's extraordinary story. 'The plenty he had seen . . . the state and bounty of Powhatan . . . especially the love of Pocahontas', this was enough; it 'so revived their dead spirits' that their 'fear was abandoned'.[15]

Newport's response was to set about organizing a suitably impressive embassy to meet this Powhatan, so that the *mamanatowick*'s military and political strength could be gauged, and perhaps a pact for peaceful cohabitation negotiated.

Powhatan, meanwhile, was making his own overtures. His scouts had tracked the *John & Francis* since it entered the Chesapeake, and he had timed Smith's release to coincide with Newport's arrival at Jamestown. He was now anxious to meet this English 'King of all the Waters'.[16] He sent delegations almost daily with supplies of food for the settlers, and gifts of venison and raccoon hides for Newport's personal use, which he could share with his 'son' Smith.

Every four or five days, the supplies would be delivered by Pocahontas herself, surrounded by attendants. Her presence in the fort caused a sensation, not least because this 'well-featured but wanton young girl' would 'get the boys forth with her into the marketplace' and persuade them to perform cartwheels, 'whom she would follow and wheel so herself, naked as she was, all the fort over'.[17]

Smith became resentful of the attention the other settlers were getting from the Indians, and loudly insisted that, as the only Englishman to have met Powhatan, he should control contact with them. He was satisfied to note that his 'feigning' to be under the command of Newport had resulted not only in Newport being recognized by Powhatan as the English *weroance*, and Smith his *quiyoughcosuck* or advisor-cum-priest, but had reduced the status of rival councillors, including President Ratcliffe, to the role of 'children, officers and servants'.

However, Ratcliffe was not prepared to tolerate this demotion. He 'so much envied' Smith's 'estimation among the Savages' that he began to compete for their attention, offering to buy commodities they were bringing to the fort at four times the official rate. To make matters worse, relief at Newport's arrival had lured the settlers into allowing the ship's crew to trade directly with the Indians. The result was rampant inflation, a pound of copper no longer enough to buy what had previously been bought for an ounce.

'Thus ambition, and sufferance, cut the throat of our trade, but confirmed their opinion of Newport's greatness.' That, at least, was Smith's conclusion, his being the only account of events to survive.[18]

The pinnace was now prepared for her first authorized journey since becoming the settlement's impromptu prison ship. Newport, Smith, Scrivener, Bartholomew Gosnold's brother Anthony, Newport's navigator Robert Tyndall, and twenty or so others climbed aboard and set sail for the bay, with the shallop in tow. They also took with them a white greyhound, and a 13-year-old boy, Thomas Savage, both brought on the *John & Francis*. Though listed as a 'labourer', young Thomas may have been the son of a distinguished family from the north of England, which also had connections to Suffolk.[19]

Heading north up the bay, they eventually reached the mouth of the Pamunkey 30 miles north-west of Cape Henry. 'The channel was good', and the ship made quick progress up the river. They passed a promontory they named 'Tyndall's Front' or 'Point', in honour of the good service offered by their pilot Robert Tyndall, and soon afterwards reached the shallow bay last seen by Smith when he was Powhatan's captive.[20]

Newport decided he would let Smith announce his arrival, while he waited in majestic isolation upon his ship.

The captain set off with a landing party of twenty or so heavily armed men in the shallop, and entered one of the creeks. He floated a mile or so into the reedy swamp before realizing he had gone the wrong way.

As if from nowhere, a party of Indians led by Powhatan's son Nantaquoud, and Namontack, a 'trusty servant', appeared on the shore, and beckoned Smith to join them. He accepted the invitation, and the soldiers did their best to disembark from the shallop without losing their boots or dignity in the mud.

Guided along a network of rickety bridges, they were 'kindly conducted' to Werowocomoco, and Powhatan's house, where forty or fifty platters of bread were laid out to greet them.

Smith went alone into the house, and was received by Powhatan, surrounded by a company of 'handsome young women'. The *mamanatowick* greeted the captain from his bedstead with a 'kind countenance', and, gesturing one of his many 'concubines' to shift up, patted a place at his side. Smith, who found the Indians' natural ease with physical intimacy uncomfortable, sat down and presented gifts, including a suit of red cloth, a hat, some jewels and the greyhound. Powhatan 'kindly accepted them, with a public confirmation of a perpetual league and friendship'.

Powhatan asked to meet the others in Smith's company, but his request that they disarm produced a moment's tension. Smith refused, arguing that asking the English to lay down their weapons was 'a ceremony our enemies desired, but never our friends'. He instead promised that 'the next day my Father [Newport] would give him a child of his, in full assurance of our loves', and not only that, but the English would conquer the Monacans, the people who lived beyond the falls, and to the north, the Pocoughtronacks, 'a fierce nation that did eat men'.[21]

'This so contented him, as immediately with attentive silence, with a loud oration he proclaimed me a *weroance* of Powhatan, and that all his subjects should so esteem us, and no man account us strangers, nor Paspaheghans, but Powhatans, and that the corn, women and country, should be to us as to his own people.'

Given Smith's still rudimentary understanding of the Powhatan language, it is unclear whether the generous donation of the fruits of the Paspahegh lands surrounding Jamestown was really intended. Smith was certainly happy to accept that interpretation, and with all the pomp and majesty he could muster, he got to his feet and took his leave.[22]

Back at the bay, he found the shallop marooned on mud, the tide having fallen. His dignity prevented him returning to Powhatan's house. But the landing party could not spend the night on the bank, the skies being 'very thick and rainy'. He was finally forced to accept the offer made by the ever-present Nantaquoud and Namontack of accommodation in a guest house.

Smith and his companions were shown into a lodge 'hung round with bows and arrows', where they settled down for the night, the Indians making fires, offering venison and preparing bedding to make their stay comfortable. A little later, Smith was summoned to come with just two of his guard to the emperor's house. Smith accepted the invitation, posting a guard at the door of the lodge to keep watch while he was away. He spent the rest of the evening engaging in 'many pretty Discourses' with Powhatan, practising the language, and renewing their 'old acquaintance' (even though it was not yet a month old). He was then led back to his lodgings by an escort carrying a torch, and the English enjoyed a peaceful night.

The following day, Newport came ashore in grand procession, a trumpet announcing his arrival. There followed a series of intensive negotiations between the 'King of all the Waters' and the *mamanatowick*, interspersed with dancing, feasting and several moments of tense disagreement, particularly over the matter of the visitors bearing arms. Newport eventually agreed that his men would lay down their weapons as a sign of goodwill, putting the English, Smith grumbled, at a tactical disadvantage.

The rift between the captain and his commander widened over the following days, Smith considering Newport too trusting of the 'political savage' Powhatan, and too dismissive of Smith's own advice.

After three days, Newport returned to his pinnace with general reassurances of goodwill, around 250 bushels of corn, and Namontack, Powhatan's trusty servant, who had been exchanged for Thomas Savage, who would stay at Werowocomoco. Smith was left to follow along in a canoe donated by the Indians as a parting gift. Once again losing his bearings and becoming 'pestered in the ooze', he had to be ignominiously rescued by six or seven of Powhatan's men, who plunged into the mud, and carried him above their heads to safety. Their reward was 'a couple of bells', which Smith was sure 'richly contented them'.

Later that day, the English sailed further upstream to Cinquoteck, Opechancanough's capital, where they were received by the chieftain and his family. After the customary feasting upon bread, beans and venison, Newport set off for Jamestown, leaving Smith with the shallop at nearby Menapucunt, to search for the source of a sample of copper 'the thickness of a shilling' shown to the English by Opechancanough.[23]

Newport reached Jamestown on 9 March 1608, and as soon as he walked into the fort, he went to see if Captain Martin had yet managed to extract gold from the soil samples that had been gathered from around the fort site. The news was grim. Despite having the help of the two refiners brought from London, Martin had found nothing. The news from Smith when he returned in the shallop a few days later was equally discouraging: he had found no evidence of copper deposits.

Before leaving London, Newport had promised Cecil that he would not return until he had found that 'which he confidently believed he had brought before', soil impregnated with precious metals. Now it was looking possible he would have to break his word, throwing the entire venture and his career into jeopardy.

He announced that every able-bodied man was to stop working on building houses or planting corn, and start digging for gold.

Smith was appalled. 'There was no talk, no hope, no work but dig gold, wash gold, refine gold, load gold,' he later complained, 'such a bruit of GOLD that one mad fellow desired to be buried in the sands, lest they should by their art make gold of his bones!' Smith was himself excluded from these 'golden consultations', Captain Martin denying him 'sight of their trials'.

Fields were left unplanted, rivers unexplored. To make matters worse, while Newport remained in Virginia, his ship's crew were tucking into supplies meant for the settlers. They were also selling off surplus luxuries, turning the ship into a 'removing tavern'. 'Never anything did more torment him than to see all necessary business neglected to fraught such a drunken ship with so much gilded dirt,' Smith wrote, referring to himself in the third person.

However, while the rest of the settlers were in the thrall of this gold fever, Smith became preoccupied with a fixation of his own. He sat down and started to pen a long, self-vindicating account of the colony's first year, a document that, to posterity, would prove as valuable as any gold.

The *True Relation of Such Occurrences and Accidents of Note as Hath Happened in Virginia, 1608* was the first of a series of colonial tracts written by Smith. Unlike his other works, it was not intended for publication, and the only version to survive was heavily edited. However, not even the most savage editorial butchery managed to remove the combination of vivid vocabulary and strangled grammar, arrogance and curiosity, intolerance and sympathy that makes Smith's literary style so distinctive. It sets the tone for a body of work that would dominate colonial discourse for centuries to come, bequeathing future historians the challenge of trying to pick out the detail of events cast into the deep shadow of the author's domineering ego.

The 'bruit of gold' is not mentioned in *True Relation*, and, in any case, Newport's efforts came to nothing. On 10 April 1608, 'having furnished him of what he thought good', Newport set sail for England, Smith and Scrivener accompanying him on the shallop as far as Cape Henry. In a much later publication, Smith noted that

Newport took with him Wingfield and Archer, 'we not having any use of Parliaments, plays, petitions, admirals, recorders, interpreters, chronologers, courts of plea, nor Justices of peace'.

With only a few weeks of the planting season left, the settlers, helped by some Indians, set about the long-delayed tasks of cutting lumber and planting corn. But they were soon interrupted. Twelve days after Newport's departure, an alarm was sounded. A ship had been spotted in the bay and was heading up the river. The English grabbed their weapons and loaded the fort's culverins, pointing the heavy barrels downriver. As the vessel approached, fears of attack became cries of relief as it was identified as the *Phoenix*, the ship that had sailed from England with Newport's *John & Francis*. Having come within a few miles of Chesapeake Bay the year before, it had been forced back by bad weather, and retreated to the West Indies for the winter.

Nelson's 'so unexpected coming did so ravish us with exceeding joy,' wrote Smith, 'that now we thought ourselves as well fitted, as our hearts could wish, both with a competent number of men, as also for all other needful provisions, till a further supply should come unto us.' In contrast to other sea captains, Captain Nelson 'had nothing but he freely imparted it, which honest dealing (being a mariner) caused us to admire him'.

'Whereupon the first thing that was concluded, was that my self and Master Scrivener should with seventy men go with the best means we could provide, to discover beyond the Falls, as in our judgments conveniently we might.' Powhatan's descriptions of the cannibals and mighty mountains, of 'brass' mines and stone buildings, had only added to the allure of the lands that lay to the west, beyond the formidable obstacle of the James river waterfalls.

Smith eagerly set about training his troops 'to march, fight, and skirmish in the woods', in preparation for the hostile reception they could expect. However, at the last minute, the expedition was cancelled. According to the *True Relation*, this was due to Nelson insisting that the settlers pay his crew's wages, and cover the costs of hiring of the ship for an extended stay. There was also a 'diversity of contrary opinions' within the fort, several men arguing that such an expedition (like the hunt for gold) would divert valuable resources.

Despite the expedition's cancellation, the *Phoenix* remained at Jamestown for a month, this time awaiting Captain Martin to complete a further series of trials. In the meantime, fifty settlers resumed the much-delayed task of clearing and planting fields. As they were doing so, it became clear that tools, in particular hatchets and 'stones', were going missing, arousing suspicions, as Smith wrote in a particularly confused passage of the *True Relation*, that there was either 'a seditious traitor to our action, or a most unconscionable deceiver of our treasures'.

Several Indians were by now living in the fort, employed by the English to help plant crops. One of them was found by Scrivener trying to steal an axe. He was chased off the island by Scrivener and a few other men. As he ran off, he discarded the axe, and turned on his pursuers, threatening to fire an arrow at any who followed him. A few days later, when Smith and Scrivener were in the fields, two warriors wearing red body paint appeared carrying cudgels, and circled Smith. Smith and Scrivener, unsure of the warriors' intent, retreated to the fort. The two Indians followed them inside, and were joined by two others, who announced they wanted to 'beat' one of their men who was living there, an Indian called Amocis.

Suspecting these were preliminaries to a coordinated attack, Smith ordered that the fort gates be shut, and arrested the four warriors. Soon after, three or four others already in the fort were found 'extraordinarily fitted with arrows, skins, and shooting gloves'. They were also arrested, and all were committed to the pinnace.

The following day, emissaries appeared at the settlement, asking for the release of the prisoners. Smith said this would only happen once all the tools he assumed the Indians had stolen were returned. The emissaries went away, and a little later it emerged that two Englishmen, who had left the fort without permission and were found by the Indians 'ranging in the woods', had been taken hostage.

The Royal Council in London had instructed the settlers to develop good relations with the Indians, and Newport had done all in his power to reinforce this policy. But Smith had argued since returning from Powhatan's custody that such a policy was impossible to pursue, as the 'naturals' were not to be trusted. Confident that the superiority of English military technology would prevail over the Indians' primitive weapons, he felt the only way to secure the land and food supplies was by force. Others were not so sure. In the forests where the Indians lurked, infantry formations, unwieldy muskets and ponderous artillery, which took time to deploy, did not necessarily offer the deadly advantage the English had assumed.

Despite these qualms, the other members of the Jamestown Council, bewildered by what they interpreted as Indian inconsistency, sanctioned reprisals. Smith undertook the commission with relish. At the first opportunity, he took a group of soldiers aboard the shallop up the James, and 'spoiled, and destroyed, what we could'.

The following morning, his actions appeared to be vindicated. The two English hostages were returned, and in response Ratcliffe released one of the Indian prisoners. As for those remaining in English custody, Smith was given permission to 'terrify them with some torture', in the hope of wringing from them a coherent account of Indian intentions.

Smith began by bringing the Indians up to the deck of the pinnace. He picked one out, and ordered him to be tied to the main mast. A squad of soldiers was then told to aim their muskets at the captive, and hold burning tapers near the firing chamber in preparation to shoot. Smith demanded the prisoner reveal Indian strategy. The Indian claimed to know nothing, but eventually it emerged that one of the other captives was a councillor of the chief of the Paspahegh, the local *weroance*. This earned the release of the first, but at the expense of the second. The councillor was seized and Smith 'affrighted' him by forcing him to lie on a makeshift rack, a torture device that had become popular for dealing with religious and political dissidents in

Elizabethan times. He then threatened to shoot him, 'which seeing, he desired me to stay, and he would confess'.

Scrivener was summoned, and to the two captains, the councillor revealed that the Paspahegh and Chickahominy chiefs had been planning to 'surprise' the English while they were at work in the fields, and seize all their tools. In a land where animal bone and bird claws were the sharpest objects available, where even cudgels and knives were made of reeds, stone or metal objects had very high value, and these two local chiefs had systematically set about trying to get hold of as many as possible from the English.

To extract the names of those who had taken part in the conspiracy, Smith ordered that the other prisoners be taken away. A shot of gunfire then sounded out over the James, convincing the councillor that his comrades were being shot, one at a time. Names were duly given, along with a confession that this was all done with Powhatan's knowledge.

Smith's brutal tactics caused some worries in the council. Captain Martin argued that they should avoid alienating Powhatan further, as it could provoke an all-out attack, which the English were in no position to repulse.

Smith swept aside such qualms, his actions justified, in his eyes, by the arrival a few days later of Pocahontas at the fort. She had brought with her Powhatan's most trusty servant, Rawhunt. He was a man 'much exceeding in deformity of person, but of a subtle wit and crafty understanding,' Smith noted – an observation which at the time must have prompted comparisons with King James's most trusty servant, Cecil. Rawhunt addressed Ratcliffe and the council, telling them that the 'injuries' suffered by the English were the result of actions by 'some rash, untoward captains,' and that Powhatan desired 'their liberties' (meaning their indulgence) 'for this time, with the assurance of his love for ever'. As a mark of good faith, he also brought from Opechancanough a shooting glove and a bracer (a binding used to tie a small shield or 'buckler' to the arm), a mark of respect from the Indian warrior to the man identified as his opposite number.

Smith's instinctive belligerence was overcome by Pocahontas's 'wit and spirit', and he willingly accepted these assurances and gifts. Having 'given the prisoners what correction he thought fit', which included a service given by Hunt in Jamestown's makeshift church, he handed them over to Pocahontas, 'for whose sake only he feigned to have saved their lives and gave them liberty'.

Summarizing the episode after Pocahontas's departure, Smith criticized the 'patient council, that nothing would move to war with the savages'. They would gladly have 'wrangled', berated him for his treatment of the prisoners, but he felt his robust actions had brought the Indians 'in such fear and obedience as his very name would sufficiently affright them, where before we had sometime peace and war twice in a day, and very seldom a week but we had some treacherous villainy or other'.

The time approached for Nelson to sail back to England. Smith gave the captain a sheaf of documents, at least some of them addressed to his friend Henry Hudson, the explorer (these inspired Hudson to set off the following year under Dutch colours to explore north of Virginia, where he found the river that now bears his name, and an island the Indians called 'Manhattan'). Smith also handed over a copy of his *True Relation*, perhaps to Robert Hunt, the preacher, whose remarkable absence from any of the records after the Jamestown fire might be explained by his decision to return to England aboard the *Phoenix*.[24]

Ratcliffe also had some letters, including one addressed to his master, Robert Cecil. The document, like so much else concerning the mysterious man who was now Virginia's president, has not survived, but later references suggest that it presented a candid summary of the faction fighting that had caused so much damage in the first months of the venture, laying much of the blame upon Smith. In particular, it mentioned a plan apparently considered by Ratcliffe himself to 'divide the country', to found a separate settlement elsewhere.[25]

Armed with these documents, and a cargo of cedar, Nelson boarded the *Phoenix* in late May to make final preparations for his departure. He moved the ship downriver to Cedar Isle, to collect some more supplies of wood, and was surprised to be joined there by Captain Martin. Martin had finally decided to yield to his many chronic ailments and return home. Smith was happy to see him go, as was Anas Todkill. Todkill had been Martin's 'man', shipped over to America at his patron's expense, but he had soon joined Smith's faction, and shared his new master's lack of sympathy for Martin's endless infirmities. He accused Martin of venturing out of the fort only twice in all the time he had been in Virginia, then only to neighbouring Paspahegh. On each of those occasions, he had found that he was 'forced to return before night . . . lest the dew should distemper him'.

Nelson set sail for England at the beginning of June 1608, just over a month after his arrival. Many of those left behind watched his ship slip away with feelings of regret that they were not accompanying him, and apprehension about what lay ahead. After just over a year of habitation, the fort was still little more than a military encampment of tents and crude wooden structures. Supplies, employment, relations with the Indians – every aspect of life remained haphazard. There were spasms of purposeful activity 'in the extremity of the heat', such as weeding around the fort to deprive the creeping Paspaheghans of cover or working on new presidential quarters commissioned by Ratcliffe, but otherwise most settlers were left to their own devices, enduring periods of listless idleness within the fort's palisade, to conserve energy and escape the relentless, sapping threat of ambush that lay in the woods beyond the walls. This provided the leisure to ask awkward questions: Why are we here? What are we looking for? What are we fighting over? When can we go home? No reassuring answers came from the three factious councillors who had been left in charge: Captains Smith and Scrivener, now allies against the third, the mysterious, prickly President Ratcliffe.[26]

TEN

The Virginian Sea

WHEN EDWARD MARIA WINGFIELD arrived back in England with Newport at the end of May 1608, the pent-up fury of Virginia's displaced president spilled out into a lengthy and valuable 'discourse' setting out his side of the story. Much of his anger was targeted at Smith, for impugning his family's honour. Smith was a mere ruffian. Wingfield had 'proved to his face' that Smith had 'begged in Ireland like a rogue without licence'. The charge was a telling, if curious one.[1]

Ever since Henry VIII had tried to bring Ireland under the control of the English Crown, 'this famous island in the Virginian Sea', as the Atlantic was now called, had confounded its domineering neighbour.[2] Theoretically, it was a kingdom of the English Crown, ruled on the monarch's behalf by an elite made up of the 'Old English', Irish-born descendants of the aristocracy installed in the twelfth century by Henry II, and the 'native' Gaelic chieftains, who in 1541 had supposedly pledged loyalty to the English Crown in return for Anglicized titles and royal protection of their lands. However, great tracts of territory that lay beyond the 'Pale', the Anglo-Irish enclave centred on Dublin, remained outside the reach of English law. When Queen Elizabeth's government tried imposing English government and a Protestant church, it provoked outright rebellion, led by Catholic factions of the Old English in the south (Fitzmaurice's Rebellion) and Gaelic chieftains in the north.[3]

The Englishmen sent in to quell the insurgency were those who launched England's first attempts at American colonization: the explorer Sir Humphrey Gilbert, Sir Walter Raleigh and Sir Richard Grenville, admiral of the Roanoke venture. Their efforts led to some of the most brutal episodes in British military history, Gilbert famously lining the path to his military tent with the heads of the day's enemy casualties, freshly decapitated.[4]

The result of their intervention was mayhem, and Elizabeth had sent a reluctant Sir Henry Sidney to clear up the mess. Faced with jealous infighting among his peers and perplexing stratagems from his enemies, Sidney had proposed a series of novel reforms with far-reaching implications for British colonial practice. One was to abolish the Gaelic tradition of 'coign and livery', a system the chieftains used to maintain private armies at the expense of their clans. He also backed schemes to develop the vast tracts of land confiscated in Ulster and Munster as English 'plantations', as a 'means to reduce [the] Realm of Ireland to civility and obedience'. In combination, he hoped that the unruly dominion could be brought under the control of a 'New English' Protestant elite of up-and-coming gentry, loyal to the Crown.[5]

The Wingfield family had been first in the queue to claim one of these plantations. In 1569, Edward Maria's uncle Jacques had been a member of a group petitioning the Queen to colonize the southern province of Munster.[6] In 1586, Edward Maria himself petitioned with others for the 'seignory' or lordship over three thousand acres, the failure of his application perhaps encouraging him to turn his attention to America as an alternative.[7]

With hindsight, such ventures appear indistinguishable from occupation, using force of arms to throw indigenous people off their land and take it over – a policy not yet officially embraced in America. At the time, however, English actions in Ireland were rationalized, if not justified, by their apologists as part of a noble enterprise. The 'native' Gaelic Irish were trapped in a primitive social and spiritual state, they argued, and it was up to the English 'to bring them from their delight of licentious barbarism unto the love of goodness and civility', which would ultimately result in peace and prosperity.

Sidney's policies were aimed at bringing about this transformation. The abolition of coign and livery would undercut the 'tyrannical' rule of the chieftains, which trapped their enslaved clansmen in the politics of the Dark Ages. Plantations would tame the land and its people, encouraging the introduction of modern methods of cultivation and trade. Unfortunately, according to the great literary champion of this venture, the poet Edmund Spenser, it was 'vain to speak of planting laws, and plotting policy' in Ireland in its current state, because its chiefs were constantly at war with one another. The sword would never leave their hands until they 'are weary of wars and brought down to extreme wretchedness', when they 'creep a little perhaps and sue for grace, till they have gotten new breath and recovered their strength again'. So first, they must be 'altogether subdued'.

This mission was to be undertaken by the English, but not because of a belief in racial superiority. The 'native' Irish were not essentially barbaric, but kept in that condition by their leaders. 'The English were at first as stout and warlike a people as ever the Irish,' Spenser had argued. But the English had escaped their undeveloped state by submitting to a more advanced imperial power: the Romans. They had been forced to endure 'many thorny and hard ways,' 'civil broils' and 'tumultuous rebellions', just as the Irish were now having to do, but they had emerged into such a state of civility 'that no nation in the world excelleth them in all goodly conversation, and all the studies of knowledge and humanity'. Civilization had been their salvation. William Strachey, soon to become an active participant in the Virginia venture, noted in response to later criticisms of English treatment of the Indians, 'had not this violence and this injury been offered unto us by the Romans . . . we might yet have lived overgrown satyrs, rude and untutored, wandering in the woods, dwelling in caves, and hunting for our dinners as the wild beasts in the forests for their prey, prostituting our daughters to strangers, sacrificing our children to our idols'. That was how an island of savages, it was argued, became the land of Shakespeare.[8]

For Spenser, Ireland's submission, like England's, was part of what one leading Protestant scholar had described as 'the very genealogy of the World', the slow, centrifugal spread of civilization from the Middle East to the edge of the world. Spenser dramatized the process in his great epic, *The Faerie Queene*, now treated as a landmark of the civilization he claimed his methods would spread. In the poem, Spenser invents an Arthurian figure called Artegall, the knight of justice, whose quest is to save the fair 'Irena', Ireland. To achieve this, he calls upon Talus, the 'yron man'

> *Who in his hand an iron flail did hold*
> *With which he threshed out falsehood, and did truth unfold.*

Together, Artegall and Talus set about trying to save Irena from

> *. . . the cruel Tyrant, which oppressed*
> *The fair Irena with his foul misdeed*
> *And kept the crown in which she should succeed.*[9]

Thus the brave and chivalrous knight, reduced to a buffoonish Petronel or Falstaff on the Elizabethan stage, was recast as the instrument of a new chivalrous mission, to plant Protestant civilization upon foreign shores.[10]

If, as Wingfield claimed, Smith had been in Ireland, it was most likely in the years leading up to his departure for Virginia. At that time, the Irish rebels were beginning to yield to the English onslaught, and there were many cashiered English captains whose military services were no longer needed, and who were forced to beg for their passage back home.[11]

In one important respect, though, Smith seems to have been touched by Ireland whether or not he actually went there. For him, the colonization of that country clearly acted as a model for America. The Virginia enterprise was the same sort of civilizing mission, and it was up to those who participated in it to 'bridle [the] brute, barbarous, and savage dispositions' of the Indians, just as Spenser argued the Irish needed to be 'brought unto that civility' enjoyed in England.[12]

The writer Robert Burton, a contemporary of Smith's, pointed out that the Germans, considered by many in England at the time as living in the most advanced state of Protestant civilization, 'were once as uncivil as they in Virginia, yet by planting of Colonies & good laws, they became from barbarous outlaws, to be full of rich and populous cities, as now they are, and most flourishing kingdoms'. Smith had gone to America, just as he may have gone to Ireland, to engineer a similar transformation. His period of captivity by Powhatan, and exposure to 'hellish' Indian conjurations, had merely confirmed the necessity of his mission and his destiny to fulfil it.[13]

It was this belief that motivated the expedition he undertook soon after Nelson had departed in the *Phoenix* back to England in early June 1608: his epic 'discovery' of America.

Smith notes with rare precision in his *Generall Historie* that 'the second of June 1608, Smith left the Fort to perform his Discovery'. He had been instructed to undertake the expedition by Newport, who wanted to locate the antimony mine used by the Indians, thought to lie somewhere to the north. Samples of the white powder had been found in the possession of Opechancanough, and tests had shown that it might yield silver. Smith was unconvinced, but he was happy to use the commission as a pretext to embark on an adventure of his own.[14]

Fifteen men were selected for the expedition: a physician, Walter Russell, seven 'gentlemen', five soldiers, and James Read, the blacksmith whose revelations had led to George Kendall's execution. The company also included Jonas Profit, a fisherman, and a fishmonger, Richard Keale, probably taken along in the hope of discovering a new Cape Cod – though, oddly, no fishing equipment was packed.

They all climbed into the cramped 3-ton shallop, and sailed in consort with Nelson's departing *Phoenix* as far as Point Comfort at the mouth of the James. There the two vessels separated, the *Phoenix*

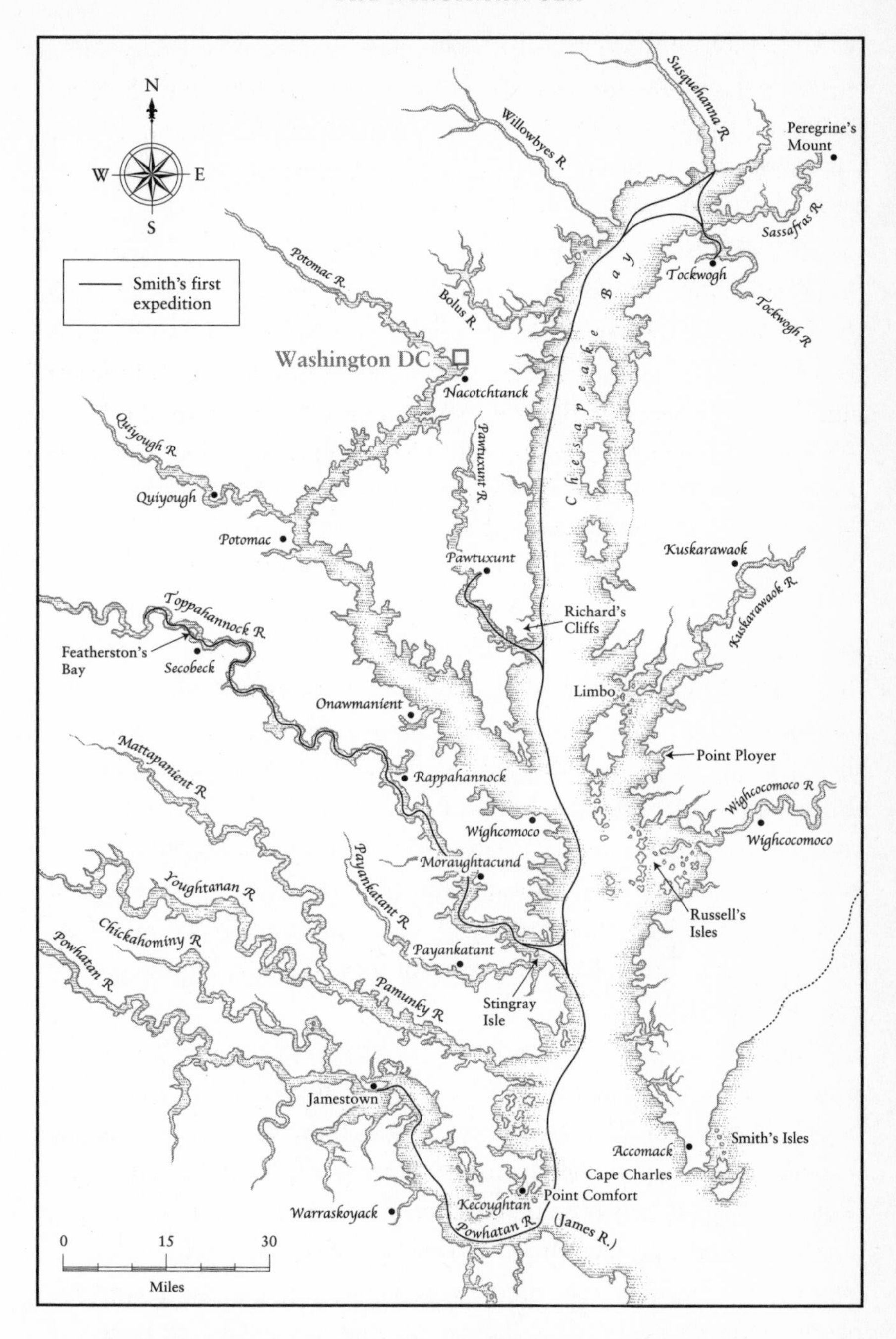

Smith's exploration of the Chesapeake Bay, June–September 1608.

heaving out into the ocean, the shallop crossing the mouth of the bay to the spit of land on its northern lip, now dubbed Cape Charles in honour of Prince Henry's younger brother. Next to it was found a cluster of islands reaching out into the ocean, which were named Smith's Isles, in recognition of the expedition leader.

The explorers landed at Cape Charles, and found themselves immediately confronted by 'two grim and stout savages', with 'long poles like javelins headed with bone'. Speaking the Powhatan language, 'they boldly demanded what we were and what we would'. Smith attempted a reply, whereupon the company was taken to the town of Accomack, Smith noting along the way the pleasantness of the countryside, with 'fertile clay soil, some small creeks, good harbours for small barks but not for ships'.

At Accomack, they were introduced to the chief, who was 'the comeliest proper, civil savage we encountered'. After refreshing them, the chief insisted upon telling his guests a story. It was about two of the village children, who had died of unknown causes. Some 'extreme passions or dreaming visions, fantasies or affection' had prompted their parents to inspect the 'dead carcasses' of their offspring. The expression on their dead faces had taken on 'such pleasant, delightful countenances', the parents believed the corpses had 'regained their vital spirits'. News of the miracle spread, and people came from all over Accomack territory to behold the apparition. Not long after, the Accomack chief told Smith, most of them died, 'being the greater part of his people'.

The chief may have told Smith the story because he had heard from tribes up the James river of English medical skills, and had been told that one of their visitors was a physician. Walter Russell, like many doctors of the time, was famous for a patent medicine, 'Russell's Drops', which he claimed could cure all manner of ailments, and he may have applied them to survivors of this mysterious disease, giving some hope to the unfortunate chief, who had witnessed his people almost wiped out by the poison that had emanated from the shining eyes of those dead children.

Smith returned to the shallop, and the company sailed on up the coast. They were caught in a summer storm, which whipped up 'ocean-like' waves. The flat-bottomed barge was not designed to withstand such conditions, and they put in at a cluster of islands in the middle of the bay, which they named 'Russell's Isles' (modern Smith and South Marsh Islands), perhaps in recognition of the physician's good work with the Indians.

The following day, the weather was gentler, and Smith took the shallop up a river in the hope of finding fresh water. They came to a village called Wighcocomoco (around the location of modern Pocomoke City, Maryland), but the people 'at first with great fury seemed to assault us, yet at last with songs and dances and much mirth became very tractable'. The English searched the town for water, but could only find enough to fill three kegs, and that muddy 'puddle' water.

Setting off in search of another supply, the shallop sailed on north-eastwards up the coast, until they found a large spring of fresh water, too hot to drink straight from the pool. By now, Smith had acquired Archer's fondness for naming features of the landscape, though without the wit. He chose to call the area Point Ployer, as it had saved him from death just as the Comte de Plouër of Brittany had in similar 'extreme extremity once relieved our captain', by taking him in when he was adrift in France.[15]

As Smith slaked his thirst, he noted in his journal that the land they had just passed was made up of narrow islands 2 miles wide and 10 or 12 long, which would be ideal for growing hay in the summer, and fishing in the winter. Beyond, the mainland was 'all covered over with wood', potentially useful as a source of timber, but less suitable as agricultural land. An image of a colonized America was beginning to form in his imagination.

Having refreshed themselves at Point Ployer, the men continued up the bay. Soon, they found themselves in the midst of another storm, this more violent than the last, 'the wind and waters so much increased with thunder, lightning, and rain that our fore-mast and sail blew overboard'. Waves crashed over the sides of the boat, inundating the shallow hold, forcing the men to bail frantically to stop the vessel from sinking, until they managed to find shelter on a small island. They were stuck there for two days by the extremity of gusts, thunder, rain, storms and ill weather, prompting the men to describe the area as 'Limbo', the region of the underworld, according to Christian tradition, that bordered hell.

Repairing their sails with their own shirts, they set sail for the mainland, and arrived at a 'convenient river' called Kuskarawaok (modern Nanticoke river, Maryland). They found themselves surrounded by people 'amazed in troops from place to place, and divers got into the tops of trees'. Arrows were fired, but Smith kept the shallop out of range, 'making all the signs of friendship we could'. The locals arrived the following day unarmed, but 'seeing there was nothing in them but villainy, we discharged a volley of muskets charged with pistol shot, whereat they all lay tumbling on the ground, creeping some one way, some another, into a great cluster of reeds hard by'.

That evening, the English landed, and found among the reeds abandoned baskets and 'much blood'. Seeing smoke coming from the opposite shore of the river, they rowed towards it, and found a cluster of huts, where they left some gifts of copper and beads. The following morning, four warriors approached the shallop in their canoe. The English treated them 'with such courtesy', and they returned later with some twenty companions. Later, others came, then yet more, until Smith estimated that there were 'two or three thousand men, women, and children . . . clustering about us', every one of them offering something to the English which they were happy to sell for a 'little bead'. These were the Nanticoke people, and they became so friendly, they were happy to fetch fresh water supplies, climb aboard the English shallop to act as hostages, 'and give us the best content'. These were 'the best merchants of all other savages', Smith decided. They were also the first to mention a feared nation of Indian warriors that lived to the north 'upon a great salt water', who came down through the bay to attack the towns and villages along the shores, the dreaded Massawomecks.[16] Smith decided to continue north in search of these ferocious people, who might hold the elusive secret of the longed-for navigable route to the Pacific.

Returning to 'Limbo', they made their way across the bay, which from nearly 30 miles at its widest point had narrowed to around 5 miles. As they approached the opposite shore, they saw a feature quite absent from the low-lying terrain around Jamestown and along the eastern shore: 'great high cliffs', which they dubbed Richard's Cliffs, perhaps after the fishmonger among them, in recognition of his London Company's support for the colonizing venture.

They proceeded northwards for a further 60 miles, rowing when their makeshift sail was limp in the wind. The land they passed was 'well watered, but very mountainous and barren, the valleys very fertile, but extreme thick of small wood so well as trees, and much frequented with wolves, bears, deer, and other wild beasts'.[17] They nosed into an inlet, and found the channel deep enough for ships. Along the rocks that lined the shoreline, they noticed just below the high-water mark deposits of a substance coloured white and red. Russell wondered whether this was evidence of clays or 'boles' used in medicine, such as bole armoniak, a red-tinted clay imported from Armenia (hence the name), and terra sigillata, also usually of a reddish colour, imported from Limnos, an island in the Aegean Sea. Both substances were prized for their astringent qualities, and used as antidotes, terra sigillata selling at around ten shillings an ounce in London's apothecary shops.[18] However, further investigation failed to turn up substantial deposits, a blow to morale that almost started a mutiny. Having been confined to the shallop for nearly two weeks, eating bread 'spoiled with wet so much that it was rotten', the men 'with continual complaints' demanded to go home.

In response, Smith stood up on one of the benches of the boat and delivered a speech. Realizing that no motive was likely to arouse their enthusiasm as much as the prospect of finding precious metals, he reminded them of Newport's commission. He quoted almost verbatim from Hakluyt the story of Sir Ralph Lane, a leader of the Roanoke venture, who had heard of a mine yielding a 'marvellous mineral', which was to be found in a land called Chaunis Temoatan, which lay somewhere towards the source of the Moratico river (modern Roanoke river, which runs through North Carolina and Virginia). Finding themselves 160 miles upriver, with just two days' provisions remaining, Lane's company had also grown restive. Their general had given them a chance to vote on whether they should continue or return home, and they chose almost unanimously to continue. If necessary they would eat the dog they had brought with them 'boiled in sassafras leaves', rather than turn back. In the face of much less challenging obstacles, 'what a shame would it be', Smith pronounced, 'for you . . . to force me to return with a month's provision, scarce able to say where we have been, nor yet heard of that we were sent to seek'.[19]

'As for your fears that I will lose myself in these unknown large waters, or be swallowed up in some stormy gust,' he told the men, 'abandon these childish fears, for worse than is passed is not likely to happen, and there is as much danger to return as to proceed.'

His words were no sooner out of his mouth than they were blown away by yet another storm. For two or three days, the shallop heaved around on the churning sea, the crew sheltering beneath a tarpaulin pulled over their heads. These 'adverse extremities added such discouragement to our discontents that three or four fell extreme sick, whose pitiful complaints caused us to return'.

Heading back along the western shore of the bay, they reached another river on 16 June. This was the mighty Potomac, mentioned to Smith by Powhatan as a possible route to a 'great sea'.

By now, the men's spirits had improved with the weather, so it was decided to enter the 'seven-mile broad' mouth of the river that opened up before them.

They came to a village near the river's mouth called Wighcocomoco (the same name as another village and river on the eastern shore of the bay, another of several examples of Indian place names being repeated, which the English found so confusing). The chief was a man called Mosco, a 'lusty savage' who presented an extraordinary sight to the English, sporting a 'thick black bush beard . . . of which he was not a little proud'. Except for the elderly, some priests and Powhatan himself, the Indians Smith had encountered around the bay were clean-shaven, taking great trouble to pluck their whiskers.[20] Mosco's exception, together with his easy manner, prompted Smith and his crew to speculate that he was 'some Frenchman's son'.

He proved to be very friendly. 'Wood and water he would fetch us,' Smith recorded, 'guide us any whither, nay, cause divers of his countrymen to help us tow against wind or tide from place to place.'

The English continued up the river, Mosco apparently accompanying them. About 30 miles upstream they were approached by two warriors in canoes. The shallop followed them 'up a little bayed creek', deep into the woods. Suddenly 'three or four hundred savages' (inflated to three or four thousand in Smith's later account) leapt out from behind the trees 'so strangely painted, grimed, and disguised, shouting, yelling, and crying as so many spirits from hell could not have showed more terrible'.

Smith ordered the soldiers to let off rounds of musket fire, aiming at the surface of the water so the bullets would skim across it. The sound echoing through the woods 'so amazed' the Indians 'as down went their bows and arrows'. Smith records little of the subsequent negotiations, but they resulted in an exchange of hostages, one of his soldiers, James Watkins, being sent 6 miles inland to the town of Onawmanient.

From his subsequent discussions with these people, Smith became convinced that the ambush had been ordered by Powhatan himself, acting on directions from the 'discontents at Jamestown'. Ratcliffe and his henchmen were apparently anxious to avenge Smith for making them 'stay in their country against their wills'.[21]

Smith was beginning to have difficulties imagining events occurring in Virginia in which he was not the principal cause or player, and the belief set in that both his compatriots and the Indians were now in league to get him.

Nevertheless, he continued in the shallop up the Potomac, until reaching the town named after it. There Mosco disembarked, leaving Smith to penetrate deeper inland, until he reached Nacotchtanck, at the navigable limit of the river, on a site overlooking modern Washington DC. They were surrounded by 'mighty rocks' shot through with promising traces of glittering substances. The English began excavating, and recovered a sample of rock or *matchqueon* flecked with 'yellow spangles as if it had been half pin-dust' or metal filings.

Returning back down the river, they stopped at the capital town of Potomac, and showed their discovery to Mosco. He negotiated with the Potomac chief for the English to be taken to a mine containing such material, up a small tributary of the Potomac called the Quiyough (possibly, modern Aquia Creek, Virginia).

They sailed 7 or 8 miles up the creek, until coming to a large rocky hill. The English landed, and found that the Indians had bored 'a great hole' into the rock face using shells and stones. The material they excavated was processed in a nearby stream to 'wash away the dross and keep the remainder', and bagged up to be sold 'all over the country to paint their bodies, faces and idols'. Smith collected as much of the material as his men could carry, and returned to the Potomac chief's house. There, after 'kindly requiting this kind king and all his kind people' with gifts, he left with Mosco.

The shallop headed back towards the bay, Mosco disembarking at Wighcocomoco. Food supplies were now almost exhausted, and Smith, for reasons he does not explain, did not attempt to get more from the Indians.

Continuing southwards in the bay, heading for the James, Smith stopped at the mouth of the Rappahannock river, and managed to strand the shallop on a sandbank. Spotting 'many fishes lurking in the reeds', he impulsively leapt from the boat to catch some. Having come without any fishing equipment – and having failed in an earlier attempt to catch fish with a frying pan – Smith drew his sword and set about trying to skewer his quarry. As he plunged his arm deep into the water, his hand was penetrated by the barbed sting of a ray, which was driven more than an inch into his wrist. 'No blood nor wound was seen, but a little blue spot. But the torment was instantly . . . extreme.' Between bouts of agonizing pain, Smith solemnly instructed the company to dig a grave for him on a nearby island, and say incantations for his imminent demise.

Meanwhile, Dr Russell injected some of his patent potion into the wound using a 'probe'. The medication, if not Smith's natural recuperative powers, proved so effective that by evening the captain was enjoying a hearty fish dinner. His recovery gave the company 'no less joy and content . . . than ease to himself', and the island upon which the empty grave had been prematurely excavated was dubbed 'Stingray Isle' to mark Smith's miraculous deliverance.

'Having neither surgeon nor surgery but that preservative oil' to treat his wound, Smith decided it was time to go home. En route, the shallop called in at Kecoughtan, the Indian town at the mouth of the James. 'The simple savages, seeing our captain hurt and another bloody by breaking his shin, [and] our numbers of bows, arrows, swords, targets [shields], mantles, and furs, would needs imagine we had been at wars.' Smith saw no reason to disabuse his hosts of this idea, and painted in words and actions a vivid picture of an heroic victory over the mysterious and mighty Massawomecks.

Convinced he had fooled the Kecoughtans, Smith set off up the James, arriving at Warraskoyack, about 16 miles upstream, the following day to find the town already abuzz with talk of his ferocious battle.

At Warraskoyack, Smith ordered that the shallop be prepared for the final leg of their journey. The crew of fifteen had covered over 500 miles of uncharted territory, most of it never before explored by Europeans. Along the way, Smith had carefully and systematically recorded the names and locations of the towns and geographical features he had encountered, for the purposes of drawing up what would be the first detailed map of Virginia. The experience had reinforced his belief that it was the settlers' responsibility to 'plant' the untamed terrain, just as others were planting Ulster and Munster as part of the 'reconquering' of Ireland.[22] As for the Indians, they could be easily overcome, as 'at all times we . . . encountered them and curbed their insolencies, they concluded with presents to purchase peace'.

On 21 July, the English set off in the shallop, having decorated it with painted streamers for their triumphant return. The guards at Jamestown's lookout post took the boat to be a Spanish frigate, and the alarm was raised, but lowered as the ragged sail and squat outline of the barge became clearer.

Smith returned to find the settlement once again on the threshold of extinction. The settlers brought by Newport with the supply had all fallen sick, Scrivener included, who was confined to his sickbed by an attack of 'calenture', severe fever. The remainder, 'some lame, some bruised', were in no fit state to do anything except 'complain of the pride and unreasonable needless cruelty of their silly President'. Ratcliffe had wasted vast resources on building his 'unnecessary palace in the woods', reducing the settlement to its customary state of misery. 'Had not we arrived,' Smith noted, he was sure the settlers would have 'strangely tormented' Ratcliffe.

According to Smith, the settlers begged him to 'take upon him the government', but he refused, perhaps mindful that Ratcliffe could not be legally deposed until the end of his term of office in two months' time. Instead he claims to have 'substituted' Ratcliffe with the delirious Scrivener as president, though how this was effected without actual mutiny, he does not say. Smith's account of the episode, the only to survive, is confusing, and made less credible by his decision, at the very moment the settlement apparently needed firm and decisive government, to embark on another exploration almost on the very next tide.

Unlike the first, the second epic 'voyage in discovering the Bay' was a freelance venture, Smith deciding to take it upon himself to track down the elusive Massawomecks, in the hope that they would lead him to the 'great salt water' upon which they lived. He took with him twelve men, eight who had accompanied the first expedition, and four who were new, including Anthony Bagnall, the camp surgeon, to replace Dr Russell.

They set off for the bay, but were forced to stop at Kecoughtan to wait out 'contrary winds'. The shallop entered the bay around 27 July, spent the night at Stingray Isle, and, making fast progress, reached the northern end of the bay the following day.[23]

Smith restlessly probed the inlets and creeks feeding into the bay, but could find no signs of habitation, until suddenly confronted by seven or eight canoes, 'full of Massawomecks'.

By now, all but five members of the crew were struck down by what may have been an infectious disease picked up at Jamestown, and were 'sick almost to death'. To hide his depleted numbers, Smith ordered the sick to stay hidden beneath the boat's tarpaulin. He put their hats on sticks propped against the shallop's deck rail, and put 'betwixt two hats a man with two pieces [guns], to make us seem many'.

The Massawomecks took some persuasion to make contact. Eventually, after much gesturing and calling, a canoe carrying two men cautiously approached, the rest following close behind. The men did not speak Powhatan, and Smith 'understood them nothing at all'. Through signs, he gathered that they had been fighting with a local tribe, the Tockwoghs. They showed Smith their 'green' (unhealed) wounds as evidence. Night fell, bringing the mimed discussion to an end.

The following morning, Smith awoke to find the canoes had gone. 'After that we never saw them.' He sailed on in search of the Tockwoghs. They were already alerted to the interlopers' presence, and in a nearby tributary (the Sassafras river) a fleet of their canoes surrounded the shallop in a 'barbarous' or wary manner. Fortunately,

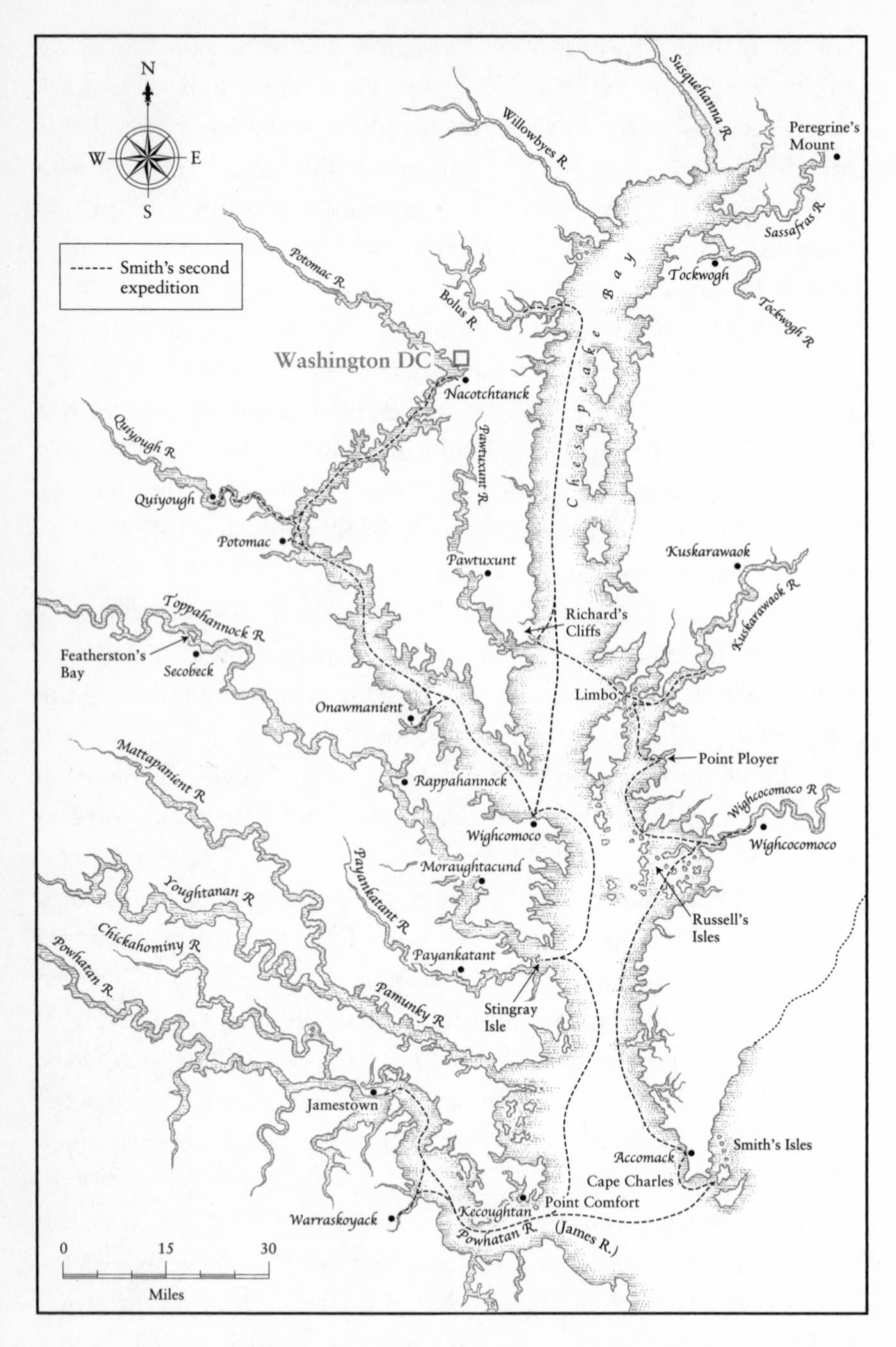

Smith's exploration of the Chesapeake Bay, June–September 1608.

one of their number spoke Powhatan, and through the interpreter Smith negotiated an invitation to their town.

The Tockwogh home was distinctly different from those Smith had encountered along the James and Pamunkey rivers. Where other Indian habitations were scattered through the woodland, the Tockwogh houses were tightly grouped, and surrounded by a palisade or screen made of wooden scaffolds neatly panelled with bark. This was a people whose territory was obviously under regular attack, a view confirmed when the Tockwoghs showed off their weapons, which, to the surprise of the English, included several made of metal. These, Smith was told, had come from the Sasquesahanock, a 'mighty people' who inhabited the main river feeding into the northern end of the bay, a two-day march above its navigable limit. They were 'mortal enemies' of the Massawomecks, Smith was told.

Smith persuaded the Tockwoghs to invite the Sasquesahanock to 'come visit us'. Three or four days later, sixty 'giant-like' warriors duly arrived on the shores of the Susquehanna river, bearing 'venison, tobacco pipes three foot in length, baskets, targets, bows, and arrows'. Smith sailed over in the shallop to greet them, and, despite gale-force winds, persuaded five of their chiefs to board the boat and cross the bay to the town of the Tockwoghs. The rest of the impressive embassy had to stay behind, the wind too strong for a canoe crossing.

The encounter between the English and Sasquesahanock developed into a strange sort of spiritual contest. Smith started to perform daily prayers accompanied by a psalm, 'which solemnity the poor savages much wondered'. The Sasquesahanock responded by demonstrating their own devotions, holding up their hands 'in a most passionate manner' to the sun, then 'in like manner' embracing Smith, deepening his conviction that he was seen by the 'poor savages' as some kind of god.

They placed a 'great painted bear's skin' on Smith's shoulders, and hung a 'great chain of white beads weighing at least six or seven pound' around his neck. They laid 'eighteen mantles made of diverse sorts of skins' together with other 'toys' at the captain's feet, 'stroking their ceremonious hands about his neck for his creation to be their governor and protector, promising their aids, victuals, or what they had to be his, if he would stay with them to defend and revenge them of the Massawomecks'.

All this was performed in a language Smith did not know, which the Tockwoghs had to translate via their own language into Powhatan, so Smith's belief that the Sasquesahanock were offering 'subjection to the English', as he put it, may have arisen from a misunderstanding if not wish-fulfilment. Like the Massawomecks, the Sasquesahanock were an Iroquois-speaking people, one of a large group reaching via the arterial watershed of the Susquehanna and its many tributaries far north, almost to Canada (which was named after the Iroquois word for settlement). The repeated references to 'a great water beyond the mountains' must after all have been an echo, muffled by layers of translation and many hundreds of miles, of the Great Lakes.

Smith also learned from these people that the French, after so many years confined by religious wars to their homeland, had returned to Canada, a sharp reminder that England was not the only European power with ambitions to claim northern America. He did not yet know it, but earlier that very month, Samuel de Champlain had founded the settlement of Quebec on the St Lawrence River. It had been claimed under a patent issued in 1603 by Henri IV of France. The Indians had already been trading with these rival colonists, who as far as Smith knew may have found a navigable route to the Pacific, given the growing evidence that it existed somewhere to the north.

In a possessive flurry, Smith set about naming as many features of the Tockwogh landscape and the surrounding bay as he could. Each of the crew members was awarded a cape, bay or point of land, and a high mountain spotted northward of Tockwogh was called Peregrine's Mount, after Peregrine, Lord Willoughby, Smith's 'most honoured good friend' and patron.

Smith now set off back towards Jamestown, stopping en route to meet the 'very tractable and more civil people' of Pawtuxunt. He also decided to attempt relations with the 'peevish' people who lived on the neighbouring Rappahannock river.

The first town they came to on the Rappahannock was Moraughtacund, where the English were delighted to find themselves 'kindly entertained'. Smith was also pleased to see his 'old friend', the 'very lusty savage' Mosco, who had come to trade with locals. Mosco warned Smith not to approach the Rappahannock, as they would kill him for having befriended the Moraughtacunds. Hostilities had apparently broken out over the 'theft' by the Moraughtacunds of three of the Rappahannock chief's women.

Smith ignored this advice, assuming that Mosco was trying to limit his contact with other groups. He sailed on up to the Rappahannock town on the opposite side of the river, where he found a reception party waiting on the bank. They indicated a nearby creek where the shallop could moor.

Hostages were exchanged, with one of Smith's trustiest soldiers, Anas Todkill, chosen to go ashore. Soon after he had landed, Todkill saw 'two or three hundred men, as he thought, behind the trees', whereupon the Indian held hostage on the shallop jumped overboard. James Watkins, who had been told to guard the hostage, shot him in the water as he tried to reach the shore.

Todkill ran back towards the water, but the air was by now thick with arrows – more than a thousand, Smith estimated – and he was forced to take cover in the grass. Several Indians attempted to catch him, but were driven back by English musket fire. Smith somehow managed to get some of his men ashore. They recovered Todkill, 'who was all bloody by some of them who were shot by us', and collected the huge number of arrows scattered around him. Smith ordered that all the arrows be broken, except for a selection offered to Mosco as a reward, together with the canoe abandoned in the creek.

For the rest of the day, the English busied themselves trying to make the shallop arrow-proof. They pushed sticks into the boat's rowlocks, to which they fastened shields, forming a barrier with gaps for muskets.

The following morning, they continued upriver, Mosco following them on the shore. As they passed a series of villages 'situated upon high white, clay cliffs', the river narrowed to a few hundred yards. Mosco asked to come aboard the shallop. A little further on it became clear why. The shallop passed a line of 'little bushes growing among the sedge'. Suddenly, a shower of arrows started thudding into the shields lining the side of the boat, 'whereat Mosco fell flat in the boat on his face, crying THE RAPPAHANNOCKS'. They had camouflaged themselves using foliage torn from nearby trees.

The English opened fire in reply, and the camouflaged warriors fell back into the sedge. By the time the shallop was half a mile or so further upriver, Smith looked back to see the forty warriors who had just attacked 'showing themselves' on the river bank, 'dancing and singing very merrily' to celebrate seeing off the interlopers.

As the English penetrated deeper inland, the reception became less hostile. Just beyond a village called Secobeck the river broadened, flowing around a series of small islands. There, Richard Featherstone, a gentleman who had arrived from England on Newport's supply ship, and accompanied Smith's previous mission, died, perhaps of wounds sustained during the encounter with the Rappahannock. He was buried on one of the islands 'with a volley of shot', and the section of river was named 'Featherstone's Bay' in his honour.

'The next day we sailed so high as our boat would float, there setting up crosses and graving our names in the trees.' A group spent an hour surveying the surrounding territory, 'digging in the earth, looking of stones, herbs, and springs', when an arrow shot past. About a hundred 'nimble Indians' suddenly appeared, 'skipping from tree to tree, letting fly their arrows so fast as they could'. The English could not load their muskets in time to respond, so Mosco let fly with the arrows presented to him after the battle with the Rappahannock. By the time he had emptied several quivers, the attackers had 'all vanished as suddenly as they approached', except one, who was found sprawling on the ground, an arrow sticking from his knee.

'Never was dog more furious against a bear than Mosco was to have beat out his brains,' according to Smith, referring to the popular London sport of bear-baiting. He had to intervene to stop Mosco killing the man. The captive was carried back to the shallop, where the surgeon, Bagnall, dressed his wounds. 'In the meantime we contented Mosco in helping him to gather up their arrows, which were an armful, whereof he gloried not a little.'

Mosco was then asked to interrogate the captive. His name was discovered to be Amoroleck, and he provided Smith with the most detailed information yet on the Indians who lived outside the Powhatan world. Amoroleck was a Mannahoack, a group living west of the Powhatan domain in the foothills of the mountains. When asked why his people had decided to attack the English, he told Mosco that 'they heard we were a people come from under the world to take their world from them'. Smith asked how many worlds the Indian knew. 'He replied he knew no more but that which was under the sky that covered him.' He spoke of his people and their 'neighbours and friends' the Monacans as living in 'the hilly countries by small rivers, living upon roots and fruits, but chiefly by hunting'. He also knew of the Massawomeck, saying they lived 'upon a great water, and had many boats and so many men that they made war with all the world'.

Amoroleck said he would introduce Smith to his chief, but on Mosco's advice – now being heeded – the English decided to turn back for the bay by cover of night. As they left, they could hear the sound of arrows 'dropping on every side the boat'.

They returned to Moraughtacund, claiming victory over the Mannahoack. From there, Smith sent a message to the Rappahannock chief, threatening to 'burn all their houses, destroy their corn, and forever hold them his enemies', unless he deliver to Smith his bow, arrow and son, and promised to become friends with the Moraughtacund.

The Rappahannock chief replied with an invitation to receive Smith at the place they had previously fought. There he presented Smith with the bow and arrow, and instead of his son the three women he had 'stolen' from the Moraughtacund. Warming to the role of colonial governor, Smith presented bead necklaces to the three women, and then 'causing' the Moraughtacund and Rappahannock chiefs, together with Mosco, to 'stand before him, bid Rappahannock take her he loved best, and Moraughtacund choose next, and to Mosco he gave the third'.

The following day, the English headed back towards the bay, taking Mosco with them. As they reached the mouth of the river, Mosco took his leave, pledging to change his name to 'Uttasantasough', a form of the word the Powhatans used to refer to foreigners, thereby electing himself an honorary Englishman. Promising his people 'ever to be our friends and to plant corn purposely for us', he left the shallop, and the English saluted his departure with a volley of musket fire. Mosco, or Uttasantasough, was never heard of again.

Smith returned to Jamestown on 7 September 1608, to find Scrivener recovered from his illness, but the fort otherwise in disarray, 'many dead, some sick, the late president [Ratcliffe] prisoner for mutiny'.

No details about this 'mutiny' by Ratcliffe have survived, but a later document hints at a far more complex story than is suggested by Smith's terse reference.

Ratcliffe, together with a man called Webbe, had become involved in an episode which resulted in them suffering (as Ratcliffe saw it), 'diverse injuries and insolences'. Webbe is likely to have been Thomas Webbe, 'gentleman', one of the original settlers. His shady background was similar to Kendall's and Ratcliffe's. He had come to government attention working for Lord Burghley, Robert Cecil's father and predecessor as chief minister. In 1591, Burghley had provided Webbe with a safe-conduct letter signed by Queen Elizabeth, so that he could go to Germany and Prague to track down Edward Kelley, a Welsh mystic who had become alchemist to the Holy Roman Emperor, Rudolf II. Burghley was anxious to see if Kelley could be lured back to England, hoping that he would produce gold on behalf of the English Crown. In the event, Webbe's efforts came to nothing, and soon afterwards he found himself facing charges for coining (dealing in counterfeit money), and was sent to the Marshalsea prison.

Later accusations alleged that Webbe, on arrival in Virginia, had tried to engineer the 'utter subversion and ruin of the colony' with one 'Prise' or Price. An intriguing but by no means definite possibility is that Webbe's accomplice was the Captain Price listed in a letter to Cecil as one of those deserving reward for helping to put down the Essex Rebellion. This Price was also connected to the men who reported Captain Gosnold's treasonous talk at an Isle of Wight dinner table to the Privy Council.[24]

However these men link up, the outline of a Cecil subterfuge is just about visible. It is not clear whether he was simply trying to put some demanding and dangerous men out of harm's way, or keep an eye on the settlement's affairs, or even sabotage the venture. Whichever, the imprint of his actions was disastrously disfiguring the government of Jamestown.

Ratcliffe's tenure as president formally expired on 10 September, and 'by the election of the council and request of the company, Captain Smith received the letters patents and took upon him the place of president'. In his own account of what followed, Smith immediately initiated a frenzy of purposeful activity. The 'palace' Ratcliffe was building somewhere in the woods was abandoned 'as a thing needless', the church was spruced up, the storehouse repaired and the palisade rebuilt to form the five-sided design preferred by military architects. Rosters were drawn up, boats were trimmed, Percy was sent off to trade for corn, while the rest of the company was made to perform regular military exercises in an open area next to the fort (punningly dubbed 'Smithfield', the name of London's meat market).

Towards the end of the month, Newport arrived with what became known as the 'second supply'. His ship was the *Mary Margaret*, and its consort the *Starr*. If the *Mary Margaret* was the ship of that name launched in Aldeburgh, Suffolk in 1604, she was capable of carrying

350 tons, making her the largest vessel to be sent by the Virginia Company so far.[25]

Newport had brought seventy more settlers with him, twenty-five of them gentlemen, including Francis West, the 22-year-old brother of the Essex rebel and Privy Councillor Thomas, Lord Delaware.

Another of the gents climbing ashore and surveying the new surroundings was Thomas Forrest. He may have been a relative of George Forrest, brought with the last supply. Just prior to his arrival, a Thomas Forrest was charged in Northumberland as accessory to the murder of one Milao Pearson using 'a lance worth 2s'. There is nothing directly linking the criminal and the settler, though the man accused of harbouring the defendants was a Thomas Errington, at whose family house John Sicklemore the Catholic priest had allegedly preached.[26] Whatever had prompted Thomas to make the crossing, he had the unique privilege of bringing with him a wife, and his wife's fourteen-year-old maid, Anne Burras, making these two the first recorded female settlers at Jamestown.

After the first Englishwomen to be seen for more than a year, the passengers who probably aroused most interest among the veteran settlers were Richard Waldo and Peter Wynn. These two, described by Smith as 'ancient Soldiers, and valiant Gentlemen', had been chosen by the Royal Council in London to join the depleted settlement council. Their credentials for this job were unknown, Smith noting that both were 'ignorant of the business'. Waldo was, and remains, a complete mystery. Captain Wynn, however, was clearly a replacement for Kendall, sent either by Cecil, or, if the charges for which Kendall was shot are to be believed, the Spanish. Like Kendall, he had served in the Low Countries under Sir William Stanley, and had continued to do so after Stanley's defection to the Spanish side, when in return for the promise of a pardon, he agreed to act as a spy for Cecil. His name also turned up in the Gunpowder Plot investigation, appearing in the correspondence of Sir Thomas Chaloner at the same time as another informer, called 'Ratlyff'.[27]

Accompanying Wynn and Waldo were eight Poles and 'Dutchmen' (Germans; only later would the word, a corruption of 'Deutsch man', be used to refer exclusively to people of the Netherlands). These men were brought for their skills in making pitch, tar and soap-ashes (used in the manufacture of soap). Little is known about them, but there is evidence that suggests one of them, Jan Bogdan, had met John Smith when the latter had fought with the armies of the Holy Roman empire.

It has also been suggested that the Poles brought along '*pilka palantowa*', a ball game which bears a striking resemblance to baseball, and which might be responsible for the game's emergence in the United States – though an English version, rounders, provided a model that was probably more readily available.[28]

Finally, there was Namontack. Powhatan's 'trusty servant' had spent two months in London, paraded around as the son of an Indian emperor. He had met King James, and apparently been taken to see the Earl of Northumberland in the Tower of London. Northumberland had given him some cheap copper jewellery, and may well have introduced him to his prison-mate Sir Walter Raleigh.[29]

Along with all those extra mouths to feed, Newport had also brought three sows and some chickens, in the hope that livestock might help relieve the settlement's chronic supply problems. Less welcome were the 'costly novelties' seen piling up on the dockside: including five sections of a huge, heavy new shallop, luxurious furnishings, such as a basin, a water-jug, bedclothes and a large bedstead, a smart new suit of clothes, and a quantity of house-building materials.

Newport's first job on his arrival was to reconstitute the council. Waldo and Wynn were sworn in, while Ratcliffe was released and reinstated.

Smith, as president, put forward a motion to exclude Ratcliffe, but was defeated. The captain now found himself all but powerless to shape the council's deliberations.

Newport read out, or circulated, a letter addressed to the council from the London Company. It contained a catalogue of brutal criticisms. The settlers were hopeless, paralysed by infighting, constantly demanding supplies when they should be self-sufficient, promising infinite wealth while delivering worthless dust. Responding to allegations made in the secret report sent by Ratcliffe to Cecil, they were even accused of trying to keep valuable discoveries to themselves, and planning to split away from the main settlement and set up private plantations.

The letter ended with an ultimatum: either the settlers load Newport's ship with £2,000 worth of goods, in particular 'Pitch and Tar, Wainscot, Clapboard, Glass, and Soap ashes', or the company would withdraw all support, leaving the settlers to their own devices.

Loudest of the outcries of indignation were Smith's, but Newport ignored them. He declared that he had two commissions to fulfil while he was in the country. First, he would launch a major expedition to explore the James river west of the falls, in the hope of finding the reported gold mines of the Appalachian mountains. He had brought the large collapsible barge from England for this purpose. Smith objected. It would be impossible, he argued, to release enough men to undertake a mission of the scale envisaged, while gathering supplies for the settlement and loading the ships with the commodities demanded by the London Company.

Newport swept aside such objections, and revealed the second mission that would be undertaken while he was in Virginia: the coronation of Powhatan. Newport himself would perform the ceremony, using a crown and scarlet cloak (scarlet being recognized in England as the colour of royalty) brought specially for the purpose. The bedstead, bedclothes and toilet ware were to be presented to Powhatan as gifts from James appropriate to the *mamanatowick*'s regal status.

Smith thought the idea insane. It amounted to handing Jamestown over to the Indians. Powhatan could be bought for a 'plain piece of copper', he protested. This 'stately kind of soliciting' would only make him 'so much overvalue himself that he respected us as much as nothing at all'. Furthermore, mounting another embassy to

Werowocomoco would waste yet more of the settlement's nearly exhausted reserves.

Newport was insistent: the coronation would go ahead. He accused Smith of being against the idea because it would expose how cruelly the captain had 'used the savages in [Newport's] absence'. As for diverting labour, he pledged (according to Smith's own recollection of the council meeting) to bring twenty bushels of corn back from the falls mission, and another twenty from Powhatan, and make up any deficit from the store in his own ship.

Whatever the motives behind Smith's objections, he cannot have been the only member of the council to have misgivings. It was such a strange idea. Making the *mamanatowick* an Earl or even Duke of Virginia perhaps made sense. This was the strategy adopted by Henry VIII in Ireland, where Gaelic chieftains such as Conn Mór O'Neill were offered Anglicized hereditary titles such as the Earldom of Tyrone, in an attempt to bring them under the control of English law. But this was a coronation. It meant acknowledging not Powhatan's subordinacy, but his sovereignty.[30]

Perhaps this was the price Newport had been forced to pay for having twice returned to England empty-handed. Without prospect of gold yielding a 20-per-cent return to the royal exchequer and profits to the investors, neither the government nor the London Company had an interest in pursuing the venture as a colonial enterprise. So, as a last desperate measure, it had been decided to concentrate on developing Jamestown as a trading post or 'factory'.

The model was the East India Company, of which Sir Thomas Smythe, treasurer with the Virginia Company, was the founder. The East India Company's royal charter explicitly excluded any colonial remit. It was set up simply to ship out gold bullion and tradeable English goods such as wool cloth to the East, and exchange them for local produce such as spices and silk. To smooth the way for these expeditions, Smythe had started to commission the building of permanent warehouses or 'factories' in the most promising localities, which handled the buying and selling of the goods through local intermediaries. This system was still in its infancy, but it was already paying dividends.

While Virginia would never offer the riches of the Orient, it was still possible that the English could profit from a presence through trade in furs, wood and other commodities. But such profits would only arise if Jamestown acted more like a factory, which existed under the local jurisdiction, than a colonial settlement, which challenged it. Such a change in approach brought numerous benefits: new trading opportunities, unimpeded access to whatever precious metals, minerals or navigable routes might eventually be found, a reliable source of local supplies to replace those brought at such vast cost from London, and most important of all, a reduction in crippling defence costs.

Newport mustered one hundred and twenty men to accompany him to Werowocomoco. Smith attempted to forestall his departure by suggesting Powhatan be crowned in Jamestown, volunteering to deliver the invitation himself. Newport agreed, and instead of one hundred and twenty men, Smith pointedly selected just five, including Namontack, all of them setting off on foot for Werowocomoco. To avoid any risk of a repeat of their earlier humiliations on the muddy banks of the Pamunkey, Smith insisted they go by foot, and cross the river in a canoe.

They arrived at night-time, and were escorted into a 'plain field', where a fire was burning. As before, Smith was invited to sit on a mat facing the fire. Presently, in the surrounding woods, there came 'such a hideous noise and shrieking that the English betook themselves to their arms'. Then Pocahontas appeared, to reassure the captain, 'willing him to kill her if any hurt were intended'. Reassured, he and his companions sat back down.

'Then presently they were presented with this antic.' Thirty young women, dressed in nothing more than body paint and a few leaves, their leader sporting 'a fair pair of buck's horns on her head and an otter's skin at her girdle', rushed from the woods 'with most hellish shouts and cries', and began to dance around the fire before the gawping men. They continued for nearly an hour 'with most excellent ill variety', some of them occasionally overcome with 'infernal passions' and falling to the ground to recover, then rising 'solemnly' to continue their dance.

Then, just as suddenly as they had appeared, they were gone.

Too distracted to notice what happened to his compatriots, the bemused Smith was led away to a house where the dancers had 'reaccommodated' themselves. They surrounded him, and began to press their naked, sweating bodies against his, one and then the next enticing and taunting him by saying, 'Love you not me?' 'Love you not me?' Smith claimed it was torment and, less convincingly, tedious.

The 'salutation' of this 'Virginian masque' ended with a feast, during which it transpired that Powhatan was away, probably on a hunt with most of the town's male population.

Powhatan appeared the following day. Smith 'redelivered him Namontack', and issued the invitation to Jamestown, to receive 'presents' brought by Newport and 'conclude their revenge against the Monacans'. *The mamanatowick* flatly refused the offer. 'If your king have sent me presents, I also am a king and this is my land,' Smith reported him as saying. 'Eight days I will stay to receive them. Your father is to come to me, not I to him nor yet to your fort, neither will I bite at such a bait. As for the Monacans, I can revenge my own injuries.'

Smith delivered the message back to Jamestown, possibly hoping the setback might persuade Newport to give up on the idea. But Newport set off immediately for Werowocomoco. He went by land with a retinue of fifty soldiers, the unwieldy presents sent ahead by barge.

The coronation ceremony is reported only by Smith, who perhaps inevitably dwells on its more farcical features. 'With much ado', Newport presented Powhatan with the gifts brought from England, 'his basin and ewer, bed and furniture'. The scarlet cloak was placed around his shoulders, but not until he was persuaded by Namontack that the vestments would not hurt him.

Newport then invited the *mamanatowick*, a tall, imposing figure, to bow down to receive his crown. This provoked a 'foul trouble', Smith was delighted to observe. Powhatan, 'neither knowing the majesty nor meaning of a crown', refused to lower his head, and Newport, presumably handicapped by having to place the heavy object with just one hand, could not reach. Eventually, the *mamanatowick*'s head was forced down by leaning heavily on his shoulders, until it was sufficiently low for Newport (or, in Smith's later recollection of the event, Newport assisted by two others) to slip the bauble over his pate.[31]

To mark the achievement, a member of the English company fired his pistol into the air, which was the signal for a guard on one of the English barges floating on the Pamunkey river to let off a volley in salute. Powhatan jumped at the noise 'in a horrible fear', but when the smoke and smell of sulphur cleared, he recovered his composure, and offered to Newport his 'old shows and mantle'.

Newport then told Powhatan of his plan to go beyond the falls and confront his enemies the Monacans, looking for help with guides and even military back-up. Powhatan flatly refused any assistance, and tried to persuade Newport to abandon the plan. His hostility evidently took Newport aback, and added to the sense of uncertainty about the occasion. Newport and Powhatan took leave of one another 'after some small complimental kindness on both sides', the old chief offering some corn (amounting to a mere seven bushels, according to Smith).

Thus, Smith was vindicated, at least by his own version of events. Far from putting English–Indian relations on a new footing, as Newport had hoped, it had made them more ambiguous.

Soon after, the admiral embarked on his expedition beyond the falls, taking 'all the council' and one hundred and twenty men, led by Waldo, Percy, Wynn, Scrivener and Francis West, leaving President Smith in the unfamiliar position of being in sole charge of the fort, with eighty or ninety men left to load the ships with the required commodities.

The restless Smith escaped at the earliest opportunity, going downriver a few miles to collect timber. With so many references to the 'goodly tall trees' that covered the Virginian landscape, the London Company's demand for clapboard (meaning high-grade timber used to make barrels, furniture and wainscoting) sounded reasonable. The land was 'generally replenished with wood of all kinds and that the fairest', Gabriel Archer had promised in his original report on the Chesapeake's natural resources, 'yea, and best that ever any of us (traveller or workman) ever saw, being fit for any use whatsoever, as ships, houses, planks, pales, boards, masts, wainscot, clapboard for pikes or elsewhat'. Smith had also noted that there were abundant oak trees 'so tall and straight, that they will bear two foot and a half square of good timber for 20 yards long'. Oaks of that quality were rare in southern England, many having been felled to make masts for the fleet that overcame the Spanish Armada.[32]

Archer and Smith had not been exaggerating. The landscape around them was covered with mature deciduous trees. But the fruits that Virginia dangled so temptingly before the settlers' eyes were easier to describe than pick, as these first attempts to develop a logging industry would prove. The problem was that the most valuable trees were the least accessible. The only way of transporting a commodity as big and heavy as logs was by water, but the river sides provided few easy pickings. Most of the trees were pines, and the quality of timber they yielded would not cover the cost of their being shipped back to England. Also, these were the areas most heavily populated by Indians, and, as Smith had pointed out, by burning and 'girding' (cutting round the trunk near the tree's root to kill it off), they had cleared much of the woodland around them for their houses and fields, 'so that a man may gallop a horse amongst these woods any way'.[33]

Hence, Smith had to go several miles downstream to find a suitable location to start logging. It was going to be hard work, so he hand-picked thirty gentlemen to accompany him. Two 'gallants', Gabriel Beadle and John Russell, excelled themselves, 'making it their delight to hear the trees thunder as they fell'. As for the rest, the skin on their 'tender fingers' was soon blistered by the axe handles, and with every third swing, Smith noted, they would emit a 'loud oath to drown the echo' of the blow. In a pious attempt to moderate their language, he counted up their curses, and in the evenings poured a can of water over them for each swear-word, 'with which every offender was so washed, himself and all, that a man should scarce hear an oath in a week'.

By the end of the week, even Smith was impressed by the work that had been accomplished. He later commented how '30 or 40 of such voluntary gentlemen would do more in a day than 100 of the rest that must be press'd to it by compulsion' (an endorsement of those of his class ignored by posterity). However, he also added that 'twenty good workmen' would probably do even better, not only because they were used to such work, but because they would be hired for that purpose. Most of the 'labourers' in the fort were the servants of the gentlemen, whose passage, employment and loyalty was owed to their masters, not the company or even the settlement.

Smith returned to Jamestown with what was presumably a useful quantity of wood, but still restless to get on, escaped almost immediately to embark on another expedition, taking eighteen men to fetch supplies from the Indians. He started with a trip up the Chickahominy, the main tributary of the James, to revisit some of the sights of his capture. He had decided to do this, he admitted, not so much to get corn, as to find an excuse to 'revenge his imprisonment and the death of his men'. He was duly obliged, the 'dogged' people he encountered treating his requests for corn 'with as much scorn and insolency as they could express'. Smith mounted an attack, and, ignoring pleas that they had little corn to spare, forced the locals to hand over a hundred bushels (according to his own estimate), and a further hundred when Percy turned up later in another barge. A trip to Werowocomoco provoked a similarly uncooperative response, though Namontack eventually managed to secure 'three or four hogsheads of corn'. Nevertheless, the Indians were becoming increasingly reluctant to provide the English with supplies.

Smith took these events as evidence that, despite the coronation, 'Powhatan's policy [was] to starve us'. This is questionable on two counts. The 'imprisonment' he sought to avenge by attacking the Chickahominians was orchestrated not by them but by Opechancanough, Powhatan's brother. The people he encountered on the banks of the Chickahominy 'lived free from Powhatan's subjection, having laws and governors within themselves', as another settler later observed. So it is unlikely they were implementing a policy of Powhatan's when they refused Smith's demands for food. Furthermore, 1608 was the third year of a drought that would prove to be the worst to hit that region of North America in eight hundred years. Not just the corn harvest, but the entire basis of the Indians' finely balanced subsistence diet was under pressure. In such circumstances, they were finding it hard enough to sustain themselves, let alone several hundred hungry English.[34]

Newport returned from the expedition into Monacan country claiming to have explored 50 or 60 miles further inland (Smith grudgingly estimated 40). 'This land is very high ground and fertile, being very full of very delicate springs of sweet water,' observed an excited Captain Wynn, in a brief letter home. He also noted that the air was 'more healthful than the place where we are seated, by reason it is not subject to such fogs and mists as we continually have'. They had visited two Monacan towns, Massinacak and Mowhemenchough, where the people spoke 'a far differing language from the subjects of Powhatan', which sounded to Wynn not unlike the Celtic language of his native North Wales.[35]

Several mineral deposits had been discovered, and William Callicut, a refiner, probably brought by Newport to replace Captain Martin, conducted tests. He claimed to find traces of silver, perhaps just enough to revive investors' fading hopes of a profitable return.

In late November, the men handed over to Newport letters to

loved ones and patrons to be taken back to London. In his capacity as president, Smith also handed over a report addressed to the Royal Council. The version of it he later published may have been doctored to exploit the benefit of hindsight, but its tone of frustration and vindication was probably authentic. Newport's escapades had wasted valuable resources, he argued, and had succeeded only in confusing relations with the Indians. Claims that the settlers had exaggerated the land's resources were unfair, while demands for more valuable commodities to be shipped home were unrealistic. The threat to cut off supplies was cruel, not least because they were being consumed by the mariners who brought them. Finally, the council may have been torn apart by faction fighting, but the cause was the very people the government had appointed to run it. That was why the 'poor counterfeited imposter' Ratcliffe was being sent back to London. If he or Archer were ever to return, Smith warned, they would 'keep us always in factions'.

'These are the causes that have kept us in Virginia from laying such a foundation that ere this might have given much better content and satisfaction,' Smith concluded. 'But as yet you must not look for any profitable returns.'

He enclosed with the report a broadsheet map of Virginia. This was an altogether more hopeful article. It laid out the extent of his discoveries around the Chesapeake, and few in London could fail to be impressed with its extent and detail. Furthermore, as Smith claimed in his report, the rivers, towns, woods, hills and creeks it charted had been explored by him 'for less charge' than Newport wasted in company expenses over a generous lunch.

One of the map's most striking features is the abundance of Indian place names, and the absence of English ones. Even the James river, dubbed by Newport in his first act of colonial appropriation, is labelled the Powhatan. It is as though Smith is turning the tables, showing that the rich and noble members of the Royal Council, rather than the settlers, were failing to exploit this wonderful discovery. On this sheet of paper was laid out a voluptuous, nubile land awaiting possession. It was theirs, but only if they had the manly resolve to take it.

Newport sailed away from Virginia at the beginning of December, his ship carrying nothing like the bounty the London Company had demanded. Besides some wood, and presumably early attempts at producing pitch, tar and glass, Smith mentioned two barrels of stones, 'and such as I take to be good iron ore at the least'.

Part Three

ELEVEN

El Dorado

Stuck in the Tower of London, Sir Walter Raleigh nurtured dreams of the golden man, or, as the Spanish knew him, *el hombre dorado*.

Raleigh had first heard of him in 1586, from a Spanish nobleman called Don Pedro Sarmiento de Gamboa, who was captured by privateers commissioned by Raleigh to prey on shipping around the Azores. Gamboa spoke of a lost kingdom somewhere in the interior of a province the Spanish called 'Trinidad y Guayana'. Each year a priest king, standing at a lake side, anointed his naked body with turpentine and rolled in gold dust, thus becoming a gilded man, *el hombre dorado*. This king was said to rule an empire called Guiana, from his capital city, Manoa. 'The empire Guiana hath more abundance of Gold than any part of Peru,' Raleigh was told, and Manoa exceeded 'any [city] of the world'.

Raleigh had led an expedition in 1595 to find this legendary empire, which succeeded only in provoking the Spanish into setting up military camps on the approaches to the highlands where the empire was believed to lie. Over the years, these camps attracted footloose soldiers, ranchers and prospectors from across the Spanish Indies, all hopeful of discovering the elusive golden man somewhere in the Guianan highlands. Their efforts were fruitless, and disillusionment drove the prospectors away. Casting around for other means to make a living, the remaining settlers discovered they could produce for themselves a commodity almost as lucrative as the elusive gold: tobacco.

There was already a thriving tobacco trade elsewhere in Spanish America, focused around Venezuela, but it aroused official disapproval. Quick profits and high prices encouraged the development of a large, freebooting black market that undermined efforts to found sustainable settlements and prevent smuggling.

In England, it was Raleigh who had popularized the smoking of this so-called 'Spanish' tobacco in the 1580s. He had ostentatiously puffed upon a silver pipe at court, which attracted the attention of Queen Elizabeth. The story goes that he bet her he could weigh the smoke rising from the pipe's bowl. She accepted the wager, and was beaten by him first weighing the tobacco, smoking it, and then subtracting the weight of the ash left behind. His feat provoked the quip from Elizabeth that she had heard of men turning gold into smoke (a dig at alchemists), but never smoke into gold. She never uttered a more prophetic word.[1]

The habit of smoking – or 'drinking', as it was known at the time – tobacco had spread quickly through the English court and beyond. Most supplies came from Spanish America, and were obtained by privateers and smugglers, whose disruption of Spanish trade had been encouraged as a patriotic act. However, with James's succession, attitudes were transformed. Anti-Spanish militants such as Raleigh were locked up in the Tower, the peace treaty was signed at Somerset House, and the King penned his famous 'Counterblast to Tobacco' in 1604, in which he condemned the 'manifold abuses of this vile custom'.[2]

With legitimate supplies now having to come from official Spanish sources, and subject to swingeing customs duties imposed by James, prices soared. By 1607, a pound of Spanish tobacco could cost 40 shillings. Raleigh's fellow inmate in the Tower, the Earl of Northumberland, was spending around £50 a year on his smoking habit, £10 more than the amount he was prepared to spend on the tutoring of his son.[3]

In 1606, Spain issued an order demanding a clampdown on smuggling activities, and banning the cultivation of all tobacco in the imperial heartlands of Venezuela and New Andalucia (modern eastern Venezuela). Desperately casting around for a new source of supply, tobacco merchants turned southwards to the neglected province of Trinidad y Guayana.

Two prominent London merchants, John Eldred, who became Sir Thomas Smythe's deputy in the Virginia Company, and Sir John Watts, led the way. Both had backed privateering missions before the peace with Spain. As late as 1607, the Spanish ambassador in London still described Watts (who that year was elected Lord Mayor of London) as 'the greatest pirate that has ever been in this kingdom'.[4] Now was a new opportunity for these resourceful entrepreneurs to profit from Spanish vulnerabilities.

The denizens of Trinidad y Guayana proved to be obliging smugglers. The place was full of 'soldiers who without spiritual or temporal yoke, run headlong into every kind of vice . . . the chosen resort of secular criminals, irregular priests and apostate friars, and in general a seminary of rascals', a Spanish official reported to Philip III. 'They tell me, on good security, that in Trinidad a great quantity of goods is smuggled and that English and Dutch ships are never lacking there,' revealed another.[5]

It was at this point, sometime in the summer or autumn of 1608, that the imprisoned Raleigh was drawn back into the business. He had shares in a ship called the *Primrose*, a vessel more substantial than its name suggested, having acted as one of Sir Francis Drake's flagships during the 1580s. Raleigh made it available to a group of merchants planning an expedition to the coast of Guiana. They were led by John Eldred, who the year before had fallen foul of the King's customs officers after importing a large quantity of tobacco sourced from Spanish America. He had objected to the swingeing 'imposts' or import duties which he had been charged for the cargo, and refused to pay. He evidently hoped that contraband tobacco from Guiana might provide a way of avoiding such charges in the future, while allowing him to continue satisfying his customers' needs.[6]

Tobacco was less likely to have been Raleigh's primary interest in the venture. He still harboured colonial ambitions for South America. An expedition mounted just before his imprisonment had established that Spain's interest in the region had lapsed. According to one report, the natives of the Orinoco had even urged Raleigh to return, now that the Spanish invaders had gone. In one of his forlorn letters to Robert Cecil, Raleigh also claimed that further tests of a sample of marcasite he had retrieved during his previous expedition had yielded traces of gold.[7]

The *Primrose*, in consort with another ship called the *Ulysses*, headed off for Trinidad in the autumn of 1608. The ships arrived safely at Port of Spain, Trinidad's main harbour, and contacted Fernando de Berrio, the governor of the Trinidad y Guayana province. Berrio invited twenty-seven of the ships' crews ashore, and promptly took them prisoner. He wrote a letter to the expedition's backers, demanding a ransom of 20,000 ducats (over £5,000), for which he would free the hostages and allow the merchants to trade for tobacco in the region. Despite the enormity of the demand, the money was raised. Raleigh was in no position to contribute, but Eldred and his rich business friends were, out of concern for their most lucrative line of business, if not the welfare of the crews.

Pewter plate and other goods of the required value were eventually gathered together to pay off the enormous ransom, and shipped out to Trinidad. Berrio took the money, and, perhaps to avenge earlier English incursions into his province, hanged all the captives, 'except the surgeon of the ship, [who] was kept alive to cure the governor's disease'. The episode became notorious in England, still being cited nearly fifty years later as an example of Spanish barbarity.[8]

Back in London, the effects on the tobacco trade were more immediate. Contraband traffic with Trinidad and Guiana came to a sudden halt, making merchants once again reliant on official sources for supply. Prices rocketed, but profit margins fell. It became clear that another source of this precious commodity was needed.

Tobacco could be grown elsewhere, in England and, as had recently been discovered, in Virginia, where the Indians called it *uppowoc* (meaning 'they puff it').[9] However, these alternatives were not nearly as sweet as the Spanish weed, and on the streets of London they commanded a fraction of the price. (One twentieth-century historian of Virginia from a tobacco-planting background compared the difference in taste between the variety originating in South America, *Nicotiana tabacum*, and that in Virginia, *Nicotiana rustica*, as the same 'as between a crab apple and an Albermarle pippin'.[10]) However, a thought began to germinate in the minds of the many merchants who shared an interest in both the tobacco trade and the Virginia venture: perhaps if one variety thrived in the soil of North America, there was no reason to believe that another from the south might not be adapted to it.

TWELVE

The Mermaid

IN JULY 1608, the newsmonger John Chamberlain wrote that a ship 'that hath been long missing' had 'newly come from Virginia'. She was the *Phoenix*. 'I hear not of any novelties or other commodities she hath brought more then sweet wood,' he added. Disappointment had been such a recurring theme of his reports concerning Virginia that he did not even bother to express it.[1]

Worse news was to come. In December 1608, the settlers sent to colonize the northern or 'second' Virginia colony returned to England, many crowded into a boat they had built for themselves while out there, which they had named *Virginia*. They had managed to build a sturdy fort at the mouth of the Sagadahoc river (Kennebec river, Maine), but, following the death of their leader, had decided they had no option but to abandon it. Their return brought the northern venture to an end, leaving Jamestown the sole representative of English interests in North America.

A few weeks later, Christopher Newport returned to London from his second supply journey. The trip had failed to cover its costs, the barrels of earth he had brought once again turning out to be worthless. As if to emphasize the desperate state of affairs, he also brought two more ringleaders of the incessant faction fighting: Captain Ratcliffe and Gabriel Archer, both full of complaints about Captain John Smith.

Robert Johnson, a prominent London merchant and a close business associate of Sir Thomas Smythe, treasurer of the Virginia Company, was one of many City businessmen being urged to help keep the project afloat. But he reflected the prevailing view when he wrote that he was 'little moved, and lightly esteemed of it'.[2]

The Spanish ambassador Pedro de Zuñiga found this all very perplexing. His monthly dispatches to Madrid had reported English ships returning from America with 'a few things of little importance'. He had noted the recent death of the Lord Chief Justice John Popham, the only government figure of any stature who had 'desired' the colonization of America, and the one 'best able to aid it'. And his contacts in the City confirmed that the venture was in 'greater straits for money than one can imagine'. Yet it continued.[3]

There was no obvious figure in the government to give it buoyancy. That same year, Robert Cecil had expanded his grip of government yet further by taking over as Lord Treasurer. This put him in charge of the royal exchequer, which, on close inspection of the books, was revealed to be in a 'chaos of confusion', the 'devouring fire' of royal expenditure having resulted in debts of £600,000. Previous attempts to douse the flames with income from royal imposts and other feudal dues had only inflamed the relationship with a restive House of Commons, which in 1607 the King had 'prorogued' or forbidden to assemble in his attempt to impose discipline. In this context the last thing James could contemplate was provoking a costly war with Spain over dubious and unprofitable claims to America.[4]

Zuñiga could only conclude from this that some freelance Protestant militants, bent on continuing Raleigh's and Drake's mission, must be aiming to 'carry on piracy from there'. Who or what else could be keeping such a fragile and leaky vessel afloat?[5]

When the Scottish King James had come to take the throne of England in 1603, it was the 9-year-old son the English had received with raptures. Prince Henry brought the clamour of youthful energy to the barren palaces of the Tudor dynasty. He also had a gravity and sense of purpose quite at odds with his tender years, characteristics that attracted adoration from courtiers and commoners alike. This aroused private jealousies, and the son adopted by the nation became estranged from his father.

Henry's response was to fill the paternal void with a court of his own.[6] Everything about this court was at odds with his father's. Where James's was opulent and hedonistic, Henry's was austere and virtuous – even swearing was forbidden, though whether this was enforced using Smith's methods of dousing offenders is unknown. Where James's circle was packed with favourites and fawners, Henry's was a 'courtly college or . . . collegiate court' of intellectuals, clerics and poet-knights. Where James sought of his advisers counsels of compromise and peace, Henry listened for idealism and boldness. And where James vacillated between treating America as a revenue-raising opportunity and a foreign-policy distraction, the prince saw only a frontier for heroic adventure.

'*Look over the strict Ocean,*' Ben Jonson had urged the prince in a poem of welcoming him to England,

. . . and think where
You may but lead us forth . . .

He had treated the invitation with boyish earnestness, sending his gunner Robert Tyndall with the first expedition of the Virginia Company, commissioning him to bring back detailed maps and reports of what was found. This interest was the reason the first identifiable landmark on the approaches to Chesapeake Bay had been named Cape Henry in his honour.[7]

By 1609, though just thirteen years old, he was in a position to extend his influence over the enterprise even further. In the face of his father's disapproval, he had struck up a relationship with Raleigh, still trapped in the Tower. 'No other king but my father would keep such a man as Sir Walter in such a cage,' he reportedly said. Henry's interest prompted Raleigh to produce a cascade of papers on maritime and colonial affairs, among them a detailed document about Virginia. Nothing is known about its contents, but according to the Spanish ambassador's spies, it had captivated some members of the Royal Council for Virginia, who had decided to 'guide themselves by it'.

Raleigh even imagined he might be able to lead an expedition. He wrote to Henry's mother, Queen Anne, begging her to speak favourably about him to Cecil. 'I have long since presumed to offer your Majesty my service in Virginia, with a short repetition of the commodity, honour and safety which the King's Majesty might reap by that Plantation.' If he were allowed to go, his wife and two sons could be held hostage in his absence, and his masters and mariners instructed to throw him overboard should he stray into Spanish waters. 'I do humbly beseech your Majesty that I may rather die in serving the King and my country than to perish here.' His pleas fell upon deaf ears.[8]

Because of the political sensitivities, no official record remains of Henry's emergence as a figurehead of a revived Virginia venture. Nevertheless, it was spotted by the ambassador's spies. In the spring of 1609, Zuñiga reported to King Philip that Henry was being promoted as a new 'Protector of Virginia, and in this way they will get deeper into this business'. Around the same time, the Venetian ambassador informed the Doge that the Prince had pledged to put his own money into the venture, 'so that he may, some day, when he comes to the crown, have a claim over the Colony'.[9]

Henry's interest electrified the debate over Virginia. What had been treated in government circles as a secret commercial venture, confined to an inner elite, took on a much more public, political complexion. Writers, politicians, lawyers and clerics began to talk about Virginia as a national enterprise, a mission to spread civilized Protestant values. These ideas surfaced first furtively in taverns, then erupted thunderously from pulpits and in print.

An early sign of their gathering confidence was the unexpected appearance at 'the sign of the Greyhound in Paul's Churchyard' of the first book on Jamestown: *A True Relation of such occurrences and accidents of note as hath happened in Virginia since the first planting of that Colony.*

It had clearly been rushed into print. According to the first copies off the presses, the book was by 'Thomas Watson', one of the tellers at the royal exchequer. This was a mistake, forcing the bookseller, William Welby, to correct the error by scribbling out the name in each copy, and replacing it with the name of the real author: Captain John Smith. Welby was also forced to stuff an extra page into the opening section, carrying an apology from the editor 'I.H.' for its haphazard appearance.[10]

'I. H.' was John Healey (I being used for J, following Latin tradition), a Catholic who became a government agent after being arrested during the Gunpowder Plot investigation. It was Healey who had informed on the activities of John Sicklemore the priest, which he claimed included an assassination plot against Cecil.

Healey had come to London soon after Smith had left for Virginia. He had started writing poetry, some of which he sent to Cecil in the hope of employment. This brought him into contact with London's thriving literary circles, and in particular Thomas Thorpe. Thorpe was the son of an innkeeper, who had connections to Catholic exiles in Spain, and would become, if he was not already, a friend of John Smith, providing a dedicatory poem to one of his books.[11]

Thorpe was one of London's most successful publishers, producing editions of the works of Jonson, Chapman and Shakespeare, as well as books on Virginia, and pamphlets featuring the speeches of prominent parliamentarians such as Richard Martin, who soon after became the Virginia Company's lawyer.[12] He was also linked to a secretive drinking club of poets, playwrights, MPs, lawyers and petty government officials known as the 'Sirenaicals', siren being the Latin word for mermaid. The term was a reference to the Mermaid Tavern in London's Bread Street, the main venue for the club's monthly meetings. Prominent Sirenaicals included Ben Jonson, John Donne, the poet who in 1609 was in the process of trying to get a job as secretary of the Virginia Company, Richard Martin and, it was rumoured, Sir Walter Raleigh, before his imprisonment by James.

'*What things we have seen, Done at the Mermaid!*' wrote Francis Beaumont, a poet and acolyte of Jonson,

> *... heard words that have been*
> *So nimble, and so full of subtle flame ...*

Smith's incendiary *True Relation* smuggled back to England aboard the *Phoenix* may well have landed on the tavern table, and from there travelled into the hands of Healey. Healey merely mentions he came across it 'by chance (as I take it, at the second or third hand)' and had been 'induced' to publish it without the author's permission 'by divers well-willers of the action, and none wishing better towards it then my self'.[13]

Despite clumsy editing which removed the most damning material and rendered some sections barely intelligible, the resulting work was sensational. It revealed to English readers for the first time the great river, from the broad, smooth channel at Kecoughtan to the 'great craggy stones' of the falls, 'where the water falleth so rudely, and with such a violence, as not any boat can possibly pass'. It told of Indians sometimes 'dancing and feasting us with strawberries, Mulberries, Bread, Fish, and other their Countrie provisions whereof we had plenty', and other times 'churlish' and full of 'jealousy'. But perhaps the most vivid material concerned neither the land nor its people, but the settlers themselves. *True Relation* revealed disturbing glimpses of the chronic infighting, poor leadership and disorder that threatened to destroy the venture before it had properly begun.[14]

The Virginia Company had not authorized the publication of the *True Relation*, which did not portray the company's management of the enterprise in a flattering light. However, the aim of publishing the work was not the destruction of the venture, but to prompt its revival. Healey tacks on a rousing conclusion arguing that, after such a dismal beginning, a bright future awaited any 'willing minds' that chose to become involved. The settlers, he claimed, were now 'contented, free from mutinies, in love one with another, and as we hope in a continual peace with the Indians'.

In his introduction, he was even more optimistic. Virginia, he wrote, 'will tend to the everlasting renown of our Nation, and to the exceeding good and benefit' of the commonwealth.

This set the tone for an outpouring of propaganda that followed within weeks of the publication of Smith's text. The emerging sense of a national mission was expressed by Michael Drayton, Prince Henry's poet, in an 'Ode to the Virginian Voyage':

You brave heroic minds,
Worthy your country's name,
That honour still pursue,
Go and subdue!
Whilst loit'ring hinds
Lurk here at home with shame.

Britons, you stay too long;
Quickly aboard bestow you,
And with a merry gale
Swell your stretch'd sail,
With vows as strong
As the winds that blow you!

Your course securely steer,
West and by south forth keep;
Rocks, lee-shores, nor shoals,
When Æolus scowls,
You need not fear,
So absolute the deep.

And cheerfully at sea
Success you still entice
To get the pearl and gold,
And ours to hold
Virginia,
Earth's only paradise!

Prince Henry's chaplain Daniel Price picked up the theme in a famous sermon delivered around this time from Paul's Cross, the stone pulpit in the northern courtyard of St Paul's Cathedral where Londoners gathered in their hundreds to hear the latest thinking on religion and politics. Price lambasted the 'lying speeches' of those who have 'injuriously vilified and traduced a great part of the glory of God, the honour of our Land, Joy of our Nation and expectation of many wise, and noble Senators of this Kingdom, I mean in the Plantation of Virginia'. 'Some idle, dull and unworthy sceptics' had persuaded Henry VII of England to ignore the pleas of Christopher Columbus to back a proposed expedition into the western Atlantic, he reminded a congregation of the pious and curious passers-by. 'We know our loss by the Spaniards.'

Having dismissed the 'sceptical humourists', he pointed out Virginia's manifold virtues: 'The Philosopher commendeth the temperature; the politician, the opportunity; the divine, the piety in converting so many thousand souls. The Virginian desireth it, and the Spaniard envieth us, and yet our own lazy, drowsy, yet barking countrymen traduce it, who should honour it.'

Drawing on the fiery language of Puritanism, he concluded with 'one word of exhortation' – in fact, several. London, he announced, had been called 'the City of God . . . the Seat of Judgment, even the seat of the house of David. Peace be within thy walls, plenteousness within thy palaces.' And yet, the Londoners crowded around the pulpit could not forget 'how manifold infections hence, as from a fountain, [have] issued out; all the tricks of deceiving, the divers lusts of filthy living, the pride of attire, the cause of oppression, gluttony in eating, surfeit in drinking, and the general disease of the fashions'. London, he cried, 'should be Jerusalem, the City of God, and it is become Murder's slaughterhouse, Theft's refuge, Oppressor's safety, Whoredom's stew, Usury's bank, Vanity's stage, abounding in all kind of filthiness and profaneness.'[15]

William Symonds was a controversial lector from Christ Church in London's Greyfriars. He was linked to Captain John Smith, sharing a patron, Robert Bertie, Lord Willoughby, and having among his former parishioners at Theddlethorpe, Lincolnshire, one Nicholas Smith, a kinsman and friend of the captain.[16] On 28 April 1609 he had delivered a sermon from Paul's Cross which was just as sulphurous as Price's. Entitled 'Virginea Britannia', it took as the text Genesis 12:

> *For the Lord had said unto Abram, get thee out of thy Country, and from thy kindred, and from thy father's house, unto the land that I will shew thee.*
>
> *And I will make thee a great nation . . .*

For Symonds, the story of Genesis revealed how God had achieved the 'plantation of heaven and earth'. He saw the settlement of Virginia as a continuation of the same mission. 'Let us be cheerful to go to the place that God will shew us to possess in peace and plenty, a Land more like the Garden of Eden.'[17]

And so the sermons continued, crowds of the curious, the furious, the pious and the disenchanted coming from across London and beyond to hear of this new civic and holy mission.[18]

A theme that surfaced repeatedly was how to treat the Indians. 'The country, they say, is possessed by owners, that rule, and govern it in their own right,' Symonds commented. 'Then with what conscience, and equity can we offer to thrust them, by violence, out of their inheritance?' His answer was pugnaciously imperialist, suited to a sermon that had as its title 'Virginea Britannia'. The destiny of a reborn 'Great Britain' was to project civilization and Protestant Christianity through the world, if necessary with military power. Had not the Lord pressed the sword of righteousness into the right hand of Cyrus, the Emperor of Persia, of King David and King Solomon, so that God's dominion would stretch 'from sea to sea, and from the river to the end of the land'?[19]

While the preachers sermonized at Paul's Cross, books on Virginia tumbled out of the presses and on to the stalls clustered in St Paul's churchyard and the surrounding lanes. Most explored the theme in similar ways to the sermonizers. In *Good Speed to Virginia*, for example, Robert Gray emphasized the venture's religious impulse. However, he also introduced a political and economic dimension. England had become overpopulated by peace and 'the blessings of the womb', he argued, and, like the children of Israel, the English were finding that 'the land is too narrow for us'. This threatened 'oppression, and diverse kinds of wrongs, mutinies, sedition, commotion and rebellion, scarcity, dearth, poverty, and sundry sorts of calamities'.[20]

The well-connected merchant Robert Johnson, a director of the Grocers' and East India Companies, wrote a more secular, practical survey, but under the stirring title *Nova Britannia*, New Britain. He outlined England's historic claim to North America, detailed Virginia's rich resources, foresaw the possibilities of cultivating wine and silk, warned of recent French successes, urged England to match the more advanced economic and mercantile methods of the Dutch. He affirmed what others had written about the Indians, arguing that settlers would 'require nothing at their hands, but a quiet residence', and promising that they would 'enjoy equal privileges'.

However, not even a matter-of-fact merchant like Johnson was immune to the sense of destiny gathering around the mission. He quoted a prophecy made by the Protestant martyr John Frith, who had been burned at Smithfield for heresy in 1533. Frith had written 'of great wonders that should be wrought by Scots and English, before the coming of Christ'. Perhaps Virginia would be the world-changing wonder.[21]

These books were in great demand. Spanish ambassador Zuñiga snapped up a copy of *Nova Britannia* more or less on the day it appeared, and packed it off to Madrid. Opportunistic publishers rushed out any book which could conceivably include 'Virginia' in the title. Hakluyt's translation of a Portuguese work about Florida appeared as *Virginia richly valued*. A translation of a work about the French exploration of Canada was called *Nova Francia: or The Description of that part of New France which is one continent with Virginia*, and provided a salutary reminder that England was not the only European power to be delving into North America. The book was dedicated to 'the bright star of the north, Henry, Prince of Great Britain'. 'Your poor Virginians do seem to implore your princely aid,' wrote the translator, the Huguenot Pierre Erondelle, 'to help them shake off the yoke of the devil, that henceforth they may be brought into the fold of Christ, and (in time) to live under your Christian government.'[22]

John Healey, having recently completed his edition of Smith's *True Relation*, produced two new works, neither with Virginia in the title, but both with relevant themes. One was *Teares for the Death of his Sonne*, from the French work by Du Plessis, the geopolitical theorist. The other was a free translation of Joseph Hall's Latin satire *Mundus alter et idem*, entitled *The discovery of a new world*. Among imaginary lands such as Gluttonia, Drinkalia and Double-Sex Isle, Healey refers to 'Womandecoia, which some mistaking both name and nation call Wingandecoia, & make it a part of Virginia', and to 'Sectariova' where 'Virginian exiles laid a plot to erect themselves a body politic', hinting at the alluring but dangerous sensual and doctrinal freedoms already thought to exist across the Atlantic. In a marginal note, Healey notes that 'Certain English Brownists' (a general term of abuse for religious radicals, named after the dissenting minister Robert Brown, living in Holland) had been 'exiled into Virginia'.[23]

Thomas Thorpe published these two works, together with another of more enduring appeal: William Shakespeare's *Sonnets*. In the flyleaf of the *Sonnets*, Thorpe inserted a dedication:

TO. THE. ONLIE. BEGETTER. OF. THESE.
INSUING. SONNETS. M^r^. W. H.
ALL. HAPPINESSE. AND. THAT. ETERNITIE.
PROMISED. BY. OUR. EVER-LIVING. POET.
WISHETH. THE. WELL-WISHING. ADVENTVRER.
IN. SETTING. FORTH. T. T.

As has been pointed out, sonnets were traditionally treated as a private form of poetical expression, which had been kept out of the public glare until Philip Sidney published his *Astrophil and Stella.* They were designed to tease, flirtatiously challenging the reader to discover scandalous secrets only an inner circle of intimates could know.[24] As a result, scholastic paparazzi have long staked out the inscription's enigmatic phrasing and punctuation, in the hope of flushing out the identity of 'Mr W. H.', who being the 'only begetter' of the sonnets may have inspired their intimate revelations about the inscrutable author. Suggestions have included William Herbert, the Earl of Pembroke, whose mother Mary was Sidney's sister and a great patron of the arts, and (his initials reversed, to disguise his identity) Henry Wriothesley, the Earl of Southampton.

One clue as to Mr W. H.'s identity may lie in the reference to him being a 'well-wishing adventurer' who was, at the time the sonnets were published, 'setting forth'. The sonnets appeared at just the moment the tide of Virginia propaganda was reaching its flood. Given Thorpe's connections to Captain Smith and Virginia, was he referring to someone who had just invested in the venture? There are several strong candidates, including Pembroke, who became a member of the Royal Council around this time. As many as four men besides Pembroke with the right initials had responded to the Lord Mayor's plea for contributions. Their claims to being W. H. are precarious, more so than one figure previously unmentioned, the outspoken MP William Hakewill. Hakewill was a lawyer and, around this time, became interested in the Virginia Company. His brother, George, was a chaplain in Prince Henry's household.[25]

Hakewill was also a Sirenaical, and his attendance of the 'wit combats' at the Mermaid provides a host of connections with the literary world. The Mermaid also provides a previously unnoticed connection with the 'well-wishing adventurers' of the Virginia venture. One Sirenaical listed a core membership of twenty-two members of the Mermaid Club. Of these, no fewer than eighteen had a documented interest in Virginia. A high proportion were also MPs, who had been effectively denied any active involvement in the project since the demise of Edward Hayes's grandiose plans 'to raise a stock for the maintaining of a Colony in Virginia' through 'the High Court of Parliament'.[26]

In the face of a sustained campaign, underwritten by the heir to the throne and promoted by a group of noisy political activists, Robert Cecil found his control over the Virginia venture beginning to slip. He was in the midst of trying to save King James from bankruptcy, and had proposed to do so with a 'Great Contract' between King and Parliament. This offered the royal exchequer a more generous annual 'subsidy' raised through taxation in return for James surrendering some of his unpopular feudal powers. Getting agreement for the contract involved some tricky diplomacy with the rebel group of MPs who had emerged with such a powerful voice in the House of Commons, leading members of which included the Sirenaicals Richard Martin and William Hakewill. Many members of this parliamentary group happened also to be involved or interested in the Virginia venture, and allowing them a small measure of reform on that front was a trivial price to pay if it enabled reform on the much greater matter of the King's finances.

The result was permission to draft a new Virginia charter, which would reflect the reforms being demanded. The man appointed to write it was Sir Edwin Sandys, the leader of the rebel MPs, who probably worked in conjunction with the company's treasurer, Sir Thomas Smythe.

Sandys's involvement in the Virginia Company before 1609 is obscure. He is not mentioned in the few official papers that have survived, and he had not been appointed to the Royal Council. But he had long harboured an interest in trade and colonization. In 1599, he had lamented how 'the Northern people' – meaning those of the British Isles, the Low Countries and Scandinavia – had not yet 'for

An Indian settlement typical of the many lining the banks of the James River, depicted in an engraving included in the *Generall Historie*.

Engravings from John Smith's *Generall Historie*, 1624.

Above An Indian chief or 'weroance'.

Left A mother carrying her child in the distinctive Indian manner.

Below A young Indian 'gentlewoman', probably the wife of a 'weroance' or chief.

Above Men hollow out a log using fire to make a canoe.

Below Indian fishing techniques, using canoes, weirs and nets.

Above Methods of food preparation, including the use of wooden hurdles to cook fish, the fore-runner of the modern barbecue.

Left and below Indian meals typically consisted of maize stewed with fish or venison.

Fire was central to Indian communal and spiritual life, being constantly tended by the women, as 'if at any time it goes out, they take it as an evil sign'.

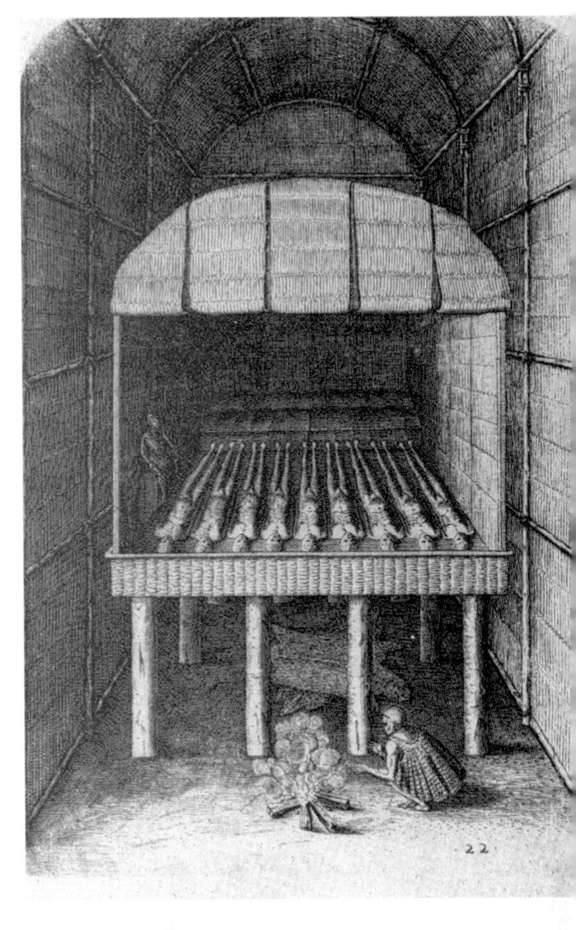

Above An Indian charnel house, where the embalmed bodies of chiefs and other nobles were interred.

Left An effigy of the Indian warrior god Okeus, sometimes called Kiwasa.

Above This portrait of Captain John Smith appears on his map of New England.

'Matoaka, also Rebecca, daughter of the great chieftain Powhatan, emperor of Virginia.' The portrait of Pocahontas, also known as Matoaka and christened Rebecca, commissioned by the Virginia Company to publicise her visit to London in 1616.

The Capture of John Smith, as portrayed in his *Generall Historie*.

Top Smith fights off his attackers, and is taken while stuck in the ooze.
Above His captors celebrate with a victory dance.
Above right A religious rite is performed before the captive Smith.
Right Smith is prepared for what he believes is his execution, watched by Powhatan.

A Coniurer.
Their
Idoll
A Preist
Their Coniuratiōn about
C: Smith 1607

C:S:
King Powhatan comands C: Smith to be slayne, his

Ralph Hamor meeting Powhatan in May 1614. The 'mamanatowick' feels around Hamor's neck for a pearl necklace, which English delegates were supposed to wear when making official visits.

The Chickahominy people, who lived 'free from Powhatan's subjection, having laws and governors within themselves', agree a peace treaty with Governor Sir Thomas Dale in April 1614.

Pocahontas being taken by the English to meet her brothers at Werowocomoco in 1614. She reportedly told them 'that if her father had loved her, he would not value her less than old swords, pieces, or axes; wherefore she would still dwell with the Englishmen, who loved her'.

Above Philip III of Spain, self-proclaimed ruler of the 'Islands of the Indies, East and West, and of the mainland of the Ocean-Sea'.

Right James I of England and VI of Scotland, whose commitment to the colonization of North America wavered from the start.

Left Sir Robert Cecil, James's chief minister, whose deformities according to a rival, made him 'void of natural affection' and mindful 'to watch and observe the weakness of others'.

Below George Percy, the Earl of Northumberland's epileptic young brother, who insisted on keeping a 'daily table for gentlemen of fashion' while in Virginia.

Above Sir George Somers, admiral of the great fleet that left London for Virginia in 1609, 'a lamb on the land, a lion at sea'.

Above right Sir Thomas Smythe, London's 'merchant prince' and prime mover of the Jamestown venture.

Below An advertisement for the Virginia Company's Great Standing Lottery, showing the company seal, a selection of prizes, the manner of the draw, and two Indians used to publicize the venture, 1615.

all their multitude and strength, had the honour of being founders or possessors of any great Empire'. England was in a strong position to achieve this, but it was yet too 'divided from all the rest of the world' by religion. The Catholic Southern countries prospered under the unifying presence of a Pope, 'seated royally and pontifically in the midst and chiefest, regarding the rich Sun in his glorious rising, and the Moon in the height of her beautiful walk: on his left hand, the Emperor [of the Holy Roman empire], the ancient remains of honour; on his right, the King of Spain, the new planet of the West; at his back, the French King, the eldest Son of the Church; all mighty Monarchs, opposed as brazen Walls against his enemies on all sides'. Only when Protestant kingdoms developed such a unity of purpose and power could they hope to compete for imperial fruits. This would happen, Sandys had dreamed, when European kingdoms, Catholic and Protestant alike, reunited around an 'honourable unity of verity', a new religious settlement.

A decade of political experience had relieved Sandys of such hopes. Instead, by 1609, the dream of unity had been replaced by the compromises of diplomacy. In March of that year, five years after England's Somerset House Treaty with Spain, the States General, representing the rebel states of the Low Countries, had signed up to a twelve-year truce with their former Spanish masters. The Northern people, Sandys calculated, were now in as good a position as they would ever be to found a great empire, and to this end, he helped produce what would be their guiding instrument: the second Virginia charter.[27]

As in 1606, early negotiations were conducted in secret. On 14 February 1609, even the well-informed John Chamberlain had no idea what was afoot, reporting in one of his regular letters to Dudley Carleton that 'news here is none at all, but that John Donne seeks to be secretary of Virginia'.[28] However, foreign spies were better informed. 'At first we always thought of sending people little by little,' one of Zuñiga's informers overheard a Privy Councillor say, 'but now we see that what we should do is establish ourselves [on a large scale] all at once, because [then] when they open their eyes in Spain they will not be able to do anything about it.'[29] The Venetian ambassador reported to the Doge that a ship was being readied to take 'eight hundred persons on board, many oxen and ponies and other things needful for developing a district near Florida'.[30]

In its opening paragraphs, the second charter allowed what Cecil had until now managed to deny: 'a further enlargement and explanation' of the company's 'grant, privilege and liberties'. From now on, its own officers were 'to manage and direct' the company's affairs. Anyone could join the company, 'whether they go in their persons to be planters there in the said plantation, or whether they go not, but do adventure their monies, goods or chattels', and to demonstrate this spirit of openness, the charter listed the six hundred and eighteen individuals, a large proportion of them MPs, and fifty-four London livery companies who had so far pledged to back the venture.[31]

The Royal Council survived, but now had fifty-two members. Sandys was to become a member, along with at least two Sirenaicals (Christopher Brook and Henry Neville). Members were henceforth to be 'nominated, chosen, continued, displaced, changed, altered and supplied, as death or other several occasions shall require' by the company itself, though new members would need their election to be ratified by one of the senior members of the Privy Council.

On the vexed issue of local government, the factious president and council were to be replaced by a single governor, appointed by the Royal Council to have sole responsibility for running the settlement's affairs. This development may have reflected the influence of Sir Thomas Smythe, who remained the Virginia Company's treasurer, and a dominant, if discreet, presence in its proceedings.

Smythe was an MP, serving alongside Sandys upon parliamentary committees investigating trade and other economic matters. But he was always careful to distance himself from controversy, and would excuse himself from speaking up in difficult debates by claiming to lack the oratorical skills. The same political pragmatism informed the reorganization of the company. The new arrangements gave the company far more control over the colony, as a single governor would not only be more manageable, but would be required to enact the company's own 'orders, laws, directions, instructions, forms and ceremonies of government and magistracy', rather than the law of England.

Though intentionally vague on the details of such laws and offices, the charter was very specific that, upon the new governor's arrival in Virginia, the authority of the existing president and council was to 'utterly cease'. Furthermore, all existing 'officers, governors and ministers formerly constituted or appointed' in the settlement were to be summarily discharged and, after being made aware of the King's 'will and pleasure', were to pledge 'that they forthwith be obedient to such governor.'

The other prominent feature of the charter was its strengthening of the powers of the settlement's officials, who were given 'full and absolute' authority 'to correct, punish, pardon, govern and rule' anyone within the 'precincts of Virginia or in the way by the seas thither and from thence'. The governor could impose martial law 'in cases of rebellion or mutiny', which allowed for summary justice and put at his discretion the use of the death penalty.[32]

The charter negotiations continued well into the spring of 1609. Meanwhile, fuelled by the outpouring of propaganda, urgent preparations got under way to raise finance and appoint personnel for a huge new mission, to be launched within weeks, in the hope that it would reach Jamestown before the settlement's collapse into anarchy or famine.

By May, it had been agreed that six hundred men would be sent in eight ships. The man appointed to lead them as the new governor of Virginia was Sir Thomas Gates, one of the patentees of the first charter. The veteran soldier had until the previous April been stationed at the Dutch frontier town of Oudewater, where he had been in the service of the States General, the body representing the rebel Dutch states. The States General, perhaps attracted to the idea of the Virginia venture as part of a great Protestant mission, had agreed to give him a leave of absence for one year 'to be in command . . . in the land of Virginia, of the peopling of that land'.[33]

A set of 'instructions, orders and constitutions' was issued to Gates 'for the Direction of the affairs of that Country for his better disposing and proceeding in the government thereof'.

They began by ordering Gates 'with the first wind to set sail for Virginia'. Upon arrival, he was to summon the council and settlers, and read out his commission, revoking the powers of the existing council, and announcing Gates's appointment as sole governor. He was to appoint a new council, which would act in a purely advisory capacity.

Gates was then to look for a new location to act as the settlement's principle city, Jamestown being only fit to act as a port 'for your Ships to ride before to arrive and unload at'. 'It is likely you shall find some convenient place to this purpose' somewhere beyond the falls, Gates was informed, 'whither no enemy with ease can approach nor with ordnance'. This intelligence was gathered from Newport's expedition into Monacan country during his last visit.

A third settlement should be established at Ocanahonan, on the Chowanoke river, towards the original site of the Roanoke. There, the instructions asserted upon unknown advice, 'you shall find four of the English alive left by Sir Walter Raleigh, which escaped from the slaughter of Powhatan [i.e. the ruler] of Roanoke'.

The instructions issued under the first charter specified that the settlers were to take 'just, kind and charitable courses' with the Indians. Those issued to Gates identified them as the enemy. 'It is clear even to reason beside our experience' that Powhatan 'loved not our neighbourhood and therefore you may no way trust him, but if you find it not best to make him your prisoner yet you must make him your tributary'. Furthermore, Gates was to free the *mamanatowick*'s subjects 'from the tyranny of Powhatan'. If it was necessary to trade with the Indians, all dealing with them should be done within the fort, and on the basis that 'you need care for nothing of theirs, but rather that you do them a Courtesy to spare such necessaries as they want as little Iron tools or copper'.

The instructions ended with a list of specifics about personnel and equipment. A new cape merchant was to be appointed, Thomas Wittingham, 'one in whose sufficiency and honesty we have great Confidence'. John Woodall, a surgeon who had been interested in Virginia since the early days, would provide a 'chest of surgery', surgical apparatus, along with 'his fellow William Wilkinson'. The company would also supply Gates with duplicate 'black boxes with divers marks, wherein are our commissions in cases of death or other vacation of the governor'. In the event of losing the governor, the boxes were to be 'succinctly' delivered to the assembled council in Virginia, so that a new governor could be installed before a further outbreak of faction fighting.

Moves were already well under way to raise the finance needed to pay for the expedition. Exploiting the new mood of religious and civic duty, approaches were made to London's livery companies, representing the city's various trades. Investing in the enterprise was 'as an action concerning God,' they were told, and a civic duty. As an extra incentive, anyone investing more than £50 was also promised automatic promotion to the reconstituted Royal Council, which would provide them with regular access to some of the most powerful people in the royal court.[34]

The response was mixed. The warden of the Company of Fishmongers, whose members had been involved in the venture since its inception, circulated the council's letter among the company's members, 'to exhort them to venture money to Virginia for Plantation'. Some of the company's officers responded positively, but 'the generality of the Company' were less eager. They had to be 'particularly, earnestly, persuaded to adventure anything', and a quarter of them flatly refused, it was disapprovingly noted.[35]

The Grocers' Company, being one of the richest in the City, merited a personal appearance by the Lord Mayor, who made a 'most worthy and pithy exhortation unto the generality' about the opportunity presented by Virginia. The company's freemen (members) were urged to line up before the company clerk 'to set down what and how much he will contribute for so honourable a service'. The exercise raised just £69 with a note being taken of the names of 'those which denied and refused to make any such contribution'.[36]

The Clothworkers had already shown an interest in colonial ventures through support of its member, Richard Hakluyt. In addition, two of its richest and most influential members, John Eldred and the piratical Sir John Watts, were centrally involved in the Virginia venture. Eldred had been appointed to the Virginia Company's old Royal Council, and both he and Watts were now made members of the new, expanded council.

Eldred was still counting the cost of the catastrophic Trinidad expedition, which had put a halt to the lucrative supply of tobacco grown in the interior of Guiana. He was also still smarting over the royal 'imposts' on imports of Spanish tobacco. Through contacts with Hakluyt, the idea may have already formed in his mind of running trials to cultivate Spanish tobacco in Virginia.[37] His enthusiasm no doubt helped persuade his fellow Clothworkers to agree to a £100 corporate investment, in addition to unrecorded sums privately ventured.[38]

The Merchant Taylors proved to be the most generous. They were famous for their extravagance, in one year lavishing £1,000 on entertaining the King. They initially offered £100 towards Virginia, with individual members of the company eventually stumping up a total of £586 13*s* 4*d*. The biggest contribution came from Ralph Hamor, who was also a director of the East India Company. Hamor ventured the princely sum of £75, and his son, Ralph junior, who was volunteered by his father to join the next expedition across the Atlantic. The Drapers, Goldsmiths, Skinners and Ironmongers all agreed to make contributions, though more in line with the Fishmongers. The Haberdashers appear to have ventured nothing.[39]

While contributions were sought from merchants and companies, individual tradesmen were being encouraged to venture their persons as well as their purse. A document was discreetly circulated around the workshops of London 'Concerning the Plantation of Virginia New Britain'. Those that bothered to put down their tools and pick it up found themselves the target of unusually desperate official pleas. 'Many noble persons, Counts, Barons, Knights, Merchants and others' were preparing to go to Virginia, it informed the reader, 'very speedily as is necessary'. This provided an opportunity for 'workmen of whatever craft they may be' to join in, be they 'blacksmiths, carpenters, coopers, shipwrights, turners . . . men who make bricks, architects, bakers, weavers, shoemakers, sawyers and those who spin wool and all others, men as well as women, who have any occupation, who wish to go out in this voyage for colonizing the country with people'. In return for investing £12 10*s* – eight months' wages for a humble carpenter or bricklayer – they were promised 'houses to live in, vegetable-gardens and orchards, and also food and clothing at the expense of the Company of that Island, and besides this, they will have a share of all the products and the profits that may result from their labour'. (There was no mention of who would pay for the cost of shipping them over, or what would happen if they chose to return to England.) Anyone who was interested was told to report as soon as possible to Sir Thomas Smythe's house in Philpot Lane.[40]

From his house in Highgate, Zuñiga beheld these events with growing horror. He claimed that eight *grandes maestros* of the main trades had been made to join the venture 'under compulsion', in the hope of attracting other artisans and craftsmen. The strategy had apparently worked, with more than forty now signed up. Earlier that month, he had also heard that Dutch 'rebels' as well as their English supporters were being actively recruited in the Low Countries, attracted by the prospects of a new adventure now that a truce with Spain had reduced their chances of active service in Holland.

The port at Woolwich was very different from Blackwall, the inconspicuous point of departure for the fleet that had left under the first Virginia charter two years earlier. This was the location of the Royal Docks, a place where many great ships and expeditions were launched.

On 8 May 1609, Prince Henry himself came to visit. He was there to discuss progress on the *Prince Royal*, an enormous three-deck warship being built for him by the crooked shipwright Phineas Pett. This gave him an opportunity to inspect the vessels of Gates's great Virginia fleet, which were being loaded and rigged for a speedy departure.[41]

A week or so later, the first passengers were beginning to assemble. Lost in the midst of the excited, anxious crowd was a man in his mid-20s called Henry Spelman, wondering what he had let himself in for. Spelman came from a distinguished family, his uncle being a famous antiquary called Sir Henry. There are other Spelmans who turn up in unexpected places in the historical record who may have been relatives. In the 1590s, a James Spelman was apparently spying on the Spanish for Raleigh, and a Thomas was an impecunious poet with links to the Sirenaicals. Henry had got into some teenagerly trouble. He volunteered no details, but it resulted in the 'displeasure' of his friends and being disinherited by his great-uncle. This prompted him to go to Philpot Lane and sign up for the Virginia venture.

Spelman found himself amidst a throng of equally disorientated and dispossessed young men. Efforts to recruit enough 'land men' or settlers had proved only partially successful. According to some accounts, the target of six hundred may have been missed by a margin of one hundred. In an effort to bolster numbers, the Lord Mayor of London had even urged magistrates to offer free passage, 'meat, drink and clothing, with an house, orchard and garden, for the meanest family, and a possession of lands to them and their posterity' to any 'inmates [that] hath not place to remove unto, but must lie in the streets'.[42]

As well as assorted vagrants and petty criminals, there were thirty 'unruly youths', their passage being paid by the parishes wanting to get rid of them. They were joined by thirty women, mostly unmarried servants shipped out to become settlers' wives. They were joined by a quantity of livestock jostling for space on the quayside: one or two stallions and fourteen or fifteen mares, a few young bulls and several cows, poultry, goats and assorted pets.[43]

Like Spelman, a majority of those who had been recruited were more concerned with escaping problems at home than embracing opportunities abroad. The departing letter of one Evelyn to his mother probably reflected the circumstances of many. He was going on 'a long and dangerous vo[yage with] other men to make me to be [able] to pay my debts', he explained, 'and to restore my decayed estate'. 'I beseech you, if I do die, that you would be good unto my poor wife and children, which, God knows, I shall leave very poor and very mean, if my friends be not good unto them, for my sins have deserved these punishments and far greater at God's hands.' Meanwhile, he wondered if his mother could clear a debt of a hundred 'marks' (about £66) owed to a Mr Stoughton. 'I am much grieved at my heart for it that my estate is so mean, that at this time I am not able to repay it.' He ended his plaintive missive with another apology, that he had not managed to bid farewell in person to her and his brother Robert, 'but that the Captain of the ship made such haste away so suddenly'.[44]

A few had religious reasons to leave the country. John Want, of Newport, near Saffron Walden, Essex, was a Puritan. His father appears to have been a 'husbandman' (owner of a small farm) who in the mid-1570s merited three citations in Quarter Sessions records for unrecorded misdemeanours. Want junior was accused of being a Brownist.[45]

The man who made the accusation was William Strachey, who, coming from Saffron Walden, knew something of Want's religious views. He, however, had no interest in or desire for religious revolution. He was a habitué of the Mermaid Tavern, and had long nurtured ambitions to join the pride of literary lions that lurked there. However, he was short of money, lacked a patron, and, now in his mid-30s, had yet to write anything of public note, except for a prophetic sonnet introducing Ben Jonson's famous political play *Sejanus His Fall*, which pondered the capriciousness of human fortune:

> *. . . as lightning comes behind the thunder*
> *From the torn cloud, yet first invades our sense,*
> *So every violent fortune, that to wonder*
> *Hoists men aloft, is a clear evidence*
> *Of a vaunt-curring blow the fates have given*
> *To his forced state . . .*[46]

Strachey was angling to become the new Virginia mission's official chronicler, in the hope of producing a book on the subject that would be as popular as those flying off the stalls in St Paul's churchyard.

John Rolfe, who had signed up to make the crossing with his heavily pregnant wife, would become one of the most famous figures in the Virginia story, but his motives for going remain opaque. He was a Norfolk farmer, classified as a 'gentleman', but probably of a modest estate. His decision to hazard not only himself but his family and future on such a dangerous enterprise was probably restlessness and ambition. He was certainly an adventurous man, perhaps escaping problems at home but certainly drawn by opportunities abroad. Given his subsequent role, it is possible that he had some undocumented contact with the tobacco merchants Eldred and Watts, which led to them pressing a sample of Trinidad tobacco seed into the palm of his hand with a suggestion that he try it in the Virginian soil.

Though they were staking their lives on being able to live there, most of the passengers would have known next to nothing about their destination. Those able to read would have found it difficult or too expensive to get hold of published material on their destination, books still being a rarity in most people's lives.

There were, however, some old hands. One of Gates's instructions had required him 'to call before you Captain John Ratcliffe and one [Thomas] Webbe', to hear their complaints of 'divers injuries and insolences'. Having done so, the governor had chosen to restore Ratcliffe to the mission's leadership, in the face of warnings from Captain John Smith. He had also decided to take Gabriel Archer and John Martin back on board.[47]

Smith's complaints in *True Relation* about Christopher Newport had also gone unheard. Newport was to be the mission's vice-admiral, a demotion probably reflecting the poor return from his previous expedition.

Newport's boss, the admiral of the fleet, in charge of getting the men and supplies across the ocean, was the 65-year-old Sir George Somers. In an age when the destructive arts of back-stabbing and invective flourished, hardly a bad word was written of Somers. Probably the worst appeared long after his death, when he was remembered for being 'a lamb on the land, so patient that few could anger him, and (as if entering a ship he had assumed a new nature) a lion at sea, so passionate, that few could please him'. He had a better claim than John Smith to modest beginnings. His mother had been a shopkeeper in the Dorset coastal town of Lyme Regis, and nothing is known for certain about his father, though a muster of the local militia dated 1539 lists a John Somer as an 'able archer'. He married modestly, too, his wife, Joan, being the daughter of a yeoman whose dowry was a candlestick. He went on to become Mayor of Lyme Regis and eventually its MP, a role he clung to, even during his long absences overseas. This, then, was the lamb. The lion first started roaming the high seas in the 1580s, and accumulated a huge quantity of prizes that transformed him into a rich man, and his crews into eager accomplices. He had invested the money in land, ships and, since 1607, the Virginia venture, which had so far earned him little but grief. Now, as the mission's admiral, he had a chance to retrieve some dividends.[48]

By mid-May the fleet was ready to make the first leg of the

journey, to Plymouth, from where it would enter the Atlantic as soon as the winds were favourable. The *Blessing* was already gone by the fifteenth, carrying Gabriel Archer, 'six mares and two horses' and an undocumented number of passengers. She was soon followed by two 'small vessels', one of them carrying Sir George Somers.

However, deep in the political background, a gargantuan power struggle was under way. Cecil was by now openly hostile towards the venture. He could see that the ambitious Prince Henry intended to use Virginia to revive the 'heroical' imperial plans of the late Elizabethan era. He was desperate to find some way to trim the mission's sails, but without provoking a direct confrontation with the Prince. His efforts eventually resulted in the issuing of an emergency order, commanding a halt to proceedings and recalling Gates to London. The remaining passengers 'assembled here [in London] for Virginia,' Zuñiga reported, 'have been detained, because the orders they carry did not seem proper'.[49]

There is no record of exactly what happened next, but the upshot was that Gates was demoted to lieutenant governor. Thomas West, Lord Delaware, was now to be 'Lord Governor and Captain General' of Virginia. As he was a member of the Privy Council, Cecil presumably hoped Delaware could be relied upon to keep an eye on the government's wider interests.[50]

The intervention threw the mission into disarray. From his house in Cow Lane, leading off London's meat market at Smithfield, Henry Spelman's uncle Sir Henry sent an urgent letter to his son John. Henry minor was among those who had been detained, and, in scholar's Latin, Sir Henry asked John to visit Richard Hakluyt at his apartment near Westminster Abbey, to see if he knew when the ships bound for 'Indiam occidentalem vz: ad Virgineam' would finally leave.[51]

In the event, Henry's ship sailed a few days after the others, and by the 29th or 30th, the fleet, now numbering nine vessels, lay at anchor in Plymouth harbour, favourable conditions inviting it into the glittering ocean beyond the harbour entrance. However, Governor

Gates was still detained in London with the commissions and sealed caskets, forcing a further delay.

The flagship of the fleet was the 300-ton *Sea Venture*, probably the large merchantman part-owned by the prominent Sirenaical Lionel Cranfield. She had just completed a series of voyages carrying cloth to Stade and Middelburg in Germany, netting for Cranfield a healthy profit of around £280 a year.[52]

Under Newport's command, the flagship was to lead her eight consorts along a new route. Another ship had already left to reconnoitre the way. Its captain was Samuel Argall, a cousin of Sir Thomas Smythe. Under conditions of great secrecy, he had been given the commission 'to attempt a direct and clear passage' to Virginia, 'by leaving the Canaries to the East, and from thence, to run in a straight western course'. This was to 'avoid all danger of quarrel with the subjects of the King of Spain', and reduce the time and cost of the journey. It was a dangerous mission, as it involved tackling 'winds and currents which have affrighted all undertakers by the North'. To help him find the way, he was accompanied by Prince Henry's gunner Robert Tyndall, acting as navigator.[53]

Gates's fleet was to follow in Argall's wake. Though theoretically quicker, the disadvantage of the new route for Gates was not only the increased danger of difficult weather and sea conditions. After leaving the Canaries, there would be nowhere to stop for emergency supplies during the 3,500-mile leg to Chesapeake Bay. The barrels of fresh water and food packed into the holds would need to last the duration, so any delay that resulted in stores being used up before departure, or spoiled before the projected date of arrival, was potentially catastrophic.

Nevertheless, the fleet continued to await the governor's arrival. During the lull, John Ratcliffe took the opportunity to go ashore, and make his way to the office of a local notary called Richard Streamer. There he made out his will. 'I, John Sicklemore alias Ratcliffe, Captain of the *Diamond* now bound for Virginia,' it began, 'give to Dorothy my wellbeloved wife all my goods and chattels, lands, leases, debts, bills.' Like the husband, next to nothing is known about the wife, except that she later married a George Warburton.

Ratcliffe also bequeathed Dorothy 'all and every part of my adventure [investment] which I have delivered unto the hands of the Treasurer for Virginia'. No other material possessions are mentioned, suggesting that Ratcliffe/Sicklemore was now a man of modest means. He made Richard Percival, one of Cecil's most senior servants and Ratcliffe's 'loving friend', his executor. Three locals were brought in to witness the captain seal the document.[54]

At around the same time, Captain William Stallenge, Cecil's agent in Plymouth and an investor in the expedition, sent an anxious letter to his master in London, asking when Gates might be expected. His 'coming hither' was 'much desired to the end the ships may be speedily dispatched', he wrote. Problems over the orders evidently remained, as 'by Sir Walter Cope's direction' Stallenge was enclosing some letters 'concerning the Virginia business, wherewith I doubt not but he will acquaint you more at large'. 'Sir George Somers hath been here this two days, and the ships, if weather . . . willing, shall be ready this next day,' he added, providing a glimpse of Somers's mounting exasperation. For unexplained reasons, the admiral had decided to leave his own ship and join Newport aboard the *Sea Venture*.

With no firm information as to when Gates would arrive, Somers may have unilaterally given the order for the rest of the fleet to set off ahead on the evening of 1 June. Gates apparently arrived later that night, carrying the sealed caskets holding the (presumably revised) instructions for the event of his death or absence.

The *Sea Venture*, with a pinnace in tow, left the following day, carrying the entire leadership of the new venture, who, thanks to the government's mysterious last-minute intervention, might not have been on good terms with one another.

The ship had caught up with the rest of the fleet by 14 June. Archer mentions seeing that 'one of Sir George Somers's pinnaces left our company and, as I take it, bore up for England' on that day, though he does not say why. This left eight ships heading off north-west from the Canaries into the deep ocean: the *Sea Venture*; the *Diamond*, carrying Ratcliffe; the *Falcon*, with John Martin; the *Blessing*, with Archer; the *Unity*, with Henry Spelman; the *Lion*; the *Swallow*, with Sir George Somers's nephew Matthew aboard; and the *Virginia*, the ketch built in America by the Northern Colonists. No opportunity presented itself for the leaders or black boxes to be redistributed among the other ships, so they all remained aboard the *Sea Venture* as it breasted into the deep ocean.

THIRTEEN

Promised Land

IN *MOBY DICK*, Herman Melville described what it was like to be lookout on a ship's masthead, broaching the tropical seas:

> In the serene weather of the tropics it is exceedingly pleasant, the mast-head: nay, to a dreamy meditative man it is delightful. There you stand, a hundred feet above the silent decks, striding along the deep, as if the masts were gigantic stilts, while beneath you and between your legs, as it were, swim the hugest monsters of the sea, even as ships once sailed between the boots of the famous Colossus at old Rhodes. There you stand, lost in the infinite series of the sea, with nothing ruffled but the waves. The tranced ship indolently rolls; the drowsy trade winds blow; everything resolves you into languor.[1]

On Monday 24 July 1609, St James's Day, the sea's agitation suddenly increased. A gust lashed the man up the masthead of the *Sea Venture*, and on the horizon, he saw a huge black-gloved cloud make a fist.

The fleet had not managed to keep to the route originally intended, and in search of suitable winds had strayed south of the Tropic of Cancer. Nevertheless, by Newport's reckoning, they were now making good progress, and were within seven or eight days of their destination.

To get even this far had been an achievement, won at great cost. The North Atlantic is a vast gyre of wind and currents, revolving clockwise, the trade winds carrying shipping east to west along its southern rim, skimming past Cape Verde and the West Indies, the

Gulf Stream carrying them back again via Labrador. In between lay the dreaded 'horse latitudes', a region of light winds and hot weather where stranded ships were forced to kill any horses they were carrying, to preserve precious water supplies from the notoriously thirsty creatures, and replenish food stocks with the meat.

Somers and his mariners had found little but 'fervent heat and loom [listless] breezes' there. Two ships, one of them the *Unity*, suffered outbreaks of 'calenture', the disease of the subtropical seas that reduced sufferers to delirium. From the relative safety of the *Blessing*'s deck rail, Archer watched the crews and passengers of these unfortunate vessels offload thirty-two bodies into the sea, while infection rampaged down below. Meanwhile, on the *Diamond*, two children were born, both barely escaping the waters of the womb before their lifeless bodies were being committed to the sea.[2]

Now the sky had gone dark, 'no less all the black night before – the clouds gathering thick upon us, and the winds singing and whistling most unusually'. Squalls disorganized the fleet's neat array, and the *Sea Venture* cast off the pinnace it had in tow for fear of collision.

'A dreadful storm . . . began to blow from out the north-east, which swelling and roaring, as it were, by fits, some hours with more violence than others, at length did beat all light from heaven, which like an hell of darkness turned black upon us, so much the more fuller of horror,' wrote Strachey, discovering in these distressing conditions his literary voice.

He had experienced storms before, 'as well upon the coast of Barbary and Algiers in the Levant, and once more distressful in the Adriatic Gulf'. The latter was endured in the 'bottom of a Candy', a small Cretan fishing vessel, 'so as I may well say, *Ego quid sit ater Adriae novi sinus & quid albus peccet Iapyx.*' (How well I know the gloomy Adriatic, and the mischief of the clear west-nor'wester.) But he had seen nothing like this.[3]

'For four and twenty hours the storm in a restless tumult had blown so exceedingly as we could not apprehend in our imaginations any possibility of greater violence. Yet did we still find it not only more terrible but more constant, fury added to fury, and one storm urging a second more outrageous than the former, whether it so wrought upon our fears or indeed met with new forces.

'It could not be said to rain. The waters like whole rivers did flood in the air. And this I did still observe, that whereas upon the land when a storm hath poured itself forth once in drifts of rain, the wind, as beaten down and vanquished therewith, not long after endureth. Here the glut of water, as if throttling the wind erewhile, was no sooner a little emptied and qualified but instantly the winds, as having gotten their mouths now free and at liberty, spake more loud, and grew more tumultuous and malignant. What shall I say? Winds and seas were as mad as fury and rage could make them.'

For Strachey, one of the most terrifying aspects of the storm's violence was the noise. No man could hear another's cries of help or instruction. 'Our clamours [were] drown'd in the winds, and the winds in thunder,' he wrote. 'Prayers might well be in the heart and lips, but drowned in the outcries of the officers, nothing heard that could give comfort, nothing seen that might encourage hope.' Not even the legendary Greek herald Stentor, whose shout was as powerful as that of fifty men, could 'express the outcries and miseries' suffered in the midst of that maelstrom.

The storm – the lashing tail of a hurricane, according to Archer – lasted nearly two whole days 'in its extremity', during which time 'the heavens look'd so black upon us that it was not possible the elevation of the Pole might be observed, nor a star by night, not sunbeam by day was to be seen'. When the sky eventually cleared and the wind abated, the fleet was scattered. From one masthead, no others could be seen. Aboard the *Blessing*, Archer and the master Captain Adams wondered what to do. In the event of the fleet being dispersed, Somers had arranged that the ships should head for the island of Barbuda, and await the others for up to seven days, before continuing on to Virginia. However, calculations made by the compass and the angle of the noonday sun, shining at last in the fresh blue sky, showed that the ship was within a few days' sailing of its destination and nowhere near the rendezvous location. With both vessel and passengers in such a battered state, and supplies running low, the decision was taken to continue on.[4]

'So we lay away directly for Virginia, finding neither current nor wind opposite,' wrote Archer. Five or six days later, they reached Cape Henry. There they found three other survivors of the storm, the *Lion*, the *Falcon* and the 'sore distressed' *Unity*, Spelman's ship. Sixty of the *Unity*'s seventy passengers were sick or dead, and all her crew were ill, except for the ship's master, his boy and 'one poor sailor'.

When news reached Jamestown of four ships entering the James, the alarm immediately went up of Spanish attack. Several weeks earlier, Captain Samuel Argall and his navigator Robert Tyndall had arrived in the Chesapeake, having in just a month successfully reconnoitred the more northerly route across the Atlantic, 'the ready way without tracing through the Torrid Zone', as Archer described it. A few days after Argall's arrival, a pinnace had been spotted entering the bay. After so many false alarms, it was assumed to be the vanguard of Gates's fleet. However, it turned out to be Spanish, and Argall, whose ship remains unidentified but was obviously a substantial man-of-war with 'two topsails and a great banner at its masthead', had to be deployed to chase it away.[5]

This time, President John Smith's scouts were determined not to be caught off guard, and news of the four ships advancing up the river stimulated a scuttling for weapons.

John Smith later boasted that, during his presidency, he had 'so determined and ordered' the settlement's defences, that he 'little feared' the Spanish. It was his 'countrymen and friends' that terrified him. So, once it had been determined that the ships coming up the James were English, he extended them a cautious welcome.

They tied up on 11 August 1609, and started to unload passengers and supplies into the fort. Smith's caution turned to dismay when Gabriel Archer and John Martin, 'graced by the title of Captains of the passengers', presented themselves, 'who as they had been troublesome at sea, began again to mar all ashore'.

Scarcely more reassuring was the condition of their charges: hundreds of battered, weak and hungry men, women, 'unruly youths', children and livestock. Having endured a two-month crossing and a hurricane, they might have been hoping to find themselves requited in the bosom of a Promised Land. Instead, they found themselves descending into an inferno of starvation and sedition.

Since taking over the presidency the previous September, Smith had endured a familiar catalogue of disappointments and disasters.[6]

Things had started off well, according to his version of events. Soon after Newport's departure for England on 3 December 1608, a series of altercations had brought the English to the brink of outright war with the Indians, but had ended thanks to Smith's firm action, which he claimed 'so amazed and affrighted both Powhatan and all his people that from all parts with presents they desired peace'.

In the absence of skirmishes and threats, the settlers had an opportunity to engage in more productive activity. Within weeks, Smith claimed remarkable progress. They had produced forty or more barrels of pitch, tar and soap ashes. They had dug a well within the fort, which yielded 'excellent sweet water'. They had constructed a glass factory on the mainland, and produced some samples. They had built twenty more houses within the fort, repaired the church, and reinforced a defensive blockhouse over the neck of land joining Jamestown island to the mainland. They had also raised more than sixty pigs from the three sows delivered by Newport with the last supply, and enjoyed the company of five hundred chickens, which had 'brought up themselves (without having any meat given them)'.

Then summer came, and familiar problems resurfaced. Checking through the barrels containing supplies built up over the previous months, the cape merchant Taverner found that half the corn had rotted away. The other half had been consumed by 'the many thousand rats (increased first from the ships)'.

Faced with yet another famine, the settlers began to agitate for a return to England. 'Thousands were their exclamations, suggestions, and devises . . . to have made it an occasion to abandon the Country,' Smith complained. His solution was to disperse the colony, in the hope they could see through the next few weeks by living off the land and waters. Sixty to eighty men were sent downriver with William Laxon, a carpenter promoted to the rank of ensign, where they were to try and live off the seafood found on the bay. Twenty were sent with George Percy to fish around Point Comfort at the mouth of the river. A further thirty or forty were sent with Francis West to see what they could find around the falls.

Such extreme measures were only possible because of the peaceful relations with the Indians, but they, too, came under strain. The previous year, Smith had sent to Werowocomoco four of the 'Dutchmen' who had arrived with Newport. There, under an agreement with Powhatan, they were to build an English-style house for the emperor.

No sooner had they gone than they defected, organizing the theft of sought-after English weapons and tools from the fort. Smith took six months to realize what was going on, as he had assumed equipment losses to be caused by petty pilfering. He tried to recover the situation by getting another foreigner, the Swiss mines expert William Volday, to 'reclaim the Dutchmen'. However, to his dismay, Smith found that the 'wicked hypocrite' Volday was actually working in league with them 'to effect their project to destroy the colony'. The Dutchmen were eventually brought back to the fort, their desertion going unpunished for reasons Smith kept to himself.

In the weeks leading up to the arrival of Archer and the others, the settlement had rallied. Most of those who had been dispersed around the country returned, and were sustained on an unexpected haul of sturgeon. Those down on the bay remained where they were, surviving well enough on an abundance of oysters. The arrival of Argall's ship had brought extra relief, as Argall allowed the settlers to buy goods from the ship's store, which was relatively full thanks to a speedy crossing.

Thus, in Smith's estimation, the settlement that Archer and the others were now to inhabit was in reasonably good shape, given the circumstances. It was, he claimed, 'strongly palisaded, containing some fifty or sixty houses'. Soldiers had been trained, fields had been planted and the Indians had, thanks to his firm treatment of them, become 'tractable'. All this had been achieved with 'but one Carpenter . . . and three others that could do little, but desired to be learners; two Blacksmiths; two sailors' and labourers who 'were for the most part footmen', gentlemen's servants, who 'never did know what a day's [hard manual] work was'. Furthermore, he claimed that few had died under his presidency, though he gave no figures on survival or population rates.

Archer had a very different view. The inhabitants he found to be 'in such distress' that many had been forced to 'disperse in the savage's towns, living upon their alms for an ounce of copper a day'. As for the eighty or so living 'twenty miles from the fort and fed upon nothing but oysters eight weeks' space', he noted that, should they return, there was no food allowance for them in the settlement's store. The Indians were in no position to help, as they too were suffering shortages.

He acknowledged that not all of the problems were Smith's fault. Insufficient food had been brought from England to alleviate the situation, because Newport 'and others' had exaggerated the quantity of provisions left by the previous supply.

A few days after the arrival of the first four ships, the *Diamond* appeared. Her mainmast had been lost during the storm, and most of her crew and passengers were 'very sick and weak'. She had no news of the *Sea Venture* and the mission's leadership.

She disgorged another of Smith's enemies: John Ratcliffe, completing the triumvirate he had specifically warned the Royal Council would be 'sufficient to keep us always in factions' if it ever returned to American shores.

A few days after the *Diamond* came the *Swallow*, also with her mainmast missing, lying low in the water due to a 'shrewd' (dangerous) leak. She, too, had heard nothing of the admiral.

'Now did we all lament much the absence of our governor, for contentions began to grow, and factions and partakings,' wrote Archer. Without the sealed caskets identifying Gates's successor, the inevitable power struggle ensued. Smith clung on as president, but, he claimed, his foes 'did their best to murder' him, 'to surprise the store, the fort, and our lodgings, to usurp the government, and make us all their servants, and slaves to our own merit'.

'To a thousand mischiefs those lewd Captains led this lewd company, wherein were many unruly gallants packed thither by their friends to escape ill destinies,' Smith railed. This way and that they would 'dispose and determine of the government, sometimes one, the next day another, today the old commission, tomorrow the new, the next day by neither ... They would rule all or ruin all; yet in charity we must endure them thus to destroy us, or by correcting their follies, have brought the world's censure upon us to have been guilty of their bloods.' It would have been better, he argued, if they had never arrived, 'and we for ever abandoned, and (as we were) left to our fortunes, for on earth was never more confusion, or misery, than their factions occasioned'.

The arguments were over the status of the new charter. Smith argued, in the absence of formal instructions telling him otherwise, that he should continue as president under the terms of the old charter. Ratcliffe pointed out that all the members of the council in Virginia alive at the time of Newport's departure (Peter Wynn, Matthew Scrivener, Robert Hunt and Richard Waldo) had since died, leaving Smith to reign as 'sole governor ... without assistants', admitting of 'no council but himself'.[7]

It was finally agreed, at least among Ratcliffe's faction, that Smith should retain the presidency for the remaining few weeks of his term, which formally ended on 10 September 1609. He would then be succeeded by Francis West, Delaware's brother, acting as provisional governor under the terms of the second charter.

According to Smith, 'it would be too tedious, too strange, and almost incredible, should I particularly relate the infinite dangers, plots, and practices' he 'daily escaped amongst this factious crew', but the upshot was an outbreak of political manoeuvring. It culminated with Smith resigning the presidency to the sickly Captain John Martin, 'who knowing his own insufficiency, and the company's scorn, and conceit of his unworthiness, within three hours resigned it again to Captain Smith'. At this point, all semblance of government began to break down.

Francis West unilaterally decided to implement the instruction given to Gates to 'find some convenient place' at the falls to establish a new colonial capital. He set off in one of the ships upriver with one hundred and twenty of the new settlers, and started to scout for locations.

In a bid to show he was still in charge, Smith followed, taking Henry Spelman with him. Somewhere towards the upper reaches of the navigable river, Smith met West, who was on his way back to Jamestown to fetch materials. West informed Smith that he had found an uninhabited area for the new settlement near the falls, on the southern bank of the river, and had ordered his men to start constructing a fort there. He then continued on his way.

Smith sailed on to the site, and decided it was 'inconsiderately' located, being 'not only subject to the river's inundation, but round environed with many intolerable inconveniences'. He thought it would be better nearer the falls, and for that purpose negotiated the purchase from the Indians of 'Powhatan's Tower', a 'Savage Fort, ready built, and prettily fortified with poles and barks of trees, sufficient to have defended them from all the Salvages in Virginia, dry-houses for lodgings and near two hundred acres of ground ready to be planted'.

As part of the deal, he 'sold' Henry Spelman to the local chief Parahunt. At the time, Spelman knew nothing of the transaction, but simply found himself left with the Indians. The chief, his new master, treated the young man well. 'He made very much of me,' Spelman later recalled, 'giving me such things as he had to win me to live with him.'[8]

Smith went back to West's fort and ordered the men to stop work. He had negotiated an excellent deal with Parahunt, he announced, with the chief promising to provide not only access to his land, but protection from attack by the Monacan, as well as corn supplies at a rate of a bushel for a square inch of copper, and an annual tribute of *pocone* or puccoon (*Lithospermum caroliniense*), a root that produced a valuable red dye.[9] But West's men, 'having bestowed cost to begin a town in another place', rejected Smith's deal, and 'an unkindness thereupon [arose] between them'.

The ensuing confrontation may have involved the drawing of weapons, and Smith, with five companions facing West's one hundred and twenty, was forced to retire. West had by now sent his ship back upriver with more materials, and Smith 'surprised' one of the boats ferrying the cargo ashore. He forced it back to West's ship, where he befriended the mariners, and persuaded them to help him commandeer the freight remaining on board, which he transferred to his own ship, thus forcing a halt to the work.

Over the next few days, Smith tried to persuade West's men to move to Powhatan's Tower, but failed. George Percy, who was not directly involved, claimed that the captain, 'perceiving both his authority and person neglected' then 'incensed and animated the savages against Captain West and his company, reporting unto them that our men had no more powder left them than would serve for one volley of shot'.[10]

Spelman had by now spent a week with the Indians and become homesick. He asked Parahunt to release him so he could go back to Jamestown, on the pretext of fetching some possessions. Parahunt agreed to this, 'and setting himself down, he clap'd his hand on the ground in token he would stay there till I returned'. Spelman made his way to Smith's ship just as the president was giving up on his attempts to bend West's men to his will. They were continuing to 'abuse themselves with their great gilded hopes of seas, mines, commodities, or victories they so madly conceived', Smith complained. He decided to return to Jamestown, taking a bemused Spelman with him.

Whether provoked by Smith or not, 'no sooner was the ship under sail but the Savages assaulted those hundred and twenty in their fort'. As the attack was under way, the ship of the accident-prone Smith became grounded a mile or so downriver. He returned by land or boat to West's fort, and by his own account quickly overcame the 'poor silly assault', which he claimed was mounted by just twelve Indians. This so 'strangely amazed' West's mutinous men that they became happy to submit themselves 'upon any terms to the President's mercy'. After punishing seven or eight ringleaders, he escorted the rest to the location he had originally determined, there being 'no place we knew so strong, so pleasant and delightful in Virginia, for which we called it Nonesuch,' after Queen Elizabeth's elegant palace to the south of London.

The removal from West's fort to Nonesuch took two days, during which time Smith's ship was refloated, and taken back by its crew to Jamestown. Hearing of the attack, West immediately set off back up the river, and arrived at the falls to find all his men relocated. He ordered them to return immediately to 'the open air at West Fort', which they did.

In fury and frustration, Smith gave up on the whole affair, and started to row back to Jamestown with his small company of men. That night, he decided to sleep on board the boat. As he lay on the deck, an autumn dew collecting on the tarpaulin pulled over his shoulders, his gunpowder bag, which was slung from his belt and lay in his lap, somehow ignited. The resulting explosion tore Smith's flesh 'from his body and thighs . . . in a most pitiful manner', removing a section of his midriff 'nine or ten inches square', including his genitals. To quench the 'tormenting fire frying him in his clothes', he threw himself overboard into the river, nearly drowning as his men struggled to pull him back into the boat.

'In this state, without either surgeon, or surgery he was to go near 100 miles' back downriver (in fact, nearer 60), before finally reaching Jamestown. According to Smith, the mutinous triumvirate of Ratcliffe, Archer and Martin now seized their chance to rid themselves of the bumptious captain. 'Seeing the President . . . near bereft of his senses', the three mutineers decided to shoot Smith in his bed. They apparently commissioned Thomas Coe and William Dyer to do the job, though the 'heart did fail him that should have given fire to that merciless pistol', and Smith survived to endure the agonies of his injuries alone in his quarters.

Meanwhile, the settlement continued its descent into chaos. The 'three busy instruments', as Percy described Ratcliffe, Archer and Martin, reverted to their earlier tactic of trying to destroy Smith with denunciations. They charged him with a range of offences: that he tried to seize from Powhatan the crown and robe he had been given by King James; that he forced settlers out of the fort to fend for themselves while he feasted on his private supply brought from England (he admitted to having 'his own private provision', but claimed to have given it all away 'to the weak and sick'); that he attempted to poison one of the Dutchmen with ratsbane; that he refused to recognize the authority of the council. One 'prophetical spirit' even charged Smith with planning to make himself king of the Indians by marrying Pocahontas – a charge Smith ridiculed, and must have now contemplated with grim irony, given his mutilated condition.

Overwhelmed by this hailstorm of allegations, and too weak to rebuff them, Smith finally submitted his office. The council was reconstituted under Ratcliffe, Martin, West and 'some few of the best and worthiest that inhabit at Jamestown', as Ratcliffe put it. Finding no one suitable for the job, the council voted to abandon the idea of appointing a governor. Instead, it would remain in power, under a president. The man elected to replace Smith was George Percy. Having been passed over so many times in the past, he was now considered the best qualified probably because he was the least objectionable.[11]

As a leader of men, Percy's style and appearance could not have contrasted more sharply with Smith's. While Smith amplified his heavy, round features with a foam of facial hair, Percy subtly delineated his long, narrow face with a pencil-thin moustache. Where Smith roared at his enemies and railed at injustices, Percy reclined upon the quiet complacency of rank. He disdained politics, and avoided unseemly displays of self-promotion. He took over the presidency out of a sense of obligation, not ambition.

Under Percy, the office of president became primarily a matter of presentation. For a man of his status and birth, it was about setting an example, rather than becoming one. Even in what he had called this 'savage kingdom,' it was not so much what he did as president that mattered, but how he did it. His presidency 'standing upon my reputation', it was essential that he 'keep continual and daily table for gentlemen of fashion,' as he put it in a letter to his brother the Earl of Northumberland. In the midst of the muck and mire, the extreme cold and heat, he insisted on wearing the finest clothes: a stiff-collared taffeta suit accessorized by a hat with a silk and gold band, shoes laced with ribbons, a pair of kid leather gloves and a garter.[12]

This did not mean that he was indifferent to the settlers' welfare. It was his patrician duty to see that the people over whom he had command were looked after. That task quickly became increasingly difficult. In a humiliating vindication of Smith's interventions, Francis West was forced to abandon his fort near the falls and return his men to Jamestown, losing eleven of them along the way. 'In charity we could not deny them to participate with us,' wrote Percy, so Daniel Tucker, an experienced captain who had been involved in the northern Virginia plantation before coming to Jamestown with the second supply, was set the task of calculating rations. After going through all the supplies remaining in the fort and aboard the ships, he reckoned there was enough for three months 'at a poor allowance of half a can of meal for a man a day'.

In a desperate attempt to get more supplies, Percy sent Captain Martin with a small troop of soldiers to Nandsamund, a small tributary of the James, to buy corn from the local people. He also dispatched Captain Ratcliffe to Cape Comfort, where he was to establish a fort at the mouth of the James, to act as a permanent lookout post for shipping, and support fishing expeditions.

At around the same time, Percy received a delegation sent by

Powhatan, bearing a gift of venison. Leading the group was Thomas Savage, the thirteen-year-old Newport had offered to the *mamanatowick* at their first meeting. As well as presenting the gift, Savage told Percy that Powhatan had removed from Werowocomoco to a royal enclosure a mile or so from Orapaks, a town about 30 miles up the Chickahominy. It was described by Smith as lying in the 'desert betwixt Chickahominy and Youghtanan', meaning the large expanse of uninhabited land between the two rivers, its name probably meaning 'deep pond'.[13]

Percy told Savage to return to Orapaks with gifts and felicitations, and to make overtures about buying more corn. Savage was loath to leave, and in the end was only persuaded when Henry Spelman agreed to accompany him.[14]

The two, together with their Indian escorts, set off on the lengthy journey up the James and Chickahominy, Spelman taking some copper and a hatchet to present to his host.

Soon after, Captain Martin reappeared at Jamestown. He was on his own. In an act considered by some to reveal shocking cowardice, he had abandoned his company of seventeen men to the charge of his lieutenant, Michael Sicklemore, 'a very honest, valiant and painful soldier' brought with the first supply.[15]

Percy sent off a search party, which eventually located the lieutenant. He reported that the men had deserted, and tried to find refuge at Kecoughtan. Their bodies were found a few days later, with their mouths stuffed full of bread. This taunt reminded Percy of the episode in Spanish Chile, when the province's governor Valdivia was captured by a group of natives, who tortured him by pouring the molten metal down his throat, crying, ' "Now glut thyself with gold, Baldivia!" – having there sought for gold as Sicklemore did here for food'.

Three weeks later, Spelman returned from Orapaks, bringing with him Powhatan's unnamed 'son and daughter'. Spelman had a message from the *mamanatowick*, 'bidding me tell them that if they would bring their ship and some copper', he would load their ship with corn. Percy eagerly embraced the offer, 'having no expectation of relief to come [from England] in so short a time'. Smith had warned that recent Indian hostility arose from their own lack of food, so such overtures were to be treated with suspicion. But, as Smith had himself previously dismissed Indian claims of food shortages as tactical, so did Percy. The president sent Spelman back to Orapaks with the reply that a pinnace would be sent on the next tide.

Ratcliffe was recalled from Cape Comfort to lead the expedition. Before departing, he penned a quick letter to his master Cecil, which he left with one of the ships' captains. He then set off for Orapaks with fifty soldiers and Powhatan's children.[16] He sailed as far up the Chickahominy as he could, then transferred to a barge for the final leg of the journey, leaving Captain William Phettiplace and twenty or so of the men to guard the ship.

As the barge approached Powhatan's royal enclosure, Ratcliffe was greeted by servants offering gifts of venison and bread from the *mamanatowick*, and the captain sent copper and beads in thanks. Ratcliffe and his crew were then escorted inland through a large cornfield to a house near Powhatan's enclosure, where they were told they could stay for the duration of the visit. Powhatan's children, meanwhile, returned to their father.

That evening, Powhatan came in person to greet the visitors, bringing with him Spelman and Thomas Savage, together with a Dutch boy named Samwell, who had been left with the Indians since Smith's debacle with the Dutch sent to work at Werowocomoco. Powhatan greeted his guests, and returned to his own quarters.

The following morning, the emperor came with Spelman and 'a company of savages', including several women, to escort Ratcliffe and his party to a nearby storehouse. There the Englishmen were shown a collection of huge baskets brimming with corn, which through Spelman Powhatan announced he was willing to trade. A price was agreed, and the captain handed over 'pieces of copper and beads and other things according to the proportion of the baskets of corn which they [had] bought'.

Powhatan took his leave, the women and Spelman following. The

English soldiers, relishing the resumption of decent rations, began to carry the corn the half-mile or so to the barge. However, they quickly discovered from the weight of the baskets that they had false bottoms, and were almost empty.

The English began to complain loudly of being cheated, 'whereat a great number of Indians, that lay lurking in the woods and corn about' began shouting 'with an oulis and whoopubb', as Spelman described it. The English made a run for the barge, carrying what corn they could. But within sight of their boat, they were ambushed by Indian warriors lying in a neighbouring cornfield. Just two of the English soldiers managed to escape the ensuing onslaught by running off into the woods.

Captain Ratcliffe was seized and brought before Powhatan at his enclosure. There was no sign of Spelman, Savage or Samwell, who, 'fearing the worst', had fled. According to Smith, Spelman had been tipped off by Pocahontas that he would be in peril if he stayed.[17] One of the English soldiers who had managed to escape the Indians' attack was hiding in the nearby undergrowth, and it was he who later reported to Percy what happened to Ratcliffe.

A fire was kindled at the foot of a tree. Ratcliffe was stripped of his clothes, and tied to the tree. Several women then approached the naked captain. They began to flay his skin with the sharp edges of mussel shells, gently teasing it away from the flesh. They then sliced through the muscle and sinews to remove the limbs and organs from his body, which were 'before his face thrown into the fire; and so for want of circumspection [he] miserably perished'.

FOURTEEN

The Astrologer

Sometime in December 1609, Jane Flud, possibly the estranged wife of the Catholic lawyer Edward Flud, shrouded against the bitter weather and prying eyes by a long cape, stepped from her front door in the City of London, and set off for the bridge. She crossed the river to Southwark, and turning westward, made her way past the Hope and Globe theatres, the bull-baiting and cock-fighting pits, the coaching inns and brothels, the fishponds stocked with fat pike, until she reached the Archbishop of Canterbury's palace, which sat on the banks of the river, facing Whitehall. Next to it lay Lambeth Marsh, an area of swampland that marked the southern limit of the conurbation, on the edge of which perched a line of riverside houses.

Behind one of these houses, just visible over the wall, was a large orchard, the bare limbs of its exotic trees reaching into a cold winter sky. Flud knocked at the front door of the house, or may have surreptitiously slipped over to a tradesman's entrance at the side. She was issued into the chamber of Dr Simon Forman, astrologer, physician, womanizer and necromancer to London's elite.

This was not her first visit. Jane had previously consulted Forman on a long list of eminent and not-so-eminent lovers: Sir Henry Wotton, Sir Thomas Walsingham, one Copell, the rector of her parish church, Sir Robert Rivington, Robin Jones, one Wilmar, the servant of her father-in-law, and the son of Lady Vane. She had also enquired about Sir Thomas Gates. Gates had professed deep feelings for her, and she wanted Forman to cast a horoscope to tell her whether he was genuine, or if another suitor, Sir Calisthenes Brook, would be more suitable.[1]

Forman's deliberations, which are not recorded, proved irrelevant, as, soon after, she revealed that she had fallen for another man, an 'untoward old fellow' in Forman's estimation, called Vincent Randall. 'She is not to be trusted,' Forman noted. 'She has a fair tongue, but with backbite and speaks evil of her best friends. She professes virtue, loyalty, chastity – yet is full of vice.'

Her affections for Gates had obviously survived her later dalliances, as on this occasion, she had come to ask Forman if he knew anything of his fate.[2] News had just reached London that, following a terrible storm somewhere in the western Atlantic, the *Sea Venture* had gone missing.

Flud was not the only anxious caller Forman had received on the matter. An Elizabeth Whitehead had visited, perhaps a week or two before, asking about her husband, who was also lost.

Forman had taken a close interest in Virginia since the start. There is no evidence that he invested: an astrologer's income, though often impressive, was perhaps too precarious for such a commitment. But he jotted down several notes about the venture, particularly concerning the wildlife and the Indians. 'There is also a kind of fly of some half inch long that flyeth upwards and down in the woods in the night with fire in their tails, like candles,' he recorded. He wondered whether they might be glow-worms, noting that 'they are seen there in the summertime, but not in the winter'. Being a keen gardener, he was interested to hear that the Indians planted 'wheat' (meaning maize, still a rare sight in England) and beans in the same mound, 'and when it groweth up, the bean doth clasp about the wheat and every wheat will have 3, 4 or 5 ears & every ear will have 300 or 400 grains, and they clear the ground in March, and set their corn in April, and reap it in August'. In his capacity as unlicensed physician, he speculated on the health of the settlers, who he had been told suffer 'a kind of burning fever' in the summer, 'swelling in their bodies and faces', which subsided during the winter.[3]

Forman's sources for this information are unknown, but one may have been Richard Staper, a backer of the second charter expedition. Forman had been to Staper's for dinner in the summer of 1604.[4]

Neither Staper nor the stars could help Forman provide the fretful Jane with any useful information on the *Sea Venture*. First news that the ship was lost had reached London with Samuel Argall, who had returned to London in October. Ambassador Zuñiga had sent a letter to Madrid the following month reporting the return of this 'fisherman's ship', as he called it.[5] Argall had departed Jamestown about a month after the first boats from Gates's fleet had limped up the James, so there was still hope that the *Sea Venture* may have since turned up in Jamestown.

Argall had also brought a letter from Gabriel Archer. Its unknown addressee did not immediately publish its devastating account of events, but it would not have taken long for news of the 'contentions . . . factions and partakings' already tearing the settlement apart to start circulating the taverns.[6]

In late November, at least two of Gates's ships had returned from Virginia, 'laden with nothing but bad reports and letters of discouragement'. One also carried the thirty 'unruly youths'. Their presence in Virginia at such a time had proved so disruptive, the decision had been made to send them straight back home. Another of the ships brought back the injured Captain John Smith, his fury barely abated by the agonies he must have suffered during the voyage.

The ill-fated fleet had suffered further casualties on the way home. Two ships, the *Diamond* and possibly the *Unity*, had perished at Ouessant, an island lying off the western tip of Brittany, France. 'And which added the more to our cross,' wrote Robert Johnson, a member of the Royal Council, 'they brought us news that the Admiral's ship, with the two Knights and Captain Newport' was still missing, 'severed in a mighty storm outward, and could not be heard of, which we therefore yielded as lost for many months together, and so that Virgin voyage (as I may term it) which went out smiling on her lovers with pleasant looks, after her weary travails, did thus return with a rent and disfigured face.'

Long-standing critics took the opportunity of this setback to 'insult and scoff' at the grandiose send-off the previous year. Returning settlers and smuggled letters added to the opprobrium. 'For those wicked Imps that put themselves a shipboard, not knowing otherwise how to live in England; or those ungracious sons that daily vexed their fathers' hearts at home, and were therefore thrust upon the voyage . . . to cover their own lewdness do fill men's ears with false reports of their miserable and perilous life in Virginia, let the imputation of misery be to their idleness, and the blood that was spilt upon their own heads that caused it.'

With the company now such a public and visible institution, the result was a catastrophic collapse of investor confidence. 'Many adventurers which had formerly well affected the business, when they saw such unexpected tragedies withdrew themselves and their monies from adventure,' lamented Johnson.[7]

In a desperate effort to recover momentum, the company rushed out a pamphlet entitled *A True and Sincere declaration of the purpose and ends of the Plantation begun in Virginia* in December 1609. It restated the mission's objectives, echoing the sermons and tracts that had sent the fleet on its way: to convert the 'pagans', to 'build up' a new nation 'for the public honour and safety of our gracious king and his estates', to transplant 'the rankness and multitude of increase in our people', to act as a 'bulwark of defence' against the Spanish.

The pamphlet also mentioned some of the more material benefits of investors' continued support. 'By recovering and possessing to themselves a fruitful land', they stood to make themselves a healthy profit, it claimed. They would corner the market in commodities which English merchants 'are now enforced to buy, and receive at the courtesy of other Princes, under the burthen of great Customs, and heavy impositions', notably tobacco. This was a bold reference to the grievances raised in the House of Commons, and clearly aligned the Virginia venture with growing agitation among MPs for freer trade.[8]

At around the same time, the company also produced a broadsheet, to be pasted up on billboards around the capital. It argued that the 'sundry false rumours and despiteful speeches' circulating the capital had been 'devised and given out by men [that] lie at home, and do gladly take all occasions to cheer themselves with the prevention of happy success in any action of public good'. Furthermore, the 'most vile and scandalous reports' coming from Virginia were put out by those of 'most lewd and bad condition' to cover their own 'misbehaviour'.

In response to this, the broadsheet audaciously announced that the Royal Council for Virginia had plans to 'instantly prepare and make ready a certain number of good ships, with all necessaries, for the right honourable Lord Delaware, who intendeth, God assisting, to be ready with all expedition' to sail for Virginia and recover the enterprise.

'Former experience' had 'too dearly taught' the company the mistake of suffering 'parents to disburden themselves of lascivious sons, masters of bad servants, and wives of ill husbands'. To ensure that such an 'idle crew' could no longer 'clog the business', all future participants were to be screened for suitability. From now on, only 'sufficient, honest and good artificers' need apply, the broadsheet providing a long list of the most sought-after trades, from brewers to ploughwrights, coopers to vine-dressers, carpenters to 'divines' (chaplains). Any with 'sufficiency' in these professions were invited to repair to Smythe's house (the Philpot Lane address by now so well known it was not even given), 'to proffer their service in this action before the number be full'.[9]

As these efforts were under way, hopes of getting news about Gates and the *Sea Venture* began to fade. Jane Flud consoled herself by marrying a wealthy widower, Sir Thomas May of Mayfield, Sussex. Meanwhile, a relieved and happy Elizabeth Whitehead returned to

Forman's house at Lambeth Marsh to report that 'by ship that came thence' she had received letters from her husband in Virginia, enclosing 'certain articles', perhaps pressed and dried samples of tobacco or sassafras, or Indian trinkets for Dr Forman. Despite the difficult conditions Whitehead was being forced to endure, there was hope that he would be back in London soon, a rich and propertied man. However, as Forman later reported, he 'came not home', his fate, like that of so many who made the journey, being unrecorded in the Virginia Company's annals.[10]

FIFTEEN

Devil's Island

SEASICKNESS, William Strachey wrote, 'worketh upon the whole frame of the body, and most loathsomely affecteth all the powers thereof. And the manner of the sickness it lays upon the body, being so unsufferable, gives not the mind any free and quiet time to use her judgment and empire'.[1]

That was the writer's condition as the *Sea Venture* tossed at the tail end of the hurricane. 'For four and twenty hours the storm in a restless tumult had blown so exceedingly as we could not apprehend in our imaginations any possibility of greater violence.'

The lower decks of the ship were now awash with water, the stresses of the violent seas 'having spewed out her oakum', the rope fibres used to caulk or seal the ship's joints. The leakage caused the hold to fill fast, and the level of the water had reached 5 foot above the ballast before it was noticed. 'This imparting no less terror than danger ran through the whole ship with much fright and amazement, startled and turned the blood, and took down the braves of the most hardy mariner of them all, insomuch as he that before happily felt not the sorrow of others now began to sorrow for himself.'

The mariners, from the master down, descended to the lower decks, and, 'creeping along the ribs' of the ship in the pitch dark, felt and listened for the gush of water. 'Many a weeping leak was this way found and hastily stop'd.' One discovered in the gunner room was stuffed with slices of beef. But still the water level continued to rise. 'It was conceived as most likely that the leak might be sprung in the bread room, whereupon the carpenter went down and rip'd up all the room, but could not find it so.'

Gates summoned the whole ship's company, numbering one hundred and forty 'besides women', and split them into three groups. One was sent to the forecastle, another to the 'waist', the ship's middle section, another to the binnacle at the stern, and each group systematically pulled up decking and partitioning in search of the elusive leak.

The ship's internals being torn apart, there was nothing left to do but man the bilge pumps, trying to evacuate the water quicker than it seeped in. But the level continued to creep up the ship's ribs. Thousands of biscuits bobbed up to the surface, confirming that the stores were now swamped. 'To me, this leakage appeared as a wound given to men that were before dead,' Strachey wrote.

A roster was drawn up to ensure all buckets and pumps were in continuous use, so the men 'might be seen to labour (I may well say) for life . . . even our governor and admiral themselves, not refusing their turn'.

For three days and four nights, 'stripped naked as men in galleys', the crews pumped and bailed, 'destitute of outward comfort and desperate of any deliverance, testifying how mutually willing they were yet by labour to keep each other from drowning, albeit each one drowned whilst he laboured'. The only fixed point in that whirling world was Sir George Somers upon the poop of the ship, 'where he sat three days and three nights together, without meals, meat, and little or no sleep', struggling to keep the ship upon an even keel.

Then, a wave rose so high it crashed down upon the poop and quarter decks, and 'so stun'd the ship in her full pace that she stirred no more than if she had been caught in a net'. A surge of water dragged the helmsman from his helm. He grabbed the handle of a whip to save himself from being sucked overboard, and was tossed 'from starboard to larboard as it was God's mercy it had not split him'. Vomiting through the portholes below, the water hurled the men at the pump from their handles, and Governor Gates, who had 'both by his speech and authority' been 'heartening every man unto his labour', was thrown to the floor, where he grovelled in his sodden uniform.

At that point, it was believed the ship was sunk. 'Like a garment or a vast cloud', the wave had 'filled her brim full for a while within from the hatches up to the spar deck'. According to one passenger, even in the midst of this chaos, some of the men managed to fetch the 'good and comfortable waters' they had hidden in their private nooks, and raise a glass to their next meeting in 'a more blessed world'.

Strachey thought the ship was already falling, and had a vision of each man 'wading out of the flood thereof, all his ambition but to climb up above hatches to die in *aperto coelo* [under an open sky], and in the company of his old friends'.

But the ship did not sink, and for another two days she drifted like flotsam, jostled 30 miles by the heaving sea without an inch of sail raised.

On the night of Thursday 27 July, Sir George Somers was sitting at his customary place on the poop. Glancing up, he noticed a glow, 'a little round light like a faint star, trembling and streaming along with a sparkling blaze half the height upon the mainmast'. He watched it at play in the rigging, running along the shrouds and yards.

Somers called out, and news of the apparition spread through the ship like the lights along the ropes. Spectral men and women rose from the bilges and the bailing stations up to the main deck. For three or four hours they were entertained, until, at the approach of dawn, the illuminations disappeared. 'The superstitious seamen make many constructions of this sea fire,' Strachey observed, 'which nevertheless is usual in storms.' Being a man of letters, well read in the latest advancements of natural philosophy, he had no time for superstition, noting that for all the 'strucken amazement' of the sailors, the mysterious fire 'did not light us any whit the more to our known way', and the *Sea Venture* continued on her haphazard journey through the unknown seas, ranging 'as do hoodwinked men' across points of the compass.

Nevertheless, as the St Elmo's fire crackled out, so the storm's violence abated, and Strachey noted how that Friday morning 'won a little clearness from the days before'. The crew and passengers had their first opportunity to survey the damage that had been done over the past three days. The booms, masts and rigging were in disarray, and the ship slouched dangerously in the water. To lighten the load, and gain an inch or two of buoyancy, 'much luggage, many a trunk and chest' were thrown overboard, 'in which I suffered no mean loss', Strachey lamented. The ship's full complement of ordnance (sixteen guns in all) and ammunition was tipped over the side.[2] Butts, hogsheads and all manner of containers were smashed open to dispose of their contents, and the passengers could only watch as the gallons of beer, oil, cider, wine and vinegar were swallowed by the now lethargic sea.

Still the ship continued to slip gently into the deep, despite the men's efforts to keep her up with their relentless pumping and bailing. By now the lower decks were so drenched, the galley had to be abandoned, it proving impossible to kindle a fire 'to dress any meat'. A discussion commenced as to whether the mainmast should be cut down and thrown overboard, to lighten the ship further. But it was too late even for that. Their debris floated around them, but their ship was about to sink.

As Friday morning ebbed, so did all hope. In the middle of the vast ocean, far from normal navigations, there was no chance of ship or harbour to save them. The whole company prepared to 'shut up hatches, and, commending our sinful souls to God, commit the ship to the mercy of the sea'.

'But see the goodness and sweet introduction of better hope by our merciful God given unto us,' Strachey wrote. 'Sir George Somers, when no man dreamed of such happiness, had discovered and cried LAND!' In the distance, from the crowded deck rail, they spied an island, and 'the very trees were seen to move with the wind upon the shore side'.

However, more than a mile offshore, the boatswain found that the sea floor rose to 4 fathoms, just 24 foot. Lying so deep in the water, the ship would soon be grounded, and, with the weight of all the water inside her hull, liable to disintegrate. Without hope of finding anchorage, Somers decided the only option was to run her ashore as near to the land as possible.

With every serviceable sail unfurled, the *Sea Venture* gathered speed, no one aboard knowing how she would come to a rest once her keel struck the seabed. Ahead lay two rocks, about three-quarters of a mile offshore. Somers steered straight for them. By good fortune, the ship reached them before the keel hit the rocky seabed, and became 'fast lodged and locked' upright between the stony groins. Before the hull split, the crew managed to ferry the passengers and some possessions ashore by boat, and by nightfall, 'about the number of one hundred and fifty' had been carried 'safe into the island'.

But which island? The latitude was just over 32 degrees north of the Equator. A check of the charts would have revealed that, across the broad waters of the Atlantic, only one miniscule archipelago was known to lie at that latitude, sitting in the middle of the Sargasso Sea, 800 miles from the American east coast: 'the dangerous and dreaded island, or rather islands, of the Bermuda'.

According to maritime lore, these were the 'Devil's Islands', 'ever esteemed and reputed a most prodigious and enchanted place, affording nothing but gusts, storms, and foul weather; which made every Navigator and Mariner to avoid them, as Scylla and Charibdis; or as they would shun the Devil himself. And no man was ever heard to make for the place, but as against their wills, they have by storms and dangerousness of the rocks, lying seven leagues into the Sea, suffered shipwreck.' They were considered the 'farthest [most isolated] of all the islands that are yet found at this day in the world'.[3]

To those who landed that July Friday, all they could see was paradise, a 'desert island' in the original sense of being unpopulated, yet 'abundantly fruitful of all fit necessaries for the sustentation and preservation of man's life'.

The weather was 'very hot and pleasant', but prone to the same 'thunder, lightning, and many scattering showers of rain' that had driven the *Sea Venture* there in the first place, which would 'fall with such force and darkness for the time as if it would never be clear again'.

The weary crew and passengers set up camp next to the east-facing sandy bay, which Somers called 'Gates's Bay', in honour of the governor. Over the coming days, they busied themselves constructing shelters from the huge leaves of the palm trees that fringed the beach, and salvaging what they could from the stricken *Sea Venture*, including the ship's dog and several pigs.

Somers ordered the clearing of a patch of land next to the beach, which he sowed with 'muskmelons, peas, onions, radish, lettuce'. Within ten days, shoots had started to poke through the garden's 'dark, red, sandy, dry, and uncapable' soil, to the company's astonishment. Gates, meanwhile, tried to plant a specimen of sugar cane recovered from the ship.

The pigs were left to scavenge in the woods around the bay, some of them taking the tender shoots of sugar cane from Somers's garden. One night, a huge boar, the descendant of hogs that had survived previous shipwrecks, came down to the garden, and was 'grovelled by the sows'. One of the men managed to placate the great snorting creature by stroking its haunch, whereupon he fastened a rope with a slip-knot around its hind leg, 'and so took him'. Soon after, groups started hunting for the boar using the ship's dog, 'for the dog would fasten on them and hold whilst the huntsmen made in'. By this method they could catch as many as fifty of the creatures during each expedition. They also managed to craft a crude, flat-bottomed 'gondola' out of the trunk of a cedar tree, which on calm days they used to catch fish and turtles in the reefs and shallows along the western shore. Some of the turtles were huge, one, 'sod [soaked], baked and roasted' providing as many as seventy-two meals.

As for fish, the supply seemed limitless: swordfish, dogfish, pilchards, mullets, rockfish, 'unscaled fish' such as lampreys and trenchers. Under every rock they would find a crayfish 'oftentimes greater than any of our best English lobsters; and likewise abundance of crabs, oysters, and whelks'. Using booms off the ship and a 'deer toil', a long net which had been brought to snare deer in Virginia, they managed to improvise a trammel, a wall of netting that could be drawn across the width of a bay. When the ends of the trammel were pulled in with ropes, the net trawled huge stretches of water, 'with which', Strachey 'boldly' recalled, 'we have taken five thousand of small and great fish at one haul'. To ensure supplies for the future, a team of two or three men was set to boiling up vats of brine, to produce salt for preserving the surplus.

So, through careful management of a bountiful supply, the hundred and fifty survivors were kept well fed through the autumn and winter, each month yielding a new abundance: a kind of 'pea' growing among the rocks 'full of many sharp subtle pricks as a thistle, which we therefore called "the prickle pear", the outside green, but being opened, of a deep murrey [blood red], full of juice like a mulberry, and just of the same substance and taste'; a palm berry as big as a damson, 'ripe and luscious'; a web-footed seabird 'the bigness of an English green plover', to be found only during the 'darkest nights of November and December', which could be caught by 'standing on the rocks or sands by the seaside, and hallooing, laughing, and making the strangest outcry that possibly they could, with the noise whereof the birds would come flocking to that place, and settle upon the very arms and head of him that so cried'. In January, thousands of these same birds would settle on outlying islands and lay eggs with 'no difference in yolk nor white from an hen's', which a single 'cockboat' (ship's boat), cruising from one island to another, could gather in sufficient quantity to feed the entire company.

'So soon as we were a little settled after our landing, with all the conveniency we might, and as the place and our many wants would give us leave', thoughts turned to finding a means of escape. Gates understood the catastrophic consequences of 'the younger and ambitious spirits of the new companies' arriving in Virginia without proper government, so commanded the swift construction of a longboat.

The distance from Bermuda was calculated to be 140 leagues – 420 miles (the actual distance is nearer 700 miles, the discrepancy arising from the lack of an accurate method of measuring longitude, which meant mariners had only the vaguest idea of the width of the Atlantic). Clearly they were going to need to build a sturdy vessel, capable of undergoing an epic ocean voyage through dangerous waters.

The shipwright Richard Frobisher (possibly a kinsman of the Elizabethan explorer Sir Martin), together with the four carpenters in the ship's company, set about building the longboat out of timbers and fittings salvaged from the *Sea Venture*. She would have been no more than 20 foot long, and when she was floated, her keel sat just 20 inches below the surface of the water.

In the hope of making her buoyant enough to survive the huge seas she might encounter, she was covered with a watertight deck and close-fitting hatches, so the crew could shelter down below during bad weather.

The risks were too great for Gates to go personally, so Henry Ravens, the *Sea Venture*'s pilot, volunteered to lead the mission, taking with him the cape merchant Thomas Whittingham and six sailors. Gates handed Ravens a letter, setting out interim arrangements for the governing of the colony. Cecil's one-time spy Captain Peter Wynn was to act as lieutenant governor, with six councillors to advise him. He also gave instructions on how Wynn and the council should deal with anyone who attempted to change 'the person . . . or form of government', identifying by name the man Gates anticipated to be the culprit. Unfortunately, that name was never recorded, leaving unresolved Gates's position in the faction fighting that was at that very moment bringing Jamestown to its knees.

If his instructions were followed, he had 'fair hopes' that the settlement would be able to manage, until such time as either a ship could be sent from Virginia to fetch him and the rest of the company from Bermuda, or Lord Delaware's fleet, which had been expected to leave London within a few months of Gates's, arrived at Jamestown.

On Monday 28 August, the longboat was launched into the bay. Ravens promised to be back with a pinnace from Virginia by the next full moon. To aid his return, he asked that beacons be set alight on nearby islands to guide him in. With all good wishes, the company saw him and his crew of seven set off on their fragile vessel, and head out to sea.

They were back by Wednesday. They had failed to find a way through the reefs surrounding the islands. They made a second attempt on Friday, pledging to have more success on this second attempt, and they were never seen again.

Several weeks later, when hope of Ravens's promised pinnace had been abandoned, Frobisher began work on another vessel, this one more substantial, capable of taking Gates together with most of the company to Virginia. Gates also promised that any who volunteered to be left behind would be given two sets of clothes, and food to last until such time as a vessel from Virginia or England came to rescue them.

By September, work was well under way on the new ship. However, a rift had opened up between Somers and Gates, probably provoked by Gates's insistence upon taking the role of governor, even though Somers, as admiral of the fleet, formally ranked above him until they reached their destination. To prevent discord degenerating into hostility, it was mutually agreed that Somers would take twenty 'of the ablest and stoutest' men to a nearby island (probably Somerset Isle, to the south), where they would set up camp for themselves and build another vessel.

Strachey remained in the Gates camp, where, in a makeshift shelter of tree branches covered in palm leaves, he continued writing his journal. 'Every morning and evening at the ringing of a bell we repaired all to public prayer,' he wrote, 'at what time the names of our whole company were called by bill, and such as were wanting were duly punished.' The Reverend Richard Buck, appointed Jamestown's new minister, led the prayers, and also delivered two sermons each Sunday, taking as his theme issues such as thankfulness and unity.

'It pleased God also to give us opportunity to perform all the other offices and rites of our Christian profession in this island.' On 26 November, the company cook, Thomas Powell, was married to Elizabeth Persons, the maid of one Mistress Horton. In early February 1610, John Rolfe's wife was delivered of a daughter, who was christened Bermuda on the 11th, Strachey joining Newport and Mistress Horton as godparents. A few weeks later, the child died.

As Frobisher's ship took shape in a woodland clearing, so did thoughts among Gates's company that life on these islands was proving much more comfortable than it had been in England, or would be in Virginia. The food was plentiful, and could be gathered without the effort of cultivation. There were no diseases, no vermin such as rats, no poisonous or predatory creatures. The risk of attack from sea was low, as the approaches to the islands were so dangerous. There were no natives to worry about. The climate was unpredictable but generally comfortable. Faced with the looming prospect of a perilous journey to Virginia, and years of servitude once they got there, many yearned to stay, to live out more gentle lives upon these soft sands, beneath these waving palms, plucking berries from the bushes and turtles from the sea.

The Brownist John Want emerged as a leading agitator. Together with seven others, including the company blacksmith and a ships' carpenter, he retired 'into the woods to make a settlement and habitation there'. They were promptly rounded up, and six of them transported to an outlying island, where they were left to fend for themselves. The smith and carpenter were allowed to stay with the company, on condition they returned to their tools.

The marooned men could soon be heard across the waters pleading for clemency, which Gates eventually granted. 'Yet could not this be any warning to others who more subtly began to shake the foundation of our quiet safety.'

Soon after, Stephen Hopkins, clerk to minister Buck, began to agitate among the men, arguing that Gates's authority had expired when they were shipwrecked on the island, and that it was 'no breach of honesty, conscience, nor religion to decline from the obedience of the governor'. Hopkins was duly arrested, and, at the tolling of the bell used to summon the company to prayer, brought before Gates in manacles, 'full of sorrow and tears, pleading simplicity [ignorance] and denial'. He was sentenced to death, but pardoned.

These 'dangers and devilish disquiets' did not subside, Gates's clemency being taken as a sign of weakness. Gates ordered a doubling of the watch, and sent out armed patrols of his trusted lieutenants.

In February 1610, the shipbuilding approached its final stages. The ships were 'breamed' or smoothed using lime made of whelk shells and stone, gaps were caulked using oakum picked out of ropes salvaged from the *Sea Venture*, and the hulls sealed with pitch and tar made from the sap of trees.

As the inevitability of departure loomed, a mood of dread about the forthcoming voyage spread through the company, culminating on 13 March with the arrest of one Henry Paine, a member of Somers's company caught stealing weapons.

Gates suspected that Somers's men, and perhaps their leader, were planning armed rebellion. After his capture, Paine refused to acknowledge Gates's authority 'in such unreverent terms as I should offend the modest ear too much to express it in his own phrase', as Strachey put it. The rebel was brought before the governor, and Gates decided to make an example of him. He was summarily condemned to be hanged. 'After he had made many confessions, he earnestly desired, being a gentleman, that he might be shot to death; and towards the evening he had his desire, the sun and his life setting together.'

Paine's execution prompted the rest of Somers's men to flee into the woods, raising the threat of further rebellion.

Strachey blamed these rifts on religious tensions, intensified by troublemakers such as the Brownist John Want and minister's clerk

Stephen Hopkins. A separatist spirit certainly seems to have been at work in the company. With the Anglican Reverend Buck confined to Gates's camp, a woodland grove, or even Somers's camp at the other end of the archipelago, would have offered a tempting refuge for dissidents.[4]

However, other tensions were at work, which emerged in the aftermath to Paine's rebellion. It appears from Strachey's garbled version of events that Somers planned to found a permanent settlement on the archipelago, and use the ship he had constructed to return to London and negotiate the necessary legal and financial support. Over the previous months, Somers had spent a great deal of time touring the islands aboard a makeshift cedar-wood barge, and had evidently concluded, like many of the 'common sort', that what Gates regarded as a temporary refuge would make a hospitable and lucrative colony. Its attractions were particularly obvious to a mariner. Somers had found that the coral reefs surrounding the archipelago formed a barrier impenetrable to ocean-going shipping. However, by spending days, even weeks carefully sounding the reefs, the admiral and his men had managed to identify two narrow gaps in the wall deep enough to admit shipping. They were well hidden, and difficult to navigate in the strong and variable winds that whirled around the islands. But armed with the charts Somers had drawn up, a friendly ship would be able to steer a course to safety, while enemy vessels would flounder on the rocks or be repulsed by artillery. In other words, Somers had found Bermuda to be 'the strongest situate in the world'.[5]

Whether Somers planned to split with Gates's company and claim the islands for himself and his men is unclear. Strachey suggests that he merely acted as an intermediary in the subsequent negotiations with the rebels in his company. Gates's eagerness to appease them suggests otherwise. Either way, it was in their names that Gates was boldly petitioned to allow those who chose to, to stay behind with the allocation of food and clothing that he had promised when construction on the escape vessels had begun. Gates replied that since there was room to take the entire company, the offer no longer stood.

Gates 'conjured Sir George, by the worthiness of his heretofore well-maintained reputation, and by the powers of his own judgment, and by the virtue of that ancient love and friendship which had these many years been settled between them', to make the rebels back down. In return, he promised that 'whatsoever they had sinisterly committed or practised hitherto against the laws of duty and honesty should not in any sort be imputed against them'.

Somers apparently co-operated in persuading his men to go, but not before secretly arranging for two, Christopher Carter and Robert Waters, to stay behind, to ensure the island's continuing possession.

The two camps reunited to watch the launch of the first of the two ships that were to carry them to Virginia. Given the circumstances of her construction, she was an impressive vessel: 80 tons draft weight, 40 foot long, 19 foot wide, with a forecastle, poop and 'great cabin'. Her beams were made of oak scavenged from the *Sea Venture*, her hull of brittle cedar wood. 'When she began to swim upon her launching, our governor called her the *Deliverance*,' Strachey noted.

A few days later, Somers brought around the ship built by his company, a smaller pinnace 29 foot long and 15 foot wide. Perhaps alluding to his state of mind regarding recent troubles and future expectations, he named her *Patience*.

Over the coming weeks, the ships were rigged and loaded, and quantities of pork and fish were salted to provide a generous supply for the voyage ahead. Gates busied himself erecting a large cross near the site where they had first landed, made out of beams from the *Sea Venture*. He nailed a coin bearing the face of James I into the centre, and added an inscription describing the circumstances that had brought them there, their efforts to reach 'Nova Britannia in America', and those that were their leaders: himself the governor, Newport the captain. The lack of any reference to Somers hints at a continuing rift between the two men, though this may have been because the admiral had planted his own memorial at the southern end of the archipelago.

April passed into May, awaiting a favourable westerly wind. Early on the morning of Thursday 10 May 1610, an 'easy gale' started to blow in the right direction, and the dangerous matter of escaping Bermuda's devilish reefs could begin. Somers climbed into the longboat, and with Newport piloted the channel that would take the two brittle hulls through the rock-strewn shoals, setting out buoys to mark the way. At 10 a.m., the two craft hoisted their sails, and gingerly navigated between the buoys. The wind suddenly dropped, and the *Deliverance* drifted, hitting a rock. 'Had it not been a soft rock, by which means she bore it before her and crushed it to pieces,' Strachey speculated, 'God knows we might have been like enough to have returned anew and dwelt there after ten months of carefulness and great labour a longer time.'

The coxswain, a man called Walsingham, used the longboat to nudge the *Deliverance* back into the channel, and the ships eventually managed to pass safely into deep water. Now all that lay ahead of them was 700 miles of ocean.

'For seven days we had the wind sometimes fair and sometimes scarce and contrary.' Twice the *Deliverance* lost sight of the *Patience*. On 17 May, 'we saw change of water, and had much rubbish swim by our ship side, whereby we knew we were not far from land'. For the following two days they took soundings, during which time the seabed rose from nearly 40 fathoms to less than 20 (120 foot). At about midnight, the ships were suddenly suffused by a 'marvellous sweet smell', and as dawn heralded Sunday 20 May, the call went up of 'land ahoy'.

Strachey clambered up to the masthead to see what he could of the land that awaited him. He saw 'two hummocks' to the south, and to the north a line of land. This, he learned from excited speculations rising up from deck, was the entrance to 'the famous Chesapeake Bay, which we have called, in honour of our young prince, Cape Henry'.

For a moment, suspended from reality as he was by the swaying mast, the approaching mainland of America appeared to him as the wild shores of ancient Italy did to the legendary explorer Aeneas. The

Trojan Aeneas had been escaping a civilization torn apart by war. He too had been shipwrecked, and saved by divine providence. And he had ended his journey with the discovery of an untamed land, where Zeus himself had prophesied he would 'overthrow the wild peoples and set up laws for men and build walls': the walls of Imperial Rome. The literary men of the Mermaid Tavern were well acquainted with Virgil's masterwork, the *Aeneid* being the classic narrative of empire building. The Sirenaical Richard Martin described Strachey as being Achates, Aeneas's trusted companion. Perhaps upon these shores, Achates would see Zeus's prophesy realized, and the civilization planted on these shores would become the new Rome.[6]

Such dreams were brought to an abrupt halt as the ship collided with the powerful tidal current flowing out of the bay, which threatened to push it back out to sea. Anchors were dropped and a change of tide awaited. By mid-morning they were off again, and soon after crossed the sandbar, Strachey underestimating the bay's gaping entrance to be 'as broad as between Queenborough and Leigh', two towns standing at the mouth of the Thames.

By the following day, they had reached Point Comfort, at the mouth of the James river. They were greeted with a loud explosion and a plume of water in their path. A cannon shot had been fired at them from a wooden palisade, overlooking the strategic point where the river channel passed close to the northern shore, which 'easily commands the mouth of the river'.

The two craft dropped anchor, and as the ropes tightened, a group of soldiers clambered into the longboat and set off towards the cannon fire. As the shore loomed, so did the thought that this fort – which had not been there when Newport had last been at these shores in December 1608 – might now be a Spanish defensive position, constructed after they had disposed of the Jamestown settlement as they did the French in Florida. A group of soldiers ran out of the fort to the water's edge. As the one party tentatively floated towards the other, each established with relief that the other was English.[7]

Gates, Somers and Strachey were welcomed ashore by Captain

James Davies. Gates's good news of his company's miraculous salvation and deliverance was met with miserable news from Davies.

Of Ravens and the longboat dispatched from Bermuda the previous year nothing had been heard. In the meantime, conditions in Virginia had gone into a sharp decline. As planned in England, three new settlements had been established: at the falls, at Nandsamund, and here at Point Comfort, where the fort under Davies's command had been named Fort Algernon, after the young nephew of Smith's replacement as president, George Percy. The settlements at the falls and Nandsamund had since been abandoned, as far as Davies knew, and all the settlers driven back to Jamestown, which was now overcome by disease and starvation. Only Fort Algernon had survived, thanks to a few pigs which they had fattened up on an unexpectedly large supply of shellfish gathered during the winter.

A few weeks ago, President Percy, just recovered from a long illness, had visited the fort to advise Davies that he would have to take in half Jamestown's survivors. Davies had protested that he had no food or accommodation for them, and was facing increased hostility from the neighbouring Kecoughtan Indians. Percy replied that he had no choice, and told Davies to prepare for an imminent influx of sick and starving men. Nothing had been heard from Jamestown since.

Gates returned to the *Deliverance* and, the following morning, ordered the ship to proceed into the mouth of the James. They had just reached as far as the Kecoughtan, half a mile or so beyond Cape Comfort, when, as Strachey put it, 'a mighty storm of thunder, lightning, and rain gave us a shrewd and fearful welcome'. The following day there was not a breath of wind. The incoming tide carried them a few miles upstream, but when it turned, they had to drop anchor and wait. For two days, they had to endure this lurching progress up the river, the inward tide raising their apprehension, the outward draining their resolve.

On Wednesday 23 May 1610 a slow tidal surge eventually deposited them next to Jamestown, where they dropped anchor. As the waters stilled, they might have stood quietly for a moment, looking and listening for signs of life from the fort. None were forthcoming.

Gates went ashore with the company to find the fort's outer defences badly damaged. Sections of the palisade were collapsing, and the perimeter gates were swinging on their hinges. Inside, Strachey's initial impression was of a surprisingly substantial settlement. There was a marketplace, a storehouse, and a guard house. In the centre was a 'pretty chapel' 60 foot long and 24 foot wide, with a communion table made of black walnut, pews, shutters, a pulpit hewn out of a large trunk, and at the west end, two bells. It was surrounded by houses, haphazardly arranged, their walls plastered using local clay strengthened with bitumen, their roofs thatched with tree bark, out of which poked 'large country chimneys'.

However, the houses were empty, their furniture and fittings broken up, the skeletal remains of their shafts and laths smoking in the fireplaces. The chapel itself was like an abandoned ship, its relics, vestments and vases discarded, the lectern empty. Gates grabbed one of the bell ropes and gave it a tug. The tolling drew a few emaciated and sluggish people from scattered houses, some naked and crawling on their knees, crying weakly, 'We are starved, we are starved.' As they approached the chapel, Reverend Buck muttered a 'zealous and sorrowful prayer'.

The mounting horror and despair were momentarily relieved by a few grateful reunions. One of Gates's lieutenants, George Yeardley, spotted his wife Temperance, who had arrived the year before aboard the *Falcon*.[8]

Gates handed a piece of paper to Strachey to read out. It was his commission, delivered a year late, appointing him lieutenant governor, awaiting the arrival of the Lord Governor Delaware, and declaring all previous commissions and orders null and void.

An exhausted President Percy walked up to Gates to deliver 'his commission, the old patent, and the council seal'. Not having sought the presidency in the first place, he happily surrendered it.

That evening, revived with some salted meat brought from Bermuda, and hopes of easier times ahead, Percy could unload himself of the 'world of miseries' that he and the settlers had endured over the previous months.

After the departure of the injured Smith in October, and Powhatan's execution of Ratcliffe, an already dire situation had deteriorated rapidly. The abandonment of the settlements at the falls and Nandsamund overwhelmed Jamestown, and the scant supplies remaining in the store were soon consumed. 'To eat, many of our men this starving time did run away unto the savages, whom we never heard of after,' Percy noted. The rest turned inwards, and began to devour the very body of the company. The horses transported from England were the first to go, then the pigs, then the chickens, then the dogs, the cats, the rats, the mice, then snakes and other 'vermin', then toadstools and fungi such as 'Jew's Ear, or what else we found growing upon the ground that would fill either mouth or belly', then anything made of leather, including shoes, then the 'flesh and excrements of man', including the corpse of a recently slain Indian, dug up from his makeshift grave and 'boiled and stewed with roots and herbs'. Some lapped up the blood 'from their weak fellows' as they bled to death.[9]

With famine came madness. Hugh Price, 'being pinched with extreme famine, in a furious distracted mood', had rushed into the central marketplace jabbering blasphemies, and cried out that no virtuous God would suffer his creatures to endure such 'miseries'. That afternoon, he went off into the woods with a 'butcher, a corpulent fat man', in a search for sustenance. Both men were picked off by Indian snipers, and Percy noted that, while the body of the fat butcher was later found untouched except for 'the savages' arrows whereby he received his death', Price's 'lean, spare' body 'was rent in pieces with wolves or other wild beasts, and his bowels torn out of his body' – divine retribution, in Percy's estimation, for his crisis of faith.

A gentleman called Henry Collins had killed his pregnant wife. He ripped the foetus from her womb, and threw it into the river. He carefully jointed and salted the rest of her remains, and secreted them round his house. When she was reported missing, his quarters were searched, and 'parts of her mangled body were discovered'. Percy, succumbing to the brutalizing conditions, ordered that Collins be strung up from the branch of a tree by his thumbs, his feet weighted down. Once he had confessed to his crime, he was 'burned for his horrible villainy'.[10]

Percy's efforts to alleviate the crisis with supplies from the Indians came to nothing. The neighbouring Paspaheghans were laying siege to the fort, picking off any who ventured beyond the blockhouse that guarded the island's causeway. The only hope was to make contact with groups beyond Powhatan's domains, in the northern reaches of the Chesapeake, or the southern realms of Chowanoke. However, the settlers no longer had the shipping to embark on such an expedition. Captain Francis West and thirty-seven of his men had disappeared with one of the company's pinnaces, apparently returning to England. The *Virginia* was at Point Comfort, and the *Discovery* had gone adrift, floating 4 miles downstream. None of the sailors dared leave the fort to retrieve it, until Percy eventually threatened them with the sword. Meanwhile, the resourceful Daniel Tucker, Percy's stalwart lieutenant, set about building a fishing boat 'with his own hands, the which was some help and a little relief unto us, and did keep us from killing one of another'.

Despite Tucker's efforts, this 'Starving Time' had taken a terrible toll. For different reasons, Percy and John Smith, who was back in England pondering his own misfortunes, later claimed just sixty were left out of the five hundred or so alive at the time of Smith's departure. Percy blamed this on the conditions he had inherited, Smith on the personnel he had left behind. 'Even in Paradise itself with these Governors, it would not have been much better,' Smith claimed.

The number of casualties was probably exaggerated, but even if accurate, did not mean that over four hundred settlers died during that period. Thirty-seven men had left the area with Francis West, and the thirty 'unruly youths' had been sent back to London. Around fifty had been lost in the months preceding Percy's presidency as a result of engagements with the Indians, so were not his direct responsibility. Taking these and other losses into account, the total under Percy's care during this period has been put at nearer three hundred and thirty.[11]

That still suggests the loss of two hundred and seventy men. However, as Percy admitted, many 'did run away'. An Indian from Florida visiting the area at the time had seen 'many women and children who went about the fields and houses of the neighbouring Indians'. The local *weroances* measured their power by the size of their population rather than territory, and treated women and children captured from enemies as the prizes or spoils of war. Any Otasantasuwak who voluntarily crossed over from the English enclave into Indian territory was likely to have been welcomed, and allowed – indeed, forced – to integrate with local society. The main limiting factor to accepting surplus numbers would have been the Indians' own food supplies, which appear to have been dangerously depleted by the ongoing drought and unusually cold winters. But even food may have taken second place to the strategic advantage of taking in outsiders, especially women of child-bearing age and their offspring.[12]

These factors may explain why the Spanish ambassador in London, who had excellent contacts in the heart of the Royal Council, put the total number of survivors at three hundred, out of seven hundred sent.[13]

Whatever the true numbers, the conditions experienced over the winter of 1608 and spring of 1609 proved catastrophic to the settlement's sustainability. For a week, Gates discussed with Percy and others ways of saving Jamestown.

Gates may have hoped that, with his two ships and the longboat, the settlement could survive by fishing until the arrival of more supplies. However, though nets were cast out and hauled in 'twenty times day and night', along the river and around the bay, they managed to catch barely enough to feed the fishermen, let alone the rest of the company.

The Indians were continuing their attacks on the English, Gates losing two men within days of his arrival, so a counter-attack was discussed. But it was realized that, even with the upwards of one hundred and twenty who had arrived from Bermuda, chances of success were slim.[14] Even if the English were victorious, the pickings would be poor, it being the 'seed time', the period of late spring when the previous year's corn harvest was all but spent, and the next only just planted.

A decision had to be taken swiftly. With the help of supplies brought from Bermuda, and 'a thin and unsavoury broth' made of mushrooms and vegetables, the entire company could last no longer than sixteen days.

Gates consulted Somers and Newport on what to do, then Percy and Martin, and finally all the 'gentlemen of the town, who knew better of the country'. The advice was conflicting. Hundreds had lost their lives since the founding of the fort in 1607. Heroic efforts had been made to keep the settlement going, and enormous amounts of money had been spent. At stake were the reputations of rich and noble men, national pride, the prospect of imperial glory, the promise of free trade, and liberation from Spanish dominance. To leave would mean abandoning not just this particle of land, but North America, which would surely be snatched by a Catholic nation at the earliest opportunity, thereby ending for ever the hope of a Protestant empire.

Nevertheless, 'it soon then appeared most fit, by a general approbation, that to preserve and save all from starving, there could be no readier course thought on than to abandon the country', Gates concluded. Everyone would cram into the *Discovery*, *Deliverance* and *Patience*, and make for Newfoundland, hoping to catch enough fish to supply them for the trip home. The ships would be dangerously overloaded for an ocean crossing, but there was always the hope of encountering one of the English fishing vessels often to be found in those waters, which could be prevailed upon to carry some of the passengers home.

Gates ordered that the fort's cannons and ordnance be buried before the fort gate. Any remaining foodstuffs or goods of value were packed into the ships. On 7 June, he told the 'taborer' or drummer Thomas Dowse to beat out a slow tattoo, as everyone lined up to be taken on to the ships. Some had wanted to burn down the fort to prevent it falling into Spanish hands, but Gates refused to allow this. 'My masters, let the town stand,' he pleaded. 'We know not but that as honest men as ourselves may come and inhabit here.' To ensure his request was carried out, he was the last to leave.

By noon, the ships were full, and the anchors were raised. With the hand weapons they had taken with them, a platoon of soldiers fired a farewell salute. The ships let the tide carry them down the river, to Mulberry Island on the approaches to Point Comfort and Fort Algernon, where they were to be joined by Captain Davies and his company in the *Virginia*.

The fleet now assembled, it headed off for home.

SIXTEEN

Deliverance

At this moment in the story of English America, some believed that God himself intervened to save the colony. As Gates's pathetic fleet set off down the river, a lone longboat was spotted coming in the opposite direction. A man stood at the prow, waving. No one could recognize him. Then they heard an English voice call out, and saw a piece of paper in his hand.

He pulled up alongside, climbed aboard the *Discovery*, and introduced himself as Captain Edward Brewster, bringing a letter from the colony's new Royal Governor, Lord Delaware. Gates was commanded to turn around and head back to Jamestown.

The letter revealed that Delaware had just arrived in Chesapeake Bay with a fleet of three ships, three hundred men, 'besides great store of victuals, munition, and other provision'. Gates immediately executed the order, and a favourable wind carried his fleet back to Jamestown, which it reached the following morning, being Sunday 10 June 1610.

Delaware followed that afternoon, and came ashore, falling to his knees as he 'made a long and silent prayer to himself'. Fifty red-cloaked halberdiers then formed a guard of honour before the main gate, and Gates, carrying Delaware's colours, led the Lord Governor into the fort, followed by his 'general of the horse' Sir Ferdinando Wainman, his 'high marshal' Sir Thomas Dale, and 'divers other Gentlemen of sort', including a familiar face, Robert Tyndall, Prince Henry's gunner.[1]

For those who saw the Virginia enterprise as a religious mission,

Delaware's miraculously timed arrival was providential. In a sermon delivered before his departure from England, the Puritan preacher William Crashaw had urged Delaware 'to go to a Land which God will shew thee', there to take the 'devil prisoner in open field, and in his own kingdom'. He should do this for the sake of Prince Henry, just as one of Delaware's ancestors had for Henry's predecessor, Edward, the famous 'Black Prince' of Wales. Through his participation, he would show that this was no mere commercial enterprise, but an imperial and religious one. 'For the time was when we were savage and uncivil, and worshipped the devil, as now they do. Then God sent some to make us civil, others to make us Christians. If such had not been sent us we had yet continued wild and uncivil, and worshippers of the devil: for our civility we were beholden to the Romans, for our religion to the Apostles and their disciples.' Delaware's task was to be civilizer and apostle, and thereby make America 'one of the most glorious Nations under the Sun'.[2]

Before departing, Delaware had ordered the sermon's immediate publication, his 'earnest desire to further the Plantation in *Virginia*' perhaps making him 'too bold with *Mr Crashaw*', whose permission to publish such a powerful and controversial political message had not been sought or given. The resulting work, entitled *A New Year's Gift to Virginia*, was dedicated not to King James or a member of the governing nobility, but to Parliament, 'L[ord] D[elaware] humbly considering the union of their interest in all endeavours for the common good'.[3]

Delaware had thus sailed to Virginia primed to see himself leading a great and noble mission. But to have arrived at this crucial moment in the venture's fortunes, at the point it was about to expire, added a divine dimension. John Smith, back in England and no longer able to reach directly into the settlement's affairs, reflected the significance of the event with an outpouring of religious rapture:

> He that shall but turn up his eye, and behold the spangled canopy of heaven, or shall but cast down his eye, and consider

> the embroidered carpet of the earth, and withal shall mark how the heavens hear the earth, and the earth the corn and oil, and they relieve the necessities of man, that man will acknowledge God's infinite providence. But he that shall further observe, how God inclineth all casual events to work the necessary help of his Saints, must needs adore the Lord's infinite goodness; never had any people more just cause, to cast themselves at the very footstool of God, and to reverence his mercy, than this distressed Colony.

For if, Smith pointed out, Gates had not arrived when he did, when the settlers were within days of dying of starvation, if he had not chosen to ignore the advice to destroy the fort after abandoning it, if he had set sail a day or two sooner, upon a course which, being towards Newfoundland, would have inevitably led to him missing Delaware's approach, then all would have been lost. 'This was the arm of the Lord of Hosts, who would have his people pass the Red Sea and Wilderness, and then to possess the land of Canaan.'[4]

Delaware's first act as governor was to command the Reverend Buck to deliver a sermon, which no doubt developed the providential theme. He then told his 'ancient' or standard-bearer to read out his commission. This confirmed his status as Lord Governor of Virginia for life, giving him 'full and absolute power and authority to correct, punish, pardon, govern and Rule, all such the subjects of his Majesty, his heirs and successors in any voyage thither, or that should at any time there inhabit in the precincts and Territory of the said Colony', in a manner 'agreeable to the Laws, Statutes, Government and Policy of this his Majesty's Realm of England'.[5]

Gates immediately surrendered his own commission, together with copies of the two Virginia patents, and the council seal handed to him by Percy. Delaware now addressed the company directly, 'laying some blames upon them for many vanities and their idleness', and 'earnestly wishing' that he would not be 'compel'd to draw the sword in justice to cut off such delinquents'. After this admonition, he made an attempt at 'heartening them' by informing them that in the fleet of three ships were provisions for four hundred, which should last a year.[6]

'Not finding as yet in the town a convenient house', Delaware decided to make his headquarters upon his ship. He was in poor health, and within a day or so of arriving at Jamestown was overcome by a 'hot and violent' fever, an illness that many of the new arrivals appear to have contracted around this time. He summoned his doctor, Laurence Bohun, a disreputable London physician who had been struck off for illegal practice. Bohun set about treating Delaware's condition with gusto, opening up a vein in the Lord Governor's arm and draining off a bowlful of blood. He also applied some 'physic', probably opium mixed with saffron and nutmeg. Delaware reported a temporary recovery, long enough for him to summon to his cabin the men who were to become his principal officers.[7]

Delaware decided to abandon any semblance of civilian government, opting to run the settlement as a military camp. Gates was to be lieutenant 'general' rather than governor, to act as his deputy. Somers would be admiral, with responsibility for the colony's shipping and naval defence. George Percy was made 'esquire', Delaware's representative in Jamestown, captaining the fifty soldiers to be stationed at the fort. Sir Ferdinando Wainman was made master of the ordnance, in charge of armaments and defences. Newport retained his title of vice-admiral.

Samuel Argall, who had been master of one of the ships in Delaware's fleet, George Yeardley and various others were appointed military captains. The tenacious John Martin was put in charge of developing iron- and steel-works, and in recognition of his stalwart service Daniel Tucker was made clerk of the stores.

The senior officers made up Delaware's council, which survived in a purely advisory capacity. All members were made to take oaths of 'faith, assistance, and secrecy'. Strachey's patience and application were rewarded with his appointment as council secretary and official recorder.

Once these arrangements were in place, the business began of assessing the state of the settlement, and ways of restoring it to sustainability. The season for planting corn had long passed, so there was no hope of an autumn harvest to provide for the coming winter, or even seed for the following spring. In addition, though Delaware had brought sufficient staples such as oatmeal, oil, butter, biscuit and salt, there was little meat, and apparently none to be had in Virginia. Livestock, once seen as providing a solution to most of the settlement's dietary needs, had turned out to be of limited use. Fields could be cleared for cattle, and hogs left to forage in the fenced-off areas of the forest, but unless they were kept under constant watch, the Indians took them. Before the onset of the Starving Time, there had been as many as five or six hundred pigs on the Isle of Hogs. Now they were all gone.

Repeated attempts at fishing had yielded very little, for reasons that mystified the settlers. Delaware confirmed this when he went with an expedition led by Robert Tyndall to the mouth of the bay, which returned with nothing but small fry.

George Somers offered to return to Bermuda in the *Patience*. Some may have doubted his motives, but he persuaded Delaware that he could easily 'fetch 6 months' provision of flesh and fish and some live hog[s], of which those islands . . . are marvellous full and well stored'.[8]

He decided to take with him Samuel Argall, 'an ingenious, active, toward, young gentleman' who had made a positive impression on his seniors since navigating a fast route across the Atlantic. He was to command the *Deliverance*.[9]

The two ships set sail on a rainy 19 June 1610, stopping at Cape Henry to take on ballast before heading out into the ocean.

Meanwhile, Delaware had to decide what to do about the Indians. He noted in a report compiled for the Royal Council in England that Jamestown was surrounded by large fields that would be suitable for corn. These had been cleared by the settlers before the Starving Time, but had since been left fallow because of fears of Indian ambushes.

So far, the new regime had had no direct contact with either the Paspaheghans or Powhatan. With only Percy's bitter reports of successive attacks to assess relations, a decision had to be taken on whether the time had come to launch an all-out attack. Gates counselled appeasement. He 'would not by any means be wrought to a violent proceeding' against the Indians, hoping that by 'a more tractable course to win them to a better condition'. Delaware apparently agreed.

On 6 July, the lieutenant governor sailed to Fort Algernon, probably in preparation for a fishing expedition. While he was there, he noticed a longboat being blown across to the south side of the river by a 'rough' northerly wind. He sent one of his men, Humphrey Blunt, in an old canoe to recover it. As Blunt crossed the wide, windswept waters, a group of Indians, probably from the nearby town of Warraskoyack, appeared on the opposite shore. Blunt was blown on to a sandbank, and the English could only watch as the Indians ran over and 'seized the poor fellow and led him up into the woods and sacrificed him'. 'Being startled by this', Gates now 'well perceived how little a fair and noble entreaty works upon a barbarous disposition, and therefore in some measure purposed to be revenged.'[10]

Delaware was still hesitant. Powhatan was powerful, and Jamestown vulnerable, its defences and armaments in disarray. Delaware had also lost his master of ordnance, Sir Ferdinando Wainman, who died around this time of unknown causes. Furthermore Powhatan was known to be hoarding a multitude of English weapons, 'by intelligence above two hundred swords, besides axes and poleaxes', accumulated from the men he had taken and killed, and from renegades.[11]

Delaware decided Gates should go immediately to Kecoughtan, the Indian town at the mouth of the river, to provoke a confrontation and test the Indians' strength.

Gates set off on 9 July, and landed on the neighbouring shore. As soon as his men had landed, he ordered his taborer Thomas Dowse 'to play and dance thereby to allure the Indians to come unto him'. The chief and his warriors were duly drawn out of the town, and, 'espying a fitting opportunity', Gates's men 'fell in upon them, put five to the sword, wounded many others, some of them being after found in the woods with such extraordinary large and mortal wounds that it seemed strange they could fly so far'.

Gates marched into the town and plundered it, coming away with 'only a few baskets of old wheat and some other of peas and beans, a little tobacco, and some few women's girdles of silk of the grass silk, not without art and much neatness finely wrought'. He then returned to Jamestown, leaving his company under the command of his lieutenant, George Yeardley. Yeardley was later joined by a Captain Holdcroft, who had been told to build two small forts on the occupied land, 'a pleasant hill and near a little rivulet' overlooking 'wood, pasture, and meadow; with apt places for vines, corn, and gardens'. For one fort, they took over two Indian houses. For the other, they erected a tent 'with some few thatch'd cabins which our people built at our coming thither'. The plan was that future settlers coming from England would stay there 'at their first landing . . . that the wearisomeness of the sea may be refreshed in this pleasing part of the country'.[12]

As these developments were under way at Kecoughtan, Delaware sent an embassy led by two gentlemen up the Chickahominy river, to meet Powhatan at Orapaks. Delaware instructed the gentlemen to register an official protest at the 'practices and outrage hitherto used toward our people, not only abroad but at our fort also'. However, they were also to inform the *mamanatowick* that Delaware 'did not suppose that these mischiefs were contrived by him or with his knowledge, but conceived them rather to be the acts of his worst and unruly people'.

To regularize relations, Delaware proposed Powhatan issue a 'universal order' to all the territories under his control, calling for attacks on the settlers to cease. He also demanded the immediate return of all English prisoners and renegades, and for the punishment of those of his own people 'whom Powhatan knew well' who had 'assaulted our men at the blockhouse', killing four guards. If he agreed to these conditions, Delaware promised that, being the 'great *weroance*' of the English, he would 'hold fair quarter and enter friendship with him as a friend to King James and his subjects'.

The emissaries returned to say that Powhatan had rejected Delaware's terms. A second attempt produced 'no other answer but that either we should depart his country or confine ourselves to Jamestown only'. According to Strachey, Powhatan also warned Delaware not to send any more messengers, unless with a coach and horses, 'for he had understood by the Indians which were in England how such was the state of great *weroances* and lords in England to ride and visit other great men'.

Having, as the English saw it, provoked this impasse, the Indian campaign of what the English regarded as relentless harassment was stepped up, parties of warriors constantly probing the blockhouse and fort defences, or setting ambushes in the surrounding woods, to strike at settlers venturing out 'to gather strawberries or to fetch fresh water'. Delaware retaliated by taking two Paspaheghans found lurking near the fort. They were manacled and dragged before the Lord Governor and his council, and one of them, 'a notable villain who had attempted upon many in our fort', was sentenced to having his right hand cut off.

Once the sentence had been carried out, the mutilated man was sent to Powhatan with an ultimatum: if all English men and weapons were not immediately returned, the other Paspaheghan would die, as would 'all such of his savages as the lord governor and captain general could by any means surprise'. Delaware would also order his soldiers to 'fire all his neighbour cornfields, towns, and villages'.

'What this will work with him we know not as yet,' wrote Strachey, reflecting anxieties in the fort as the aggression was ratcheted up.

These were among the final words of Strachey's account or 'reportory' of his first momentous year away from England – a masterpiece meant for publication, but destined to be confined by the Virginia Company's internal censorship rules to private circulation by the 'excellent noble' but unnamed lady to whom it was addressed. He handed it to Gates who, with Dale and Newport, was due to return post-haste to England in the *Blessing* and the *Hercules*, to arrange more supplies.[13]

Delaware also had some documents to send, including his official report on the state of the colony, and a letter for Cecil. 'If God restore me to health', he promised the Earl of Salisbury, he would 'return something valuable unto the adventurers'. He also mentioned Dr Bohun's good work in dealing with an epidemic of 'strange fluxes and agues' that had afflicted many other newcomers, and the need for more doctors and medicines to treat it. Hinting at the political tussles that had led to Delaware's appointment to the Virginia mission, the Lord Governor begged Lord Treasurer Cecil to become 'a favourer and a furtherer herein unto us, and make it your own cause, since it is undertaken for God's glory and our country's good'.[14]

Around 12 July, Delaware and a company of soldiers accompanied Gates, Newport and Dale on the *Blessing* and the *Hercules* to Point Comfort, where Delaware set up camp at Algernon Fort.

To avenge the killing of Humphrey Blunt, a squad was sent over the river to Warraskoyack, where they captured the *weroance*, called Sasenticum, together with one of his 'chief men' and his son Kainta. Delaware decided that Kainta would be sent back to England with Gates. The fate of the other two captives is unknown.[15]

The *Blessing* and *Hercules* were loaded with cedar, clapboard, black walnut and more soil samples, hoped to contain iron ore. The ships departed on 15 July 1610.

Over the following weeks, Delaware drew up plans for a decisive assault on the neighbouring town of Paspahegh. Percy was appointed to lead an expedition of seventy soldiers plus Kemps, who was to act as guide. Kemps was one of 'two most exact villains' captured by Captain Smith the previous year during a skirmish with the Paspahegh. Kept a 'fettered' prisoner in the English fort, he had been released, but apparently he 'so well liked our companies' he had chosen to stay at Jamestown, to teach his captors 'how to order and plant our fields', though more probably to act as a spy.[16]

Percy set off in two boats on the morning of 9 August, and that night reached the confluence of the James and Chickahominy, 3 miles from Paspahegh. The men went ashore and, led by a handcuffed Kemps, set off for the town. Percy soon realized Kemps was leading them astray. The captain thrashed the 'subtle savage', and threatened to cut off his head, 'whereupon the slave altered his course and brought us the right way near unto the town'.

When the glow of the town's fire or the trail of its smoke became visible through the trees, Percy signalled his men to halt, and whispered their orders. The houses being scattered around a woodland clearing, the soldiers were divided into squads, each to surround a cluster of houses to ensure that none of the slumbering inhabitants escaped. Once they were all in position, Captain William West would fire off his pistol, to signal the attack.

'And then we fell in upon them,' Percy recalled in his memoir of the event. Fifteen or sixteen were put to the sword, and almost all the rest fled. 'Whereupon I caused my drum to beat, and drew all my soldiers to the colours.' One of Percy's lieutenants returned with the town's 'queen', her children, and one other prisoner. Scolded by his captain for sparing them, the lieutenant replied that now they were in Percy's custody, 'I might do with them what I pleased'. Percy ordered that the prisoner's head be cut off. He then told his men to put the town to the torch, and to ransack the surrounding gardens and fields, laden with ripe fruit and corn.

Having witnessed this destruction and carnage, the queen and her cowering children were taken to the river, to be shipped back to Jamestown as hostages. Aboard the boats, several of the soldiers started to complain of the captives' soft treatment. Percy summoned a council, at which 'it was agreed upon to put the children to death, the which was effected by throwing them overboard and shooting out their brains in the water'.

Two miles downstream, Percy gave Captain Davies permission to take most of the soldiers ashore, perhaps to allow them to cool off. No sooner had they landed, than arrows began to rain out of the dark sky.

Davies launched a counter-attack, his men charging into the woodland towards their attackers' hiding place. Their fury carried them several miles inland, until they came across a 'spacious temple, clean and neatly kept', possibly the main temple of the Quiyough-cohannock. They destroyed it, along with surrounding houses, before returning to the boats.

Percy decided that the killing spree had gone far enough, and headed straight back for Jamestown. Delaware was still confined by illness aboard his ship, and so it was there that the queen was presented to the 'lord general' as a trophy of the night's business. The general was 'joyful of our safe return'. But, according to Davies, he felt 'discontent because the queen was spared'. It was suggested that, being a pagan, she should be burned like a witch. Percy replied that 'having seen so much bloodshed that day, now in my cold blood I desired to see no more'. He turned his back upon Davies, who with two soldiers took her ashore, and killed her with their swords.[17]

Percy recorded the details of these horrific events in an account written two years later for his brother, the Earl of Northumberland. They are in a manuscript intended to vindicate his period as the settlement's president, in response to John Smith's very public and trenchant criticisms of Percy's role as governor. He makes no effort to disguise the savagery that took place that night under his command, even though he knew the treatment of the captive Indian queen and her children to be a violation of the 'law of arms' – a war crime, in more modern terms.[18] Rather, he indicates how, in the aftermath of the Starving Time, relations with the Indians passed a threshold. A struggle for territory had become a clash of civilizations, a pre-sumption of cohabitation had become a need for domination.

Nothing more was said of the incident, which was quickly submerged by more pressing concerns. A few weeks later, towards the end of August 1610, Samuel Argall returned from his expedition to report that, after an abortive attempt to reach Bermuda, he had lost contact with Sir George Somers's ship while fishing around Sagadahoc (modern Kennebec river). Making his way back to Virginia, gathering more fish as he went, he had taken a detour around a large inlet north of the Chesapeake, which he had dutifully dubbed Delaware Bay. The fact that Somers had not turned up at Jamestown, however, aroused fears that this stalwart veteran of the venture was lost on the reefs around Bermuda, or had sailed to London to claim the islands for himself.[19]

Argall had brought back with him a good supply of cod and halibut, temporarily alleviating the continuing shortage of meat in the settlers' diet. After the supplies had been unloaded, Delaware sent him straight to Warraskoyack to continue the policy of attrition begun by Gates.

Argall arrived at the town to find it abandoned, the inhabitants having been forewarned by 'their neighbours' harm' of the more aggressive tactics now being adopted by the English. Argall commanded the town be burned down, and the corn in nearby fields destroyed.

But no sooner were the Indians driven away, than they were back. 'The savages, still continuing their malice against us, sent some as spies to our fort,' Percy noted. One was captured, and after one of his hands was amputated, he was 'sent unto his fellows to give them warning for attempting the like'. But it had little effect.

With Jamestown thus effectively still under siege, Delaware decided upon another strategy. He ordered Argall to sail to the north of the bay, to see if he could find Indians there who might be prepared to form an alliance with the English to challenge Powhatan's grip over the region.

Argall set off, following the course pioneered by Captain Smith. He eventually reached the Potomac river, where he encountered a chief called Iopassus. The *weroance* claimed to be a brother of the unidentified overlord of the Potomac people, a 'great king' who was, Iopassus boasted, 'as great as Powhatan'. He also knew something of the English, as Henry Spelman, the young man left with Powhatan, had come to live with him following the execution of Captain Ratcliffe.

Iopassus greeted Argall as a 'brother', and, in return for gestures of friendly consort and some sheets of copper, offered to return Spelman to the captain's care and throw open the doors of his granary. Argall was only too happy to oblige, delighted that he had found a chief who might be encouraged into rivalry with the 'subtle' Powhatan.

'After many days of acquaintance with him', Argall invited Iopassus aboard his ship to deliver Spelman back into his countrymen's care. The chief accepted the invitation, eager, as many Indians were, to find out what it was like aboard these large wooden wind-powered vessels.

Argall entertained the chief in his cabin where, it being 'about Christmas', there was a roaring fire in the grate. In honour of the season, Argall asked one of his officers to read out a passage from the Bible. Iopassus, who may have understood a word or two of the text, 'gave an attent ear, and looked with a very wish'd eye' upon the reader. Argall noticed his curiosity. Taking the bible, he 'turned to the picture of the creation of the world in the beginning of the book', depicting Adam and Eve, the Tree of Knowledge and the serpent, and showed it to the chief, asking Spelman to explain what he was looking at. This prompted Iopassus to explain his own people's creation story, of the great hare Ahone, of the men and women that he kept in a bag in his hut, of the four jealous gods of the north, south, east and west who tried to steal the bag, of the moment Ahone 'opened the great bag wherein the men and the women were, and placed them upon the earth, a man and a woman in one country and a man and a woman in another country. And so the world took his first beginning of mankind.'

Part Four

SEVENTEEN

A Pallid Anonymous Creature

IN 1611, PHILIP III OF SPAIN, King of Castile and León, Aragon, Portugal, Sicily, Naples and Jerusalem, of the Islands of the Indies, East and West, and of the mainland of the Ocean-Sea, Archduke of Austria, Duke of Burgundy, Lorraine, Brabant, Prince of Swabia, Marquis of the Holy Roman empire, and the most Powerful Man in the World, sat in the Escorial, trying to ignore the clamour of his distracting domains. But even in his gloomy chamber deep in the palace, this 'pallid anonymous creature', as the historian John Elliott has described him, could not escape the swirl of events outside: the re-establishing of Catholic dominance across Europe, the battle against the forces of Islam spreading through Africa, the 'corsairs' or pirates emerging from Ottoman ports to prowl the Mediterranean, the protection of Spanish and Portuguese dominions in the New World and the treasures they yielded, the protection of trade routes to the East Indies and China.[1]

The Escorial, part royal residence, part monastical retreat, had been built by his father Philip II near Madrid, Spain's new capital. The building was a huge structure of grey stone designed to have the layout of a gridiron, because that was the device used to martyr the saint to whom it was dedicated: St Lawrence. Inside, a pervading mood of piety was relieved only by Philip II's weakness for the works of the great Flemish painter Hieronymus Bosch, who lingered over the grotesque and diabolical aspects of human transgression in such works as the *Seven Deadly Sins*, which had as its central image the eye of God, the iris reflecting scenes of gluttony and murder, and the

Garden of Earthly Delights, a fantastical vision of humanity deformed by sinful pleasures.

Following his father's death in 1598, Philip III had dutifully buried him in the Royal Pantheon and fled, opting to base his court in gentler Valladolid, the old capital of Spain. However, his nobles had invested too much of their money and time establishing themselves at Madrid, and he was eventually prevailed upon to return. It was there that he set up his peculiarly baroque system of government, built around *privados*, favourites, led by the Duke of Lerma, and *Juntas*, or small, informal cabinets of confidants.

The empire Philip ruled was without equal in the history of the world. Its European core had been inherited almost fully formed by his father, and it was a matter of some pride that it had been forged not through violent conquest, but for the most part by treaties, marriages and election. Now it stretched across the world, from the Philippines (named after his father) in the east to Peru in the west, and produced unimaginable wealth. Income from gold and silver bullion imported from South America peaked at 36 million pesos in 1595 (at least £4 million in English money, far greater than England's total national income). Of this, as much as a third was reaching Philip's exchequer, making him rich beyond the dreams of James I or any other European royal.[2]

However, this fabulous wealth encrusted a very fragile economy. Like James I, Philip III ran an expensive government. The gifts, sinecures and privileges he had to lavish upon his *privados* to sustain their loyalty was costing him more than 200,000 ducats a year by 1602. At around the same time, concerned administrators pointed out that, while his father's annual expenditure on maintaining his court had amounted to around 400,000 ducats a year, he was spending twice that or more. Meanwhile, the influx of bullion was producing rampant inflation. As the English settlers had discovered on a miniature scale in Virginia with copper, the more precious metal there was in circulation, the less precious it became.[3]

In 1607, Philip was forced to default on repaying his debts, effectively rendering his kingdom bankrupt. His vast empire may have been responsible for bringing in millions of pesos, but it was costing millions more to maintain and protect. Spain's domestic economy had remained more or less unreformed. Rather than becoming enriched by new trades and exchanges built around the importation of goods from across the world, it was still largely reliant on the efforts of sheep farmers. Meanwhile, the Low Countries, with their Protestant urban enclaves constantly slithering from Philip's grasp, went from strength to strength. While he struggled to make money out of an empire, Dutch merchants were somehow managing to make an empire out of money. The Dutch East India Company, founded in 1602, already stretched its tendrils into the Pacific, snaring valuable commodities and trading deals from under Spanish and Portuguese noses.

The difficulty was the Spanish empire itself. It ruled Spain as much as Spain ruled it. Some have argued that it was barely Spanish at all. The vast armies Philip commanded across Europe were mostly made up of foreigners. He had ports and settlements scattered throughout the seven seas, but they were often run not by his loyal subjects, but by local and often disloyal agents. In fact, Spain's imperial possessions and lines of command were so dispersed, even its monarch had difficulties keeping track of them. On one occasion, Philip's father mislaid a whole country, even though it had been named after him. 'I think I have some maps,' he wrote to his secretary, when asked to endorse a plan to send an expedition to the Philippines, 'and I tried to find them when I was in Madrid the other day. When I go back there I shall look again.'[4]

This threw up problems of unimaginable complexity: having to cope with negotiations in Brussels one moment, demands for religious freedom in Bohemia another, piracy in the South China Seas after that. Philip III's response was to create a huge and expensive bureaucracy to handle the paperwork, and wherever possible to simplify foreign and imperial relations by making them peaceful. Policy was thus realigned with expediency, and *Pax hispanica* became the order of his reign.

Philip's father, 'withered and feeble', had begun the process just before his death, by signing the Peace of Vervins with France's King Henry IV, a major rapprochement given Henri's Protestant upbringing in the Kingdom of Navarre, which straddled the French/Spanish border.[5]

When James had ascended to the English throne, Philip III found the Scottish King even more eager for peace, and the result was a speedy and satisfying negotiation that culminated with the Somerset House Treaty of 1604. Notwithstanding a fluke victory over the Armada in 1588, England had failed to establish naval superiority. By agreeing to peace terms, Spanish dominance of the oceans, and the right to pass freely through the Channel and North Atlantic, now seemed secure.

The Somerset House Treaty had to some extent eased the situation in the Low Countries, as radical English Protestants were no longer allowed to aid the rebels, as they had done in Elizabeth's time. This simplified negotiations with the Dutch, resulting in the Twelve Year Truce of 1609. Though not a final settlement, it provided a chance to consolidate positions, with the satisfying side effect of igniting religious disputes between heretical sects within the so-called United Provinces.

Thus, Philip saw himself as the architect of a pan-European settlement that had brought peace to the continent, as well as peace of mind to himself.

There was, however, a disturbance on the fringes, an irksome distraction of which Philip would hear about each and every month, sometimes more often, from his ambassador in London: Virginia.

Since the first news of English plans to found a settlement in North America had reached Madrid in 1606, Philip's military advisers had been discussing the implications. On 14 March 1607, when details of the venture were still unclear, the Council for War in the Indies, responsible for the defence of Spain's American possessions, recommended that 'all necessary force should be employed to hinder this project' as Virginia had been 'discovered by the crown of Castile and lies within its demarcation'.[6]

In August 1608, the King's Council of State had met to consider the issue, having consulted Sir William Stanley, the renegade English commander who had taken his regiment over to the Spanish side during the Dutch rebellion.[7] Stanley was known for his aggressive stance towards England, and no doubt inspired the council's verdict, that 'this matter of Virginia is not to be remedied by any negotiation, but by force, punishing those who have gone there'.

The King, however, had misgivings, and to put off having to make a decision, told the governor of Florida at St Augustine, Pedro de Ybarra, to send a ship up the Florida coast under the command of Captain Francisco Fernandez de Ecija, who had reconnoitred those waters in 1605. Ecija did as he was commanded, reaching the 'Bay of Jacan' (the Chesapeake) on 25 July 1609. He spotted a ship (Captain Argall's) anchored in the bay, 'and since it was of much greater tonnage' than Ecija's, carrying 'two topsails and a great banner at its masthead', he decided to withdraw, and return to St Augustine, where he produced a report of his voyage together with a 'pilot book . . . stating the sea-marks and the character of the harbours and bays that there are from Florida to the place to which they went'.[8]

Ecija's mission proved that the English had established a military presence in Virginia, and provided the navigational intelligence needed to mount a mission against it, but still the King refused to act. By the autumn of 1610, Don Alonso de Velasco, who had replaced Zuñiga as Philip's ambassador in London, was beside himself with frustration at the King's procrastination. If only His Majesty would send a few ships, 'I think this plan might be brought to naught with great facility,' he urged. But Philip remained unmoved.

Then, sometime in late 1610, the King received a report concerning one Francis Maguel or Magner which produced a shift in attitude.

Maguel was an Irish sailor from the King's Head Tavern, Ratcliffe, just outside London. He had gone to Virginia with the first settlers, and appears to have returned to London with Newport the following June. His subsequent movements resulted in a visit to Madrid in the summer of 1610, where he met a 'Captayne Ralfe'. This was the Irishman John Rafe, a famous figure in Irish history who in September 1607 had captained the French ship that took the Gaelic leader Hugh O'Neill, Earl of Tyrone, from Ireland to France in what came to be known as the 'Flight of the Earls'. Rafe showed a keen interest in Virginia, asking Maguel 'where the English colony had planted themselves', of the 'manner, situation and height [latitude] of that place and what hurt he thought it could do unto the King of Spain'. Rafe also wanted to know of a 'good pilot' who could guide an Armada to 'go and surprise Virginia', perhaps hoping to recruit Maguel to the role.[9]

Maguel's replies were noted down and translated into Spanish by Florence Conry, Hugh O'Neill's Franciscan chaplain. Conry then presented the report to Philip.[10]

Maguel revealed an enterprise poised to dominate North America. While Spanish attempts to colonize Florida had withered, Virginia, Maguel suggested, was thriving. He mentioned the coronation of Powhatan, and the emperor's promise to convert his people to 'the god of the English' (not being the Catholic God). He spoke of 'many mines of iron and of copper' which the English were keeping secret until their fortifications were complete. He had personally carried a sample of metal ore weighing 80 pounds back to England, and claimed to have found several ounces of gold, silver and copper.

Maguel also mentioned English optimism about finding a passage to the Pacific. 'The English desire nothing else so much as to make themselves masters of the South Sea in order to secure their share of the riches of the Indies and to cut off the trade of the King of Spain,' he explained. Revealing that he had retained good contacts with the venture, he also mentioned Argall's discovery the previous year of a fast northern route, which had opened the way to shipping twenty or thirty thousand settlers to the region over the next few years.[11]

Maguel's testimony, combined with increasingly shrill calls for action from his exasperated Council of War, finally prompted Philip to take action. The Armada that his military commanders had wanted was still considered too costly. So the King gave orders that a well-armed 'caravel', a fast, light ship, be prepared in Lisbon to 'reconnoitre the port and land called Virginia which is on the coast of Florida'. It would be under the command of one Don Diego de Molina, with Francis Lembry, a 'confidential' Englishman, acting as pilot.[12]

Don Diego de Molina arrived at Havana, Cuba, on 24 May 1611. There he was handed his commission, signed by Governor Pereda of Cuba, to search the coast of Florida for a Spanish ship called *Plantation* of 300 tons, which had been carrying a consignment of ammunition from Cartagena. She had lost her rudder and broken a mast, Molina was informed, and fallen out of contact with the rest of her convoy somewhere north of St Augustine. Molina was told to sail up the east coast of North America until he found her, and recover whatever cargo he could if she was no longer seaworthy. This was to be the cover story for the expedition, its true purpose of spying on the English in Virginia to be kept an absolute secret from the crew and anyone he encountered.

Molina set off on 2 June, with a longboat or 'sloop' in tow. He reached St Augustine six days later, where he and his crew spent the following eight days, during which time 'their purpose was not discovered or even suspected'.

San Augustin, originally founded in 1565 to launch the attack on the Huguenot settlement of Fort Caroline, was a small, neglected outpost still reliant on supplies from Havana and Spain for its survival. Like Jamestown, it was planted on a small island, and fortified with 'planks and thick beams', which rotted 'with the dampness of the land, and because the land is salty'. It was surrounded by impenetrable woodland, which could not be cultivated 'except in one part where it is sandy, where they sew corn'. Around the time of Molina's visit, there were fewer than twenty permanent settlers tending the cornfield. They had fifty cows on a neighbouring island, which 'do not flourish' and are 'occasionally killed when the governor wants it'. Since the turn of the century, there had been repeated efforts by the colonial authorities in Havana to close the settlement down. According to one report, there was 'no reason for it, either to defend against pirates or for settlement'. Another objected to the 'deception involved in keeping [St Augustine] occupied with so many people and at such cost'. Only with the arrival of the English in Virginia in 1607 were plans for its abandonment finally dropped.[13]

Molina left the outpost on 8 June. His caravel was quickly carried upon the Gulf Stream to the 'Bay of Virginia' (at Cape Henry), which Molina calculated to be about 200 leagues' or 600 miles' sailing distance from St Augustine. Molina measured the latitude, which he found to be 37°10′ (actually 36°55′).

He took the caravel into the bay, and had started taking soundings when a ship was spotted lying at anchor near the shore. Molina announced to his crew that it was the *Plantation*. However, as they approached her, they noticed on the adjacent land 'an earthwork, like trenches', manned by sixty or seventy men. Soon after, a cannon was fired, and in response Molina ordered one of his cannons to fire back, though without a cannon ball.

Despite the hostile reception, Molina announced that he, Lembry and one of the sailors, Marco Antonio Pereos, would go ashore. They were taken by Pedro, the ship's master, aboard the sloop.

As they approached the beach, one of the sloop's oarsmen spotted footprints in the sand, and swore they were made by English or Flemish shoes. Molina told him to 'keep quiet and say nothing to him, because there were no enemies who would do them harm'. As soon as they were within reach, he leapt out of the boat, and waded towards the shore, followed by Lembry and Pereos.

When Molina reached the beach, three or four detachments of men appeared, disarmed Molina and his two companions, and led them away. After an hour, the English soldiers returned to the beach, and called to Pedro to bring the sloop ashore. Pedro refused. Lembry was then thrust forward, who gestured frantically, 'striking with his hand outward and crossing his arms, declaring and making them understand that he was a prisoner'.

Pedro told one of his men to swim to the shore. He was not allowed to speak to Lembry, but returned to the boat with one of Molina's captors, carrying him on his shoulders through the surf. Tipped unceremoniously into the sloop, the man introduced himself as John Clark, an English mariner. If Pedro agreed to go ashore with four or five men, Clark said he would guide the caravel to safe anchorage near the fort. Pedro refused, and ordered his men to start rowing for the caravel. Clark tried to leap overboard, but was restrained and muffled. By the time they reached the ship, it was late in the day, so Clark was kept prisoner overnight while Pedro pondered his options.

The following day, Pedro took Clark in the sloop to negotiate an exchange of prisoners. However, as he approached the shore, he saw some English soldiers lying in ambush, and held back. Clark once again tried to escape, but was thwarted. In a final desperate attempt to secure his release, he revealed to Pedro that he was Virginia's chief pilot, and was of such importance the English were bound to agree to an exchange. At this point, James Davies, captain of the fort, appeared on the shore, and across the surf shouted to Pedro that he would not release any of his Spanish prisoners. Through Clark, Pedro threatened to attack if they were not handed back immediately. Having heard this, Davies replied from the shore 'with great anger that they might go to the devil. At this time it was seen that they took away Francisco Lembry with much violence.' As Lembry was being manhandled, he managed to gesture to Pedro 'that they should push out to sea, crossing his arms and hastening [them] to get away'.

Pedro sailed the sloop back to the caravel. Seeing an English boat heading out for them, he raised anchor and set off out of the bay, taking Clark with him and leaving Molina, Pereos and Lembry behind.

Pedro was back in Havana by 20 July, barely two weeks after he and Molina had originally set off on the mission. He reported to Governor Pereda the loss of Molina, and handed over John Clark, who was put into solitary confinement.

Three days later, Clark was interrogated, one John Lak, an English prisoner, acting as interpreter.

Clark described himself as a 35-year-old pilot from London who had been in Virginia nearly eighteen months. Asked about English defences in Virginia, he reported that there were four forts along the James river. Three were at its mouth, barely a musket shot apart. The other was 60 miles upstream, and was called Jamestown. Clark said there were as many as one thousand English people now living in Virginia, six hundred bearing arms, the remainder, 'woman, boys, and old men'. Most were 'outcasts', accustomed to living by 'piracy' (by which he probably meant begging or pilfering). Attempts had been made to find gold and silver, but they had yielded nothing and, Clark pointed out, 'the Indians bring them none of those metals'. The soil produced 'no other fruit but maize and nuts', and game was confined 'very far inland'. There is 'no intercourse with the Indians, because at one time it is war, at another time it is peace'.[14]

Clark's testimony caused considerable excitement and concern. Governor Pereda, together with the chief pilot in Havana, attended some of the interrogation, bringing maps and compasses so they could chart the exact location of Jamestown and the forts guarding the entrance to the river. Soon after, Luis de Velasco, Marques de Salinas, Viceroy of New Spain, arrived from Spain, and Clark was told to repeat his entire testimony. It summoned up in the mind of Velasco the same troubling image as it did to Pereda. Just four years ago the English had planted a single fort. Now three more had sprung up, and Clark's mention of a further four ships arriving from London within the month suggested yet more, spreading across the lands north of Florida like some rampant weed.

In the light of the information, Pereda ruled 'that the English pilot should remain at Havana on account of his safety there, without any communication with any one, and especially of his own nation'.[15] He also impounded Molina's caravel in Havana, which he would use to mount an attack on Virginia, for which he would pay out of his own pocket.

Back in Spain, the Council of War discussed options, the consensus favouring a force of as many as four thousand men 'to turn the English out' of Virginia. To reassure the King that an attack would not provoke war, they reminded him that the English government considered the venture to be a private one, which the settlers had undertaken at their own risk.[16]

EIGHTEEN

Strange Fish

On the evening of 21 June, 1611, a bedraggled, flustered Lord Delaware arrived 'by stealth' at Westminster Palace, begging for a private audience with King James. His request being denied, he returned to his horse and rode at speed to Kensington, to the opulent home of Sir Walter Cope. Cope was shocked to see him. He was supposed to be in Virginia.

Delaware breathlessly tried to explain himself. Having contracted a horrible disease in Virginia, he had decided to seek refuge in the West Indies, in the hope that the change in climate would allow him to recover. But on his way to the Caribbean, his ship had been blown off course, and he had ended up at the Azores, in the eastern Atlantic. Since he was now closer to Europe than America, he had decided it would be better to sail straight back to England.

Cope observed that Delaware's physical condition was remarkably good. 'I have no touch of my disease remaining on me,' Delaware admitted, saying that it was thanks to the therapeutic effects of a few days he had spent on the Azores island of Fayal.

Whether Cope believed Delaware's story or not, no royal audience was forthcoming. Cope respectfully, or perhaps sarcastically, suggested His Lordship 'attend of [his] health' before seeking to talk to the King or Cecil. Fearing a rebuff, Delaware sent a note to Cecil the following morning, repeating his request to see the King, and the explanation he had given Cope for his untimely return. The reply, if one was sent, does not survive.

News of Delaware's return soon reached the ears of Virginia's promoters and investors, and it provoked 'coldness and irresolution'. One investor probably reflected the opinion of most when he wrote that he had begun to 'conceive great hope' of the enterprise until 'my L. Delaware's unseasonable return'. It made matters no better to learn that the man Delaware had left in charge in Virginia was George Percy, who had presided over the Starving Time.[1]

On 25 July, just four days after his return to London, Delaware was summoned before a 'general assembly' of the Virginia Company's managers and investors to account for himself.

Contriving as much lordly indignation as he could at those 'who spare not to censure me in point of duty', Delaware recapitulated his excuses, explaining that he had been forced away from America by a 'hot and violent ague' which would have killed him within twenty days had he not left. He detailed the debilitating symptoms he had endured 'lest any man should misdeem that under the general name and common excuse of sickness I went about to cloak either sloth or fear'. There had been 'flux' (diarrhoea), which 'surprised me and kept me many days'; cramps, which 'assaulted my weak body with strong pains'; gout, which rendered him 'unable to stir or to use any manner of exercise', and finally scurvy. After 'long consultation', he had decided to leave for the West Indies, but being blown off course ended up in the Azores, where he started to recover, thanks to a 'fresh diet, and especially of oranges and lemons, an undoubted remedy and medicine for that disease which lastly and so long had afflicted me ... which ease as soon as I found, I resolved, although my body remained still feeble and weak, to return back to my charge in Virginia again.' But once more he was advised (he does not say by whom) 'not to hazard myself before I had perfectly recovered my strength', hence his presence before their lordships and gentlemen of the company that day.

He assured his audience that he had left the settlement in good shape, most of the two hundred or so living at Jamestown in reasonable health, with at least ten months' provisions in the store-house, supplemented by provisions gathered by Samuel Argall from his trading expeditions to the Potomac. Though Powhatan 'still remains our enemy', the settlers were too well-defended for him to do them any harm.

As for himself, notwithstanding the fact that he had 'suffered beyond any other', he remained 'willing and ready to lay all I am worth upon the adventure of the action'.

No sooner were the words out of Delaware's mouth than Sir Thomas Smythe had them taken down by the company's secretary, Samuel Calvert. Two weeks later, they were published under the humiliating title *A Short Relation made by the Lord De-La-Warre to the lords and others of the Council of Virginia, touching his unexpected return home.*[2]

The appearance of the *Short Relation* failed to calm nerves among Virginia's beleaguered backers, which were already set on edge by external factors. The Great Contract, Cecil's strategy for saving the King from looming bankruptcy, had failed to find support in the House of Commons, leading to acrimonious exchanges, and James once again dismissing Parliament. To complicate matters, Cecil, the man upon whom James had come to rely to deal with such crises, was ill, the deformities that others saw as the source of his political potency now beginning to sap his strength. This may have contributed to a rash of poorly judged money-raising schemes introduced in the early months of 1611, the most notorious and demeaning of which was James's decision to start advertising the sale of aristocratic titles. Rumour spread of baronies being offered at £5,000 apiece, and a new semi-aristocratic order of baronet, aimed at the gentry and costing £1,000. 'I thought this dignity would not be much affected at so dear a rate,' a correspondent caustically noted in June, 'but now I find the power of ambitious wives passes my understanding. Twenty patents are already passed, and many more are suitors.' Adding to the unedifying spectacle of the King's desperate scrabble for money was his continuing use of monopolies and impositions, which left some with the impression that no aspect of commercial or public life was safe from his grasping hands.[3]

By the summer of 1611, it was clear that such developments were having an impact on the Virginia Company's own finances. Many investors were threatening to default on share options they had pledged to buy, against which sums of money had already been borrowed. There were also growing worries about the Spanish. By the end of October, news had reached England of the Molina mission, and the capture of the English pilot John Clark. It could only be assumed that the Spanish now had all the navigational information they needed to mount a full-scale attack on the fledgling colony.

On 4 November, Sir John Digby, King James's ambassador at Madrid, reported to Cecil that he had heard 'by a Frenchman, and an Irishman' that Clark had been secretly brought to Seville. He had also intercepted a letter from the Spanish King to the London ambassador Velasco, which raised some hopes that Molina's mission had not been hostile, his caravel having entered the Chesapeake 'in search of a ship which sailed from the Port of Cartagena of the Indies with certain artillery' – the cover story Molina had told his crew and captors. Philip instructed Velasco to 'express to the said king [James] the just resentment which I feel at the seizure of these men', and demand that they be set free immediately.

The information was evidently meant to fall into the hands of the ailing Cecil, and whether he fell for this subterfuge is unclear. He did not apparently have access to the contents of a coded section of the letter, which instructed Velasco not to reveal that one of those captured was the Englishman Francis Lembry, 'who by my orders went to reconnoitre those ports'.

Cecil's response was the familiar and effective one of offering compromise in private while affecting outrage in public. The Spanish ambassador was summoned before the Privy Council 'where it was roundly told him what criminal wrongs and injustice our nation was still offered in Spain, with this conclusion, that if there was not present redress, the king was fully minded to recall his ambassador'. But, as critics noted, just a few days before, Velasco was allowed a private audience with King James, which had been granted without the Privy Council's knowledge.[4]

Meanwhile, Cecil quietly ordered Digby to negotiate the swapping of captives, Clark to be exchanged for Molina and his two associates. To ease the way, the ambassador was authorized to pledge 'our Renouncing the Plantation of Virginia'.[5]

On 1 November 1611, All Saints' Day, King James was invited to the Banqueting Hall, Whitehall, to enjoy a new entertainment by the playwright William Shakespeare. It was to be performed by Shakespeare's players, known as the King's Men. Like Virginia, the troupe was the creation of a royal charter, permitting it to 'use and exercise the art and faculty of playing comedies, tragedies, histories, interludes . . . for the recreation of our loving subjects'.[6] Such art and faculty was to be stretched to the limit in the staging of Shakespeare's latest work, which in terms of production values as well as subject matter was ambitious even by his standards. It was called *The Tempest*.[7]

Natural light being weak so late in the year, and the action depending so heavily upon a sense of magic and dislocation, Shakespeare's company may have attempted to stage the play using a revolutionary new method pioneered in Italy. Conventionally, performances took place during the afternoon, illuminated by daylight. However, in 1598, the dramatist Angelo Ingegneri had started to stage his plays in a darkened space, relying on artificial rather than natural lighting to illuminate the action. English stage designers were certainly aware of such methods, and *The Tempest* provided an ideal opportunity to try them out.

Such techniques would have contributed to the impact of the play's unforgettable opening scene. Sitting in complete darkness, awaiting the performance to begin, the audience would have been overwhelmed by an outburst of flashing squibs, rolling cannon balls, cascades of quartz crystals and even sprays of water, recreating the experience of being aboard a ship during a violent storm.[8]

In the midst of the din, a boatswain entered, staggering from side to side. He shouted orders to his crew, but was distracted by panicking passengers: Alonso, King of Naples, his son and heir Ferdinand, and his unruly entourage.

The boatswain's efforts to save the ship were to no avail. 'All lost! To prayers, to prayers! All lost!' the mariners cried out, and the action disappeared back into the darkness.

Later, the ship's passengers awoke to find themselves cast away upon the shore of an enchanted island, the setting for the rest of the play. What followed took the audience into a world of wizards, spirits, savages and spells, of rape, romance, vengeance and hope. The island, it transpired, was located somewhere between North Africa and Italy, in the middle of the Mediterranean. This, as the audience well knew, was the latitude dividing the Catholic and Ottoman empires, Christendom and Islam. Alonso's ship had been on its way back from Tunis, where his daughter had just married the Muslim king, a match that even now his courtiers carped at. ''Twas a sweet marriage,' said one, sarcastically, 'and we prosper well in our return.' King Alonso wondered if the storm was a punishment for crossing a religious boundary, as having already lost a daughter, it now seemed he had lost his son Ferdinand, who had not been seen since the storm's fury took the ship.[9]

Sitting in the audience, James would have been alert to the thinly disguised parallels with his own dynastic challenges. His negotiations to marry his son Prince Henry to a Spanish princess had revived in his court the polarities between pro- and anti-Spanish factions that had torn at Queen Elizabeth's government. The tempest was also significant, suggestive of a ship of state all at sea.[10]

Any concerns, however, would have been driven swiftly off, as attention was drawn from the survivors' past follies to their present predicament. As they began to explore the island, they found it to be an extraordinary place, populated by strange and wonderful characters: a powerful magician called Prospero, once the Duke of Milan but usurped of his title by his brother Antonio; Miranda, Prospero's daughter, who since infancy had been stranded on the island with her father; Caliban, the monster offspring of a witch who was forced to be Prospero's slave; and Ariel, an androgynous nymph who performed Prospero's magic.

The magic disorientates the castaways, but also the audience. Prospero, like his slave Caliban and spirit Ariel, is a polymorphous character, sometimes a wizard, sometimes an angry or exhausted old man. There are strong reminders of Dr John Dee, the famous magus of Mortlake, whose work on navigation and geography had inspired England's first forays to America. In 1604, Shakespeare's troupe had stayed at Mortlake, a few yards from where Dee had his famous library. At that time the ageing scholar was also marooned among his books, banished from King James's court for being 'a *Conjurer*, or *Caller*, or *Invocator* of devils'.[11]

As the play proceeded, the island itself started to shift, apparently moving from the Mediterranean into the mid-Atlantic. The opening scene had already invoked memories of the famous tempest that wrecked the *Sea Venture*, which had been the talk of the taverns since news of the miraculous deliverance of Gates and his company to Virginia had reached London. The play later revealed that the storm had been conjured up by Ariel, who had haunted the ship in the form of the same St Elmo's fire that Strachey saw playing upon the rigging of the *Sea Venture*.

When the courtiers get to the island and are confronted with Caliban, there are unmistakable reminders of the Indians brought back from the Americas in recent years, who were 'shewed up and down London for money as a wonder'.[12] 'What have we here?' asks the jester, Trinculo, confronted with Caliban. 'A strange fish!' The fish is a clue. Caliban later reveals that he is knowledgeable of the fishing methods used by the Indians of Virginia. His appearance and demeanour also recall a line in Captain John Smith's *True Relation*, published 1608, where Powhatan's 'most trusty messenger, called

Rawhunt' is described as being 'much exceeding in deformity of person, but of a subtle wit and crafty understanding'.

Ariel is the agent who works this transformation of the island into an outpost of the New World, capable of fetching a shrouding dew from 'the still-vexed Bermudas' to swathe the castaways from reality. And for a moment, the refuge becomes a place of escape. The isle of devils becomes the home of free spirits, as it had proved to be for the rebel castaways who had tried to stay on Bermuda rather than leave with Gates. 'Had I plantation of this isle, my lord . . .' speculates Alonso's faithful old councillor, Gonzalo, it would be a commonwealth without 'magistrates . . . riches, poverty . . . contract, succession'

> all men idle, all;
> And women too, but innocent and pure;
> No sovereignty . . .

As these musings develop into a dangerously republican fantasy, the old retainer is mocked by Sebastian, the King's cynical, treacherous brother, and Antonio, Prospero's usurper, but Gonzalo will not be deflected.

> . . . All things in common nature should produce
> Without sweat or endeavour. Treason, felony,
> Sword, pike, knife, gun, or need of any engine,
> Would I not have; but nature should bring forth,
> Of it own kind, all foison, all abundance,
> To feed my innocent people.

To any Virginia Company investors sitting in the audience, this exchange encapsulated the struggle between Utopianism and realism that had characterized their troubled venture since its inception. They might have also recalled Seagull's drunken promise in *Eastward Hoe* of a New World 'without sergeants, or courtiers, or lawyers, or intelligencers'.

But like James, Alonso is deaf to such a dream, and refuses to silence Gonzalo's teasing tormentors. 'Thou dost talk nothing to me,' he says, when Gonzalo appeals to the King to quieten the mockery.

In parallel with the antics and arguments of Alonso's courtiers runs the story of his son Ferdinand, who is saved from the sea by Ariel and brought to Prospero's cavern. There he meets and falls in love with Miranda, a romance that culminates with their betrothal, and Prospero's triumph over his old adversaries. This epiphany moves Miranda, whose only human company until now has been her father, to utter the play's most famous lines:

> O, wonder!
> How many goodly creatures are there here!
> How beauteous mankind is! O brave new world
> That has such people in't!

''Tis new to thee,' Prospero reminds her.[13]

Thus, at the climax of the play, experience is confronted by hope, naivety is momentarily overwhelmed by innocence – concentrating into a moment of high drama the familiar struggle between expectation and disappointment of colonial adventure.

William Strachey was in London at the time of the *Tempest* performance at the Banqueting Hall, and he must have realized, if he did not know, that Shakespeare had been inspired by his 'True Reportory of the wrack and redemption of Sir Thomas Gates, knight, upon and from the Islands of the Bermudas'. There are fragments scattered through *The Tempest* of Strachey's vivid account, which have led most Shakespeare scholars to accept it as a source.[14]

How Shakespeare got his hands on Strachey's unpublished manuscript (it did not appear in print until 1625) is not known, but a possible source is Mary, Countess of Pembroke, whose son the Earl had recently been appointed to the Royal Council. Strachey, ambitious to forge a literary career, would have been happy for her to circulate the work among the Sirenaicals at the Mermaid, whence it might have fallen into Shakespeare's hands.[15]

Although the most eloquent, Strachey's was not the only work offering inspiration, nor publicizing the magical qualities of Bermuda.

Two others had appeared since first reports of the *Sea Venture* being washed up on the islands' shores reached England. One was *A Discovery of the Barmudas, Otherwise called the Ile of Divels*, written by one of the survivors of the wreck, Silvester Jourdan, dated 13 October 1610. The other was the Virginia Company's own propaganda tract, the *True Declaration*. This included a passage on Bermuda that powerfully invokes, if only to dismiss, the magical theme:

> These Islands of the Bermudas, have ever been accounted as an enchanted pile of rocks, and a desert [i.e. deserted] inhabitation for Devils; but all the Fairies of the rocks were but flocks of birds, and all the Devils that haunted the woods, were but herds of swine.

The *True Declaration* went on to conclude that even these natural phenomena were miraculous, as none of the islands of the West Indies were populated with such a plentiful stock of animals when they were first discovered by the Spanish. 'An accident, I take it, that cannot be paralleled by any History,' the author observed, suggesting that the passengers on the *Sea Venture*, like those aboard Alonso's ship, had been drawn to the island by supernatural forces, exercised in this case not by a wizard, but God himself.[16]

The discovery of Bermuda's magical allure proved to be conveniently timed. Matthew Somers, Sir George's nephew, had just arrived in England from Bermuda aboard the *Patience*. His uncle, he reported, had reached the island to find the two men left there still alive. But the old admiral died 'of a surfeit in eating of a pig' soon after. Matthew had buried his uncle's heart and entrails on the island, and brought his body back for burial in their native county of Dorset. He was now in London, pursuing Somers's vision of creating a settlement upon the island. *The Tempest* provided a poignant context in which to promote its wonders, and its miraculous role in the saving of the

Virginia venture. Abundant fauna and flora, a hospitable climate, natural defences and the absence of natives allowed young Matthew to delineate a refuge every bit as enchanting as Prospero's.

To start with, there was some hard-headed scepticism among investors about Bermuda's commercial prospects. As they had lost so much money in Virginia, there seemed no reason to assume that they would make it back in Bermuda. However, 'it came to be apprehended by some of the Virginia Company, how beneficial it might be, and helpful to the Plantation in Virginia'. It could act both as a plantation in its own right, as well as an entrepôt for trade between Europe and America, even a refuge if Virginia came under attack.

But there was a problem. The Virginia Company did not have any rights to the islands, because the area covered by its 1609 charter extended only 100 miles from the Chesapeake shore. So a plan was formulated, probably under the joint management of Smythe and Sandys, to apply for a new charter, extending those rights to 300 leagues (900 miles), which would cover Bermuda and any other islands that might subsequently be discovered in the western Atlantic.

As before, Sir Edwin Sandys seems to have led the drafting of the document. This provided an opportunity for his supporters in the House of Commons, who now made up a substantial segment of the company's investors, to shape the Virginia project to their own agenda. What they wanted was wider public participation and freer trade. Having been denied by King James a chance to put these policies into practice through Parliament, they now saw an opportunity to implement them through the Virginia venture. In particular, 'the Monopolizings of trade into a few men's hands' that was such a feature of the closed, cosy world of city wheeling and dealing could be challenged by creating a company that was truly open to all with the money or trade skills to make it prosper.[17]

Using this as the model, Sandys crafted a new kind of joint stock company. It would henceforth be under the control not of an unaccountable council of government appointees, but 'one great, general and solemn assembly' of all shareholders and participants, which would take place four times a year. This assembly would be known as the 'Great and General Court of the Counsels and Company of Adventurers for Virginia', and would effectively take over the King's powers of oversight. It would elect the members of the Royal Council, appoint the company's officials (including the treasurer, the most senior director), and pass the regulations that would govern the settlement. By even modern standards of corporate governance, it was a thoroughly democratic structure, all shareholders having an equal voice at the Great and General Court, no matter the size of their investment. Furthermore, anyone could become a shareholder. Unlike other London companies engaged in foreign trade, they were not even required to take the Oath of Supremacy, usually an obstacle for Catholics.[18]

A further innovation was the introduction of a public lottery. This method of raising public finance had first been tried in England in Elizabeth's time, with the 'Very Rich Lottery' of 1567, conducted in the City of London to raise money to repair sea ports, which offered as its first prize £5,000 worth of tapestries and gilded plate. By reviving it for the Virginia project, the company would raise the public profile of the venture, with ordinary citizens acquiring a stake (if not a share) in its welfare.

In the earlier years of the venture, a document as bold as this third Virginia charter would never have got past Cecil. But the ailing Lord Treasurer was now too ill, and too preoccupied with other matters of state, to intervene. After passing through the peristaltic process of government approval in the early weeks of 1612, it emerged in March in the form its promoters had originally intended, barely trimmed by bureaucratic and political wrangling.

Meanwhile, the company found itself heading towards bankruptcy. More than £3,000 were needed to fund yet another supply, and all the usual avenues for raising such a large sum had been exhausted. A stopgap solution was needed, and Smythe controversially decided to start selling off company assets. Delaware's ship, which had ignominiously brought him back from Virginia, was the first item to go, sold probably around January 1612 for between £600 and £800. This barely covered a third of the money required, so other items were sought. A check of their inventory revealed that they had nothing much else of value, so Smythe resorted to selling an asset which the company did not yet strictly possess: Bermuda.

In a complex arrangement, an inner circle of directors paid £2,000 into the Virginia Company's coffers in return for the rights to settle Bermuda, which would be done by an 'under-company' or subsidiary.

The under-company's new owners had early anxieties about its role and identity. They 'have changed their name twice within this month', John Chamberlain commented to his friend Dudley Carleton, the company 'being first christened Virginiola as a member of that plantation, but now lastly resolved to be called Somer Island as well in respect of the continual temporal air, as in remembrance of Sir George Somers that died there.'[19] It also changed the way investors were rewarded. Whereas a holding in the Virginia Company meant a share in the overall profits made by its plantation's operations, those in the subsidiary bought a portion of the land itself. One of the beneficiaries, the surgeon John Woodall, found this to be the most alluring aspect of the deal, boasting to a friend that he had acquired a 'hundredth part of the whole island'. Others were furious at being left out, one Virginia Company investor later citing the deal as evidence that the directors kept to themselves 'all the benefit made by the several voyages to Virginia aforesaid and all the several sums aforesaid without yielding any account'.[20]

The Virginia Company's Third Charter received the royal seal on 12 March 1612. A month later, the first group of between fifty and sixty settlers departed for the 'Somer Isles' in a ship called the *Plough*, under the command of Richard Moore, a ship's carpenter.

With funds still tight, the Virginia Company arranged for its first public lottery to be held in May. There were to be £5,000 worth of

prizes, the winner to receive a set of gilded plate worth £1,000. To publicize the draw, a pavilion was constructed next to the main entrance to St Paul's cathedral, where the glittering prizes could be admired, and tickets could be bought for as little as a shilling each.

The event became big news, even inspiring popular ballads such as 'London's Lottery', sung to the tune of 'Lully Gallant'. It summoned citizens and countrymen of the provinces to London, where . . .

The Merchants of Virginia now,
Hath nobly took in hand,
The greatest golden Lottery,
That ere was in this Land.

A gallant house will furnish forth
With Gold and Silver Plate,
There stands prepared with Prizes now,
Set forth in greatest state.
To London, worthy Gentlemen,
Go venture there your chance:
Good luck stands now in readiness,
Your fortunes to advance . . .

Who knows not England once was like
A Wilderness and savage place,
Till government and use of men,
That wildness did deface:
And so Virginia may in time,
Be made like England now;
Where long-loved peace and plenty both,
Sit smiling on her brow.[21]

Such songs aroused interest across the capital, even among London's growing number of Puritan congregations. Several churches bought tickets to support the company's evangelic mission. The congregation of St Mary Woolchurch at the Stocks Market (the location of the official pillory) bought 10 shillings' worth, while the communicants of St Mary Colechurch at Old Jewry staked a very substantial £6.[22]

A sequel to Robert Johnson's 1609 work *Nova Britannia* also appeared, entitled *The New Life of Virginia*. It provided the first public information on the state of the colony since Delaware's return to England, and was predictably optimistic. It reported that Sir Thomas Gates had returned after getting more supplies from England, and was now installed as the settlement's acting governor. There, he had implemented a new set of 'Laws Martial, Divine and Moral', which promised to impose some discipline among the unruly settlers, putting an end to 'all wrongful dealing' and 'injurious violence against the Indians'.

Johnson claimed that there were now seven hundred people in Virginia 'of sundry arts and professions', most in reasonable health. They were taking up residence in the new city of Henrico, 80 miles upstream from Jamestown (at the site of modern Virginia's state capital, Richmond), 'a place of higher ground, strong and defensible by nature, a good air, wholesome and clear (unlike the marish seat at Jamestown) with fresh and plenty of water springs, much fair and open grounds freed from woods, and wood enough at hand'. There they had built 'competent and decent houses, the first storey all of bricks, that every man have his lodging and dwelling place apart by himself, with a sufficient quantity of ground allotted thereto for his own use'. There was even a hospital, 'with fourscore lodgings (and beds already sent to furnish them) for the sick and lame, with keepers to attend them for their comfort and recovery.'[23]

Such enthusiasm bemused critics, who were astonished at the project's tenacity. Commenting on a report to Philip of Spain about doubts surrounding Delaware, ambassador Velasco pointed out that 'those who are interested in this Colony' nevertheless continued to 'push this enterprise very earnestly'. The reason, he speculated, was that 'the Prince of Wales lends them very warmly his support'.[24]

Just days before the lottery draw was about to commence, Robert Cecil died, forcing its postponement. The Lord Treasurer had succumbed on 24 May, while returning from the spa town of Bath, where he had hoped the waters would revive him. 'My audit is made,' said the man charged with managing the King's accounts, as he breathed his last.[25]

The news did not provoke the expected outpouring of grief in the city. 'My Lord Treasurer died upon Sunday last at Marlborough in returning from the Bath homewards, much against his own desires,' noted Anthony Withers, a London merchant writing to his friend William Trumbull, King James's agent in Brussels. Withers passed quickly on to speculating on Cecil's successor, and noted that 'the lottery for Virginia is put off for a month'.[26]

The draw eventually began on 29 June, and continued to run over several days. It was quite a spectacle, attended each day by 'knights, esquires and sundry grave discreet citizens'. The winner of the grand prize was a tailor called Thomas Sharpliss, and the glittering plate he had won was conveyed to his house 'in a very stately manner'. For their £6 stake, the people of St Mary Colechurch won the more modest prize of two spoons, valued at £1.

The pomp and ceremony of the lottery raised the venture's profile, but not the mood of its investors. The lottery's 'benefit will scarce pay old debts', Samuel Calvert, the Virginia Company's secretary, informed William Trumbull. 'There is another intended.'

Calvert's gloom reflected a wider malaise. 'All our state is in several motions,' he wrote, provoked by the return to London of Velasco's predecessor as Spanish ambassador, Don Pedro de Zuñiga. It was feared Zuñiga had come to negotiate Prince Henry's Spanish match. 'Don Pedro begins to plot his business', Calvert observed disapprovingly, noting that Queen Anne, James's Catholic wife, 'hovers about London, to whose court the Spanish hath access at pleasure'.[27]

John Chamberlain echoed Calvert's mood. In a letter to his friend Dudley Carleton written on 9 July 1612, he noted the attention James was lavishing on Zuñiga, feasting him at Windsor Castle one day, entertaining him at a royal retreat the next. 'It is generally looked for' that Zuñiga would 'expostulate about our planting in Virginia', Chamberlain reported, 'wherein there will need no great contestation, seeing it is to be feared that the action will fall to the ground by itself, by the extreme idleness of our nation, which (notwithstanding any cost or diligence used to support them) will rather die and starve than be brought to any labour or industry to maintain themselves'.[28]

Early in October 1612 the 19-year-old Henry, Prince of Wales, developed a strange illness, which made him lethargic and feverish, suffering a persistent headache. On the 12th, a Monday, he suffered 'a great looseness, his belly opening 25 times, avoiding a great deal of choler, phlegm and putrefied matter.' Despite this, he bravely continued with his official duties. On Sunday the 25th, after a 'violent play at tennis', he went to listen to a sermon, as was his habit, which happened to be on the subject of mortification. Later that day, the onset of fever forced him to retire to St James's Palace, his official London residence. The following morning, he rose, but suffered a continual 'drought' or thirst, and a headache even more severe than before, which 'broths and jellies' prescribed by his physicians did nothing to alleviate.

Over the following days, physicians gathered around the prince's sickbed, unsure what to do, awaiting some 'crisis' that would disclose the true nature of his disease. Meanwhile, more symptoms developed: he complained of a singing in his ears, his tongue turned black, his eyes became so sensitive he could not endure even candlelight. Then, on the evening of Thursday 29 October, there appeared 'a fatal sign': a glow hanging over St James's Palace, 'bearing the colours and show of a rainbow'.

Over the next two days, the patient continued deteriorating, the physicians vacillating. Finally, on Sunday 1 November, it was decided to let his blood, a vein in his arm being cut open, and 8 ounces drained off. 'The blood being cold, was seen of all to be thin, corrupt and putrid, with a choleric and bluish water above.' He briefly rallied, but the following day relapsed. The night of 2 November, there 'came upon him greater alienation of brain, ravings and idle speeches'. The next morning his hair was shaved off, and hot cups were placed on his scalp, which drew blood through the skin as they cooled.

On Wednesday 4 November, after a night 'in which he began to toss and tumble, to sing in his sleep', a cockerel was brought into the room, cut in half along its spine, and pressed against the soles of the prince's feet, 'but in vain'.

By Friday, his 'retentive power was gone, the spirits subdued, the seat of reason overcome, and nature spent'. Two or three times he seemed to be 'quite gone', unleashing a 'great shouting, weeping and crying in the chamber, court and adjoining streets', which revived him. That afternoon, a bottle of cordial water arrived from the Tower of London, sent by Sir Walter Raleigh from his cell, 'which whether or not to give him, they did a while deliberate'. After it had been tasted and 'proved', a spoonful was trickled between the Prince's chapped lips, producing a flicker of life, but no improvement.

He died at a quarter to eight that evening.[29]

The outpouring of literature prompted by Henry's death far exceeded that for Queen Elizabeth, and, as the Prince's modern biographer Sir Roy Strong put it, 'in theme it matched that which mourned the passing of another quintessential perfect Protestant knight, Sir Philip Sidney'. Sir Thomas Dale, Gates's deputy, summed up its impact upon the American adventure, when the news finally reached him in Virginia:

> He was the great captain of our Israel, the hope to have builded up this heavenly New Jerusalem. He interred (I think) the whole frame of this business [when he] fell into his grave. For most men's forward (at least seeming so) desires are quenched, and Virginia stands in desperate hazard.[30]

NINETEEN

The Good Husband

THE *TREASURER* was a swift man-of-war, owned by a family of Essex gentry equal to its name, the Riches. Sir Robert, the patriarch, was so rich that, in a few years' time, he could afford to buy the title Earl of Warwick from James at a cost of £10,000. He would not have achieved such a title otherwise, as he was notoriously independent in his religious views, Richard Bancroft, the Archbishop of Canterbury, bidding him to 'go amongst his puritans'.[1]

Sir Robert first became officially linked with Virginia when he signed up to the Third Charter. However, he, or his eldest son and namesake, only ever bought a single share. This suggests that the Riches were not interested in Virginia's commercial prospects, but had other designs. In particular, both Sir Robert and his son were leading members of what became known as the 'war party', the group of nobles and senior gentry in government who were agitating to end James's appeasement of the Spanish. They saw in Virginia exactly the sort of role Zuñiga had envisaged in his letters urging King Philip to destroy the colony before it grew too strong: a base for 'piracy' in the Spanish Main. It was probably with this aim in mind that Rich had bought the *Treasurer* and made it available to the Virginia Company.

The man he chose to captain it was the 32-year-old Samuel Argall. Argall had returned to London with Delaware in March 1611, having been given command of the Lord Governor's ship. He had since been involved in extending and refining the Virginia Company's official map of North America, which was pinned to a wall in Smythe's Philpot Lane office.[2]

Argall hoisted the *Treasurer*'s sails and set off for Virginia on 23 July 1612. She was equipped for a military rather than a supply mission, and carried sixty-two crew and soldiers, none of whom were expected to become settlers.

He reached Point Comfort on 17 September. Arriving at Jamestown a day or two later, he was pleased to find 'both the country and people in far better estate' than recent reports had suggested. Don Diego de Molina and Francis Lembry were to be found moping around the fort, having endured captivity for over a year. Their shipmate Pereos had died earlier in the year, of starvation, according to Molina.

As John Rolfe later put it, a form of 'absolute government' was now in place. Sir Thomas Gates was based in Jamestown as overall governor, his deputy Sir Thomas Dale at Henrico. A set of draconian 'Laws Divine, Moral and Martial' now regulated the colony's affairs, and Gates and Dale were implementing them rigorously, particularly Dale.

The laws covered every aspect of the settlers' lives, from their religious conduct to their military duties. Death was dealt out for just about every infringement imaginable: sodomy, adultery, fornication (sex with an unmarried partner), sacrilege, theft, slander, desertion. There was death for making too much noise 'where silence, secrecy, and covert is to be required', and death for slaughtering without permission any 'Bull, Cow, Calf, Mare, Horse, Colt, Goat, Swine, Cock, Hen, Chicken, Dog, Turkey, or any tame Cattle, or Poultry, of what condition soever'. Lesser punishments were handed out for lesser offences. The use of 'disgraceful words' was outlawed 'upon pain of being tied head and feet together,' a month of night guard duty, and a lifetime ban on owning property or holding office. And anyone found 'swaggering' would be made to 'ask forgiveness upon his knees, of the officers, and rest of the Guard, before the Captain of the watch at that time' for the first offence, and for a second, to be 'committed to the Galleys for one year', which effectively meant being a slave.

In future years, the regime created under these laws would be accused of being 'most cruel and tyrannous'. At the time, harshness was considered necessary. Ralph Hamor, son of the merchant tailor and one of the Bermuda castaways, considered such treatment as the only way of dealing with 'dangerous, incurable members' of the colony. Choosing words that via the English Bill of Rights would form such a significant phrase in the US Constitution, he acknowledged that the laws meted out what might seem 'cruel, unusual' punishments, but argued that worse had been used in France. John Rolfe also backed the regime, at least as a temporary measure, arguing that it ensured that men 'spent not their time idly nor unprofitably'.[3]

While Gates implemented the laws in Jamestown, Dale was still at work building Henrico. The city of brick homes and hospitals Robert Johnson described in his *New Life for Virginia*, turned out to be little more than a few rickety huts surrounded by a fence. Several settlers had run off to join the Indians, and others, faced with starvation, began to steal. Dale showed little sympathy. 'Some for stealing to satisfy their hunger were hanged, and one chained to a tree till he starved to death . . . Many famished in holes and other poor cabins in the ground, not respected because sickness had disabled them for labour, nor was there sufficient for them that were more able to work.'[4]

In the face of such setbacks, Henrico had yet to acquire a proper church. But it did have its own minister, Alexander Whitaker, who delivered his inaugural sermon from a makeshift pulpit in the summer of 1612.

The son of a famous theologian, the 27-year-old Whitaker was highly educated, fluent in Greek and Latin, and boasted a master's degree from Cambridge. He was keen to use this learning to plant a new and thriving Protestant church in Virginia, and upon arrival at Henrico had been quick to set about the business. He supervised the building of a temporary chapel, staked out a choice hundred-acre plot or 'glebe' of church land, and ordered the construction of a 'fair framed parsonage', which he dubbed Rock Hall.

He also tried his hand at some personal planting. 'Being a stranger to that business', and 'having not a body inured to such labour', he struggled, and was eventually forced to call upon the help of three men to assist. Nevertheless, he still congratulated himself on setting enough corn '*horis succisinis unius septimanae*' (in the idle hours of a single week) to keep him in bread for a year.

He was not so impressed by his parishioners. Many were 'most horribly plagued . . . with famine, death, the sword, etc'. This was due to their 'intolerable' sinfulness, he decided, rather than a failure in government. 'I marvel more that God did not sweep them away all at once than that in such manner He did punish them,' he noted in a letter to his friend in London, the Reverend William Crashaw. Dale was inclined to agree. Writing to an unidentified member of the Virginia Council in London, he expressed some hope for the thirty-six planters he had nominated the settlement's 'farmers', but as for those hired as soldiers and workmen, 'oh sir, my heart bleeds'. 'Did I not carry a severe hand over them, they would starve one the other by breaking open houses and chest[s] to steal a pottle of corn from their poor brother, and when they have stolen that, the poor man must starve.'[5]

But the shortages of food were not entirely the settlers' fault. Gates had brought several hundred cattle and some pigs from England, to replenish the livestock lost during the Starving Time. There had been high hopes that the grassy uplands surrounding Henrico would provide ideal grazing conditions for them. But as soon as they were let out, the Indians took them, forcing the English to keep them in protected compounds, where they had to be fed with corn out of the common store. 'For howsoever we could well defend ourselves – towns and seats – from any assault of the natives,' Rolfe observed, 'yet our cattle and corn lay too open to their courtesies and too subject to their mercies.'[6]

So, while relations with the Indians remained so hostile, the settlers still had to rely on external sources for food, either from England, or from Indians outside Powhatan territory.

Argall's prime objective upon arriving in Virginia was to address this problem by re-establishing trade with the friendly Indians of the Potomac. Preparations began soon after his arrival, with the repair and maintenance of the settlement's fleet of boats, which he had found 'decayed for lack of pitch and tar'. In the meantime, he joined Dale in a number of raids on nearby Indian towns 'for their corn', during one of which, on the banks of the Nandsamund river, Dale 'escaped killing very narrowly'. Argall also led scouting expeditions to Smith's Isle, near Cape Charles, hoping to catch fish and find a safe route for fishing boats along the eastern shore to Delaware Bay.

Argall finally set off for the Potomac on 1 December 1612, and arrived a few days later. As he was sailing up the river, his 'brother' Chief Iopassus appeared upon the banks. The *weroance* had just returned from a hunting expedition, and came aboard the ship, 'seeming to be very glad of my coming, and told me that all the Indians there were my very great friends, and that they had good store of corn for me'. Iopassus continued with Argall upriver to the stretch of river near Passapatanzy, where a collapsible shallop was assembled to fetch the corn from the town's granary.

Argall sailed back to Jamestown, arriving on 1 January with over a thousand bushels of corn, three hundred of which Gates allowed him to retain to feed his own men, the rest being placed in the company store.

The following March, Argall set off again for the Potomac, this time with the intention of furthering Captain Smith's explorations. Picking up guides along the way, he penetrated nearly 200 miles upriver. Leaving his ship, he headed west towards the mountains, and came across a herd of bison, which his guides demonstrated 'are very easy to be killed in regard they are heavy, slow, and not so wild as other beasts of the wilderness'. Carcasses were butchered and cooked over a fire, the meat proving to be 'very good and wholesome'.

Heading still further inland, Argall found mineral deposits, and took samples to send back to London, 'and likewise a strange kind of earth, the virtue whereof I know not, but the Indians eat it for physic, alleging that it cureth the sickness and pain of the belly'.[7] Having covered so much territory, with supplies running low, he decided it was time to return to the ship and head back for Jamestown.

As he was making his way down the Potomac, he heard that Powhatan's daughter Pocahontas 'was with the Great King Patowomeck'.

The Indian princess was now in her mid-teens. She had not been in contact with the English for some years – not since John Smith's departure, by Smith's own account, having chosen thereafter to forsake further contact with the English and live quietly among the Potomac people.[8]

According to other sources, she had gone to the Potomac not to pine for Smith, but because she and some friends were 'employed thither as shopkeepers to a fair to exchange some of her father's commodities for theirs'.[9]

Argall sailed to Passapatanzy, where he had heard she was currently lodged, and moored the *Treasurer* nearby. Iopassus came aboard, and Argall told him bluntly that 'if he did not betray Pocahontas into my hands, we would be no longer brothers nor friends'. Iopassus pointed out that, if he were to do such a thing, 'Powhatan would make wars upon him and his people'. Argall promised to come to his defence if this proved to be the case.

Iopassus returned to Passapatanzy to talk the matter over with his brother, the 'great king', and his councillors. They had discussions lasting several hours, during which the great king was persuaded that Argall's proposal presented an opportunity to weaken Powhatan's dominance. Word was sent to Argall that the plan would go ahead.

Iopassus then announced to his people that he was to return to the ship to have further discussions with the English captain, and got his wife to 'feign a great and longing desire to go aboard' with him. He publicly and very sternly denied this request, prompting her to weep ('as who knows not that women can command tears!' Ralph

Hamor, who accompanied Argall, caustically noted). As expected, this provoked the strong-willed Pocahontas to spring to her friend's defence, and announce that she wanted to go aboard too. Iopassus chivalrously relented, if only to 'please' the princess, who was 'desirous to renew her familiarity with the English'. 'So forthwith aboard they went [and] the best cheer that could be made was seasonably provided.'

The party was entertained to supper in the captain's cabin, Iopassus reportedly spending much of the meal 'treading upon Captain Argall's foot' to signify how well their scheme was going. The guests were afterwards invited to spend the night in the ship's gun-room, Iopassus briefly drawn to one side so that he could receive a reward, which Hamor claimed to be 'a small copper kettle and some other less valuable toys so highly by him esteemed that doubtless he would have betrayed his own father for them'.

The following morning, Pocahontas awoke early, 'being most possessed with fear and desire of return' to Passapatanzy. It was at this moment that she found out she had been lured into a trap. Argall announced he would 'reserve' her upon the ship, only surrendering her to her father if he delivered up 'eight of our Englishmen, many swords, pieces, and other tools, which he had at several times by treacherous murdering our men taken from them'. 'Whereat she began to be exceeding pensive and discontented, yet ignorant of the dealing of Iopassus, who in outward appearance was no less discontented that he should be the means of her captivity.'

Pocahontas was distraught, and 'much ado there was to persuade her to be patient, which with extraordinary courteous usage, by little and little, was wrought in her'.

The date of Pocahontas's capture was 13 April 1614, and Argall returned to Jamestown a day or two later. Governor Gates immediately dispatched a messenger to Powhatan with the news of his daughter's abduction and demands for the return of prisoners and weapons. Argall, meanwhile, disappeared in the *Treasurer* to Point Comfort, with orders to fit the ship for a fishing expedition around Cape Cod, to gather supplies for the starving men at Henrico.

Gates waited in Jamestown for a reply.

From beyond the curtain of trees, no cries of fury were heard. No wars were unleashed upon the Potomac people, no threats made to the English. In fact, for three months, there was no response at all. The English, searching for an explanation, attributed this to political dithering, Powhatan apparently being unable to take any decision 'without long advice and deliberation with his council'.

Powhatan was now in his mid- or even late seventies. As a figure, he could still strike 'awe and sufficient wonder' in the English who met him. Strachey had been so overwhelmed by his presence that he beheld 'an infused kind of divineness' in his eyes and countenance. But, perhaps befitting a man touched by divinity, there was also a growing world-weariness.[10]

Back in January of 1609, during a biting winter that built ice sheets half a mile across the wide course of the river James, Smith had visited Powhatan at Werowocomoco, and in a curiously intimate exchange glimpsed a powerful man enfeebled by age and exhausted by events. Stretched out upon his ceremonial bedspread, swathed in thick furs and tobacco and wood smoke, Powhatan revealed to the captain that he had 'seen the death of all my people thrice, and not one living of those three generations, but myself'. Smith had no idea what he meant, and the comment has been left to gather speculation ever since: that he had witnessed epidemics of European diseases sweeping through the Indian population, or great losses in the bloody battles that had led to the formation of his empire, or, most likely, the passing of three generations. 'I know the difference of peace and war better than any in my country,' the septuagenarian *mamanatowick* had told the younger captain, 'but now I am old, and ere long must die.' He had no stomach for further battles. He did not want to live in a state of enmity with the English, where 'if a twig but break, every one cry there comes Captain Smith, then must I fly I know not whither, and thus with miserable fear end my miserable life, leaving my pleasures to such youths as you'.[11]

Smith suspected Powhatan of trying to lure him into a false sense of security, but the *mamanatowick* may have been more candid than the captain realized. Powhatan's empire, the Tsenacomoco, appears to have been shaken by some sort of seismic power shift soon after Smith's departure, undetected by the English, but indicated by the removal of Powhatan's court from Werowocomoco, a prominently located, well-defended seat overlooking the wide river, to Orapaks, a small outpost lost in the forests.

Orapaks brought Powhatan closer to the source of his power: the falls. Across those rocks cascaded the precious metals and stones of the mountains that had underpinned his empire and authority. Now the English threatened to seize and overwhelm the source, not only by flooding it with their own trinkets, but by taking from him the surrounding land, which the Henrico plantation was threatening to swallow up. Even Alexander Whitaker had sensed the importance the Powhatans attached to this place, reporting their attempts to cast spells upon the encampment being constructed by Dale.

William Strachey mentioned a prophecy sweeping through the towns of the Tsenacomoco around this time, attributed to Okeus, according to which the Powhatan people would twice overthrow 'such strangers as should invade their territories or labour to settle a plantation amongst them'. But a third time, 'they themselves should fall into [the strangers'] subjection and under their conquest'. Perhaps Powhatan was preparing to prevent this eventuality, or escape it, or even submit to it.[12]

Eventually, the long-expected reply regarding Pocahontas arrived. Seven men were spotted paddling down the river towards the jetty near Jamestown's riverside gate. They were some of the prisoners Gates had demanded Powhatan return, and they had brought with them some corn, and a selection of weapons: three broken muskets, a 'broad axe' and a long whipsaw. They also delivered to the governor a message from Powhatan that 'whensoever we pleased to deliver his daughter, he would give us in satisfaction of his injuries done to us – and for the rest of our pieces broken and stol'n from him – 500 bushels of corn, and be forever friends with us'.

Gates refused this offer, claiming that the few weapons returned did not come close to accounting for the large number that had been stolen. 'This answer, as it seemed, pleased him not very well,' observed Ralph Hamor, as nothing more was heard from the *mamanatowick* for the rest of that year.

Meanwhile, Pocahontas joined Molina and Lembry as a prisoner of the English settlers. The condition of her captivity is unknown, and in the early days must have been fairly restrictive, in expectation of an attempt by the Indians to take her back. She may even have been confined to one of the ships, possibly the *Elizabeth*, which had arrived soon after her abduction, having dropped off supplies at the fledgling settlement on Bermuda en route. But the expected attack did not materialize, and by the time the *Elizabeth* set sail for London, Pocahontas may well have had free rein of the fort.

The *Elizabeth* was the first ship bound for England for nearly a year, and she had much to take with her. There was the news of Pocahontas's capture. The governor Sir Thomas Gates was also aboard, his release from Dutch service having expired. The ship also carried the first samples of a series of agricultural tests that had been conducted by John Rolfe.

During the spring of 1612, Rolfe had planted a packet of seeds given to him by one of the tobacco merchants in London. They had flourished, producing Virginia's first crop of 'best tobacco, of Trinidad and the Orinoco'. It spread through the fort like a rampant weed, growing in 'the market-place, and streets, and all other spare places'. The plants had 'large, sharp' leaves that had grown '2 or 3 yards from the ground' – not as large as they would have been in their native South America, but with foliage far superior to the indigenous variety, with its 'biting taste'. Cured over a fire until they were as brown as shoe leather, the leaves had given off a rich aroma, and smoked in a pipe, produced a 'pleasant, sweet, and strong' taste, according to Ralph Hamor. With the loss of the Trinidad and Orinoco trade, there were hopes that this might provide an alternative, fetching a good price in London's apothecary shops.[13]

With Gates's departure, Dale became acting governor, and one of his first acts was to move Pocahontas to Henrico, where she would be under the care of Alexander Whitaker at Rock Hall. Whitaker's job was to convert her to the Protestant religion, a process that involved not just religious indoctrination, but teaching her to read and write, so she could study the newly published King James Bible, and take notes for her studies.

The education was not all one way. Regular contact with the Indian princess produced something of a conversion in Whitaker, his attitude to the 'savages', initially superstitious and histrionic, approaching respect. He noted approvingly how orderly their society was compared to that of the settlers. They lived under a 'civil government . . . which they strictly observe'. They were not so grasping of each other's property. They were obedient to authority. They were loyal. He commented how Chaopock, the chief of the Quiyoughcohannock, and his disgraced brother Pipisco had remained friendly towards the English since their arrival, welcoming the strangers into their houses since the settler William White had run to them in 1607.[14]

For Whitaker, this did not mean that Indian culture should be cherished or preserved. But it might mean the Indians were sufficiently close to a civilized state to make religious conversion a practical step. Success with Pocahontas would provide high-profile evidence of this, and Whitaker set about the task with gusto.

As part of her education, Pocahontas was expected to mingle with the 'better sort' of Englishmen and women, to observe their manners and ways. This brought her in contact with the widower John Rolfe, who had emerged as a model not only of Protestant piety, but also of enterprise and industry. Their contact resulted in Rolfe falling 'in love with Pocahontas and she with him', and some time in the early spring of 1614 they decided they would get married.

The issue of intermarriage was fraught. Sex with Indians was forbidden, though banned only as part of a general prohibition against rape and 'fornication', sex out of wedlock. Raleigh himself had noted how Spanish invaders 'used' Indians 'for the satisfying of their own lusts', and had argued that such practices should be banned, not least because they destroyed any hopes of cooperation from the natives.[15]

On the other hand, America was seen, indeed advertised, as a place of sexual opportunity. This was revealed in a moment of drunken candour by the captain in *Eastward Hoe*, when he said, 'Come boys, Virginia longs till we share the rest of her maidenhead.'[16] For some of the male settlers, this wantonness was confirmed by the carefree manner and skimpy clothes of the Indian women. Even casual encounters between the two sexes were seen as dangerous, and officially discouraged.[17]

There were also religious objections to relations with the Indians. The sons of Abraham 'keep them to themselves', one preacher had reminded a group of settlers about to embark for Virginia. Invoking passages in Deuteronomy in which God ordered the Israelites to seize the land of Canaan for themselves, he warned them that 'they may not marry nor give in marriage to the heathen, that are uncircumcised . . . The breaking of this rule, may break the neck of all good success of this Voyage.'[18]

This rule troubled Rolfe. His relationship with Pocahontas had also aroused some comment among his fellow settlers, some disapproving, some ribald, and he felt a need to justify himself.

He did so by writing an open letter addressed to Governor Dale. He wanted to 'sweep and make clean the way wherein I walk from all suspicions and doubts'. He was aware of the 'frailty of mankind, his proneness to evil, his indulgency of wicked thoughts', and admitted that such weaknesses had provoked 'a mighty war in my meditations'. He also acknowledged the Biblical prohibition against 'strange wives', but his 'reading and conference with honest and religious persons' had given him 'no small encouragement' that a match with Pocahontas would be lawful. This was because his love of Pocahontas was founded not upon the 'unbridled desire of carnal affection', but a desire to act 'for the good of this plantation, for the honour of our country, for the glory of God, for my own salvation, and for the converting to the true knowledge of God and Jesus Christ an unbelieving creature'.[19]

Rolfe's letter proved to be highly influential, and was quickly dispatched to England for publication by the Virginia Company. The arguments he gave for taking Pocahontas as a wife acted as arguments for England taking Virginia as a colony, showing it to be an act of lawful possession rather than rape.

When Dale first received Rolfe's letter, the young girl had been with the English nearly a year. During that time no word had come from her father or her people enquiring of her welfare, or offering terms for her return. If she now married Rolfe, she could no longer be used as a hostage. So, before Rolfe's relationship was allowed to develop any further, Dale resolved that the time had come 'to move [the Indians] to fight for her'.[20]

In a fleet of frigates and boats, led by Argall in the *Treasurer*, he set off with Pocahontas, Rolfe and one hundred and fifty soldiers for Werowocomoco (which the English still believed to be Powhatan's capital). Cruising up the Pamunkey river for a day or two, he eventually saw a small group of warriors standing on the bank. Dale went ashore with some of his soldiers, making sure that Pocahontas was visible on the deck of the *Treasurer*.

He announced to the warriors that he had come to deliver Princess Pocahontas back to her father. In return, he expected all previous demands for the return of weapons and renegades to be fully met. Dale especially insisted upon the return of one Simons, 'who had thrice played the renegade', and was accused of practising 'lies and villainy' among the Indians. If these conditions were not met, his men would 'fight with them, burn their houses, take away their canoes, break down their fishing weirs, and do them what other damages we could'.

The warriors said they would need time to communicate these demands to their 'king'. Dale agreed, and men were exchanged as a sign of goodwill. The next day, the warriors returned to inform Dale that Powhatan was three days away (presumably at Orapaks) and Opechancanough, Powhatan's military chief, was nearby and could negotiate on the *mamanatowick*'s behalf. This offer was refused. The afternoon hours passed with no further developments, and Dale became impatient. He ordered a troop of his men to climb into one of the boats, and set off for the shore. Suspecting an attack, the Indians unleashed a shower of arrows. 'We were not behindhand with them,' Dale later boasted. His men landed, 'killed some, hurt others, marched into the land, burnt their houses, took their corn, and quartered all night ashore'.

The next day, the soldiers returned to their ships, and Dale ordered that they continue upstream. As they sailed, Indians lined the banks, putting up a 'great bravado', challenging the English to fight, 'bragging, as well they might, that we had ever had the worst of them in that river'. As the channel forced the ships nearer the river bank, the Indians fired more arrows, one hitting a man in the forehead, 'which might have hazarded his life without the present help of a skilful surgeon'.

Dale continued until he came to Werowocomoco.[21] He went ashore with a company of men, taking Pocahontas with him, where he was awaited by the same delegation that had negotiated with him further downriver. They were now accompanied by four hundred warriors.

Despite the presence of so many troops, the Indians now seemed to be in a more conciliatory mood. The attacks had been perpetrated by stragglers, they said, who were unaware of the ongoing negotiations. They promised him that 'they themselves would be right glad of our love, and would endeavour to help us to what we came for'.

It was eventually agreed that they would send 'two or three men once more to their king to know his resolution, which, if not answerable to our requests, in the morning, if nothing else but blood would then satisfy us, they would fight with us and thereby determine our quarrel'. The two armies spent the rest of the day and the following night in each other's apprehensive company upon the shores of the river.

The next morning, two of Pocahontas's brothers appeared. English accounts of their reunion with their sister suggest an emotional and awkward confrontation. Fearing that she had been badly treated, they 'much rejoiced' at her being in good health. Pocahontas, however, was furious. She told her brothers 'that if her father had loved her, he would not value her less than old swords, pieces, or axes; wherefore she would still dwell with the Englishmen, who loved her'. Perhaps taken aback by her anger, they promised that they would 'undoubtedly persuade their father to redeem her'.

The two brothers were then taken aboard the *Treasurer*, and Dale selected two men to go and negotiate direct with Powhatan. One was Robert Sparks, who had had previous dealings with the Indians during an expedition with Argall. The other was John Rolfe. They returned a day later with the news that they been denied an audience with Powhatan, but instead had been taken to Opechancanough. He now claimed to be Powhatan's 'successor – one who hath already the command of all the people'. He would use his 'best endeavours to further our just requests', he promised. They also learned that Powhatan had agreed to Pocahontas becoming Dale's 'child', to 'ever dwell' with the English governor.

Dale had assumed that Powhatan was still in ultimate command of his empire. Now it seemed there had been a fundamental shift in power. Writing in England, John Smith claimed that Opitchapam not Opechancanough, was supposed to be Powhatan's successor. But Opitchapam was 'far short of the parts of Opechancanough'. Some even believed that it was 'fear of Opechancanough' that was driving Powhatan into retirement.[22]

In recognition of this fundamental, if confusing change, Dale realized he had to shift his position. Powhatan had apparently surrendered his daughter to the English. Dale therefore decided that, despite having rejected them only a day or two before, he would have to accept Opechancanough's assurances, and withdraw.[23]

The English sailed back to Jamestown, and Pocahontas was returned to Henrico. There is no record of her response to these events. She may have felt her father had betrayed her, or that she had been sacrificed as part of some greater plan, or that political considerations were overcome by her personal feelings towards Rolfe. Whatever her reaction, she now embraced a new life with her captors.

She was taken to Henrico church, and, leaning over a pail or trough, allowed Reverend Whitaker to wash away her former identity as princess of the Tsenacomoco, daughter of Powhatan and child of Ahone with a trickle of water. She emerged baptized Rebecca, after the wife of Isaac, son of Abraham. The choice of name was portentous, rich with hopes and anxieties about her new role. In Genesis, Rebecca becomes pregnant with twins, and God tells her, 'Two nations are in thy womb, and two manner of people shall be separated from thy bowels, and the one people shall be stronger than the other people.' The Biblical Rebecca had given birth to Esau, a 'cunning hunter', and Jacob, 'a plain man, dwelling in tents', two brothers who would be perpetually engaged in a struggle for primacy one over the other.[24]

After the christening came the banns of marriage, the public declaration before the congregation of Henrico of John Rolfe's betrothal to 'Rebecca'. Powhatan heard the news, and gave his 'sudden consent thereunto'. A few days later, Pocahontas's 'old uncle' Opachisco arrived with his two sons to offer her father's blessings.

The wedding took place in Henrico church on 5 April, 1614. The bringing together of the English planter and the Indian princess was hailed as a turning point in the settlement's fortunes. If this daughter of the Powhatan people could be converted to Christian understanding and joined in civil partnership, then so could the country that gave birth to her. Virginia, as the English saw it, had survived the onslaughts of the Spanish empire and Catholic Church with her maidenhood intact. She had now surrendered it to the English nation and the Protestant cause.

Within days of the wedding, two representatives of the

Chickahominy people arrived at Jamestown offering peace. Shortly afterwards, Dale went with Argall up the Chickahominy river to negotiate terms. According to the English, these turned out to be extremely favourable, with the council that ruled the Chickahominy territory agreeing to pledge allegiance to King James, to rename themselves 'Tassantassee' (an English variant of the Indian word for foreigners), and at the beginning of each harvest to provide as tribute two bushels of corn for each warrior, and three hundred fighting men in the event of Spanish attack.

'For their diligence', Dale offered each member of the Chickahominy ruling council a red coat, a copper chain and a portrait of King James, and a promise that henceforth they would be considered English nobles.

The negotiations concluded, a councillor then gave a speech to his people, informing them of their new obligations to the English, which he urged them to observe so they would no longer be under the subjection of Powhatan, 'whom they held a Tyrant'.[25]

With this new treaty in place, Dale decided to send Ralph Hamor to Powhatan to discover the truth of the inscrutable *mamanatowick*'s retirement.

Hamor departed on 15 May 1614 with Thomas Savage, the young man who had lived some years with the Powhatans, and was now acting as an interpreter. Two Indian guides had come from Powhatan to escort them.[26]

They went by foot to the Pamunkey river, where a canoe was summoned to take them across. The *mamanatowick* himself was waiting for them on the opposite bank.

The old chief greeted Savage warmly, saying he had been 'a stranger to me these four years', and reminding him that he was 'my child by the donative of Captain Newport in lieu of one of my subjects, Namontack'. He wondered where Namontack had gone, as he had not seen him since 'though many ships have arrived here'. The English had evidently not told Powhatan of Namontack's conversion to Christianity while in England. Nor had he heard of Namontack slaying his Indian companion Machumps, and making 'a hole to bury him', and because it was too short cutting off his legs and laying them by him, 'which murder he concealed till he was in Virginia', according to John Smith.[27]

Powhatan turned to Hamor, and felt around the Englishman's shirt collar. He wanted to know what had happened to a pearl necklace he had given 'my brother Sir Thomas Dale for a present at his first arrival'. It was provided, he said, on the basis that any Englishman who visited him must wear it, 'otherwise I had order from him to bind him and send him home again'. It emerged that Dale had made arrangements for the necklace to be given to Hamor before his departure, but the governor's page had failed to carry them through.

Powhatan was in no mood to cavil, and since Hamor had been brought by two of his own men, was happy to admit him to his house, 'not full a stone's cast from the waterside'. The *mamanatowick* sat upon the edge of his bed, with at each arm 'a comely and personable young woman, not twenty years old the eldest, which they call his "queens" '. Hamor and Savage sat at his feet on a mat.

A pipe was brought, from which Powhatan 'drank' thick wafts of smoke, before passing it to Hamor, who did likewise.

Powhatan asked after his brother Dale, and after 'his daughter's welfare, her marriage, his unknown son' John Rolfe, wanting to know 'how they liked, lived, and loved together'. Hamor claimed that Pocahontas was 'so well content that she would not change her life to return and live with him, whereat he laughed heartily, and said he was very glad of it'.

'Now proceed,' the great chief commanded. Hamor said he had come on private business, so Powhatan ordered everyone to leave, except for the 'two queens' sitting at either side, and one of his councillors, Pepaschecher, who had acted as one of the escorts bringing Hamor from Jamestown.

Savage proceeded to deliver a well-rehearsed speech, saying they had come about Powhatan's youngest daughter, an 11-year-old whose 'exquisite perfection' had become 'famous through all your territories'. Dale had requested that Powhatan grant permission for her to go back to Jamestown with Hamor, because Pocahontas wanted to see her. Also, 'if fame hath not been prodigal' and exaggerated her beauty, Dale intended eventually to make her 'his nearest companion, wife, and bedfellow' (an offer he was in no position to make, as he was already married). The English and Indians were now 'friendly and firmly united together and made one people', Savage explained, quoting his governor, and in the 'band of love' between Dale and Powhatan's daughter would be formed a 'natural union between us'.

As Savage spoke, Hamor laid before the chief a series of gifts as 'testimony' of Dale's intentions: two large pieces of copper, five strings of white and blue beads, five wooden combs, ten fish-hooks, and a pair of knives.

Surveying the presents, Powhatan noted that they were not as generous as 'formerly Captain Newport, whom I very well love, was accustomed to gratify me with'. He also told Hamor that he had just 'sold' his daughter to another chief for two bushels of beads. Hamor suggested Powhatan buy her back, as the English would offer 'treble the price of his daughter in beads, copper, hatchets, and many other things more useful for him'. Powhatan refused. He did not hold it 'brotherly' for the English 'to desire to bereave me of two of my children at once'.

In any case, further pledges of Indian friendship were not required, Powhatan said. Dale need not fear any injury from his people. 'There have been too many of his men and mine killed, and by my occasion there shall never be more,' he told Hamor. He was feeling old and wanted to end his days in peace. His people would not threaten the English any more, nor would he. 'I will remove myself farther from you,' he announced.

That night, Hamor and Savage were offered a house to lodge in, but found 'fleas began so to torment us', that they were more comfortable on a mat, spread beneath a broad oak.

The following day, Hamor was approached by a man saying he

was William Parker, one of the renegades Dale had demanded back during the negotiations over Pocahontas. Parker had become 'so like both in complexion and habit to the Indians' that Hamor 'only knew him by his tongue to be an Englishman'. He now wanted to return to Jamestown.

When Hamor mentioned this to Powhatan later that morning, the chief flew into a fury. The English had taken one of his daughters, but would not allow him one of their men. If Hamor insisted, Parker could go back with him, but Powhatan would not provide guides for their homeward journey. Hamor pointed out that, if anything were to happen to him as a result, 'his brother, our king [James], might have just occasion to distrust his love'. At this, Powhatan left the room, and would not speak to Hamor for the rest of the afternoon.

That evening, at supper, Powhatan's mood was restored, and at midnight, he came to Hamor where he slept and woke him up. He said he would allow his counsellor Pepaschecher to accompany the English party home. Hamor was also to take a list of items Powhatan wanted: more copper, a shaving knife, an iron 'froe' (a tool for cleaving wood), a grinding stone, two bone combs 'such as Captain Newport had given him (the wooden ones his own men can make)', fish-hooks, a cat and a dog. In return, Powhatan promised the English hides.

Powhatan was insistent that Hamor remembered 'every particular' of his list, and to ensure that he did so, asked him to write it down. To show Hamor what he meant, he brought out a notebook given to him by Captain John Smith when Smith was his prisoner. Hamor asked the chief if he could have it, but Powhatan would not hand it over, as it 'did him much good to show it to strangers which came unto him'. So Hamor used his own notebook instead.

The following morning, after a breakfast of 'good boiled turkey', Hamor prepared for his departure. Powhatan gave him some food for the journey home, and to each member of the English party 'an excellent buck's skin very well dressed and white as snow'. Hamor was given two further skins, to be given to Powhatan's 'son and daughter each of them one'.

He escorted Hamor and his companions to the waterside. Before they left, he told Hamor that he hoped Dale was happy to continue their accord. If he was not, the *mamanatowick* would 'go three days' journey farther from him, and never see the Englishman more'.

This was, indeed, the last formal meeting Powhatan would have with the English, and it left the impression of a leader whose former greatness was fading, if it had not already dispersed, of a man who no longer had the strength within himself or among his people to prevent the English presence in the midst of his lands from spreading.

By the summer of 1614, English activity along the banks of the river James was feverish. Letters sent back to London mentioned several new 'cities' taking shape along the upper reaches of the James. Henrico, according to Hamor, was now a town with three streets of 'well framed houses', with the foundations already laid for a 'stately' new brick church, which was projected to be 100 foot long and 50 foot wide. Five 'blockhouses' or fortified farmhouses stood on the banks of the river, while a 2-mile-long 'pale' or fence stretched between the river banks protected it from intruders.

On the opposite shore was 'Coxendale', an area of land about 12 miles square set aside for pigs to forage, protected by three fortified outposts called Faith, Hope and Charity.

A few miles downriver (around modern Hopewell) was Bermuda city, considered the 'most hopeful habitation'. It was sited on Appamattuck territory, which the English claimed to have seized the previous year 'to revenge the treacherous injury of those people done unto us'. The surrounding land had been divided up into five 'hundreds' (the old English term for an administrative region of a county or shire), 'the most part champion and exceeding good corn ground'.

To develop these hundreds, and exploit the Indian peace, Dale introduced a version of the policy adopted for the settlement of the Bermuda islands. Until now, all the settlers had worked collectively for the Virginia Company, constructing defences and public buildings, clearing and planting the land for the growing of food for the common store and commodities for sale, a system later wrongly characterized as 'communist' ('corporatist' might be a better word). But, as Captain John Smith later noted, under this system, 'glad was he [who] could slip from his labour, or slumber over his task', hard work bringing no individual improvement or comfort to a life already full of peril and pain.[28]

Dale proposed that the 'ancient planters' who had come to Virginia under the First Charter (before Gate's fleet of 1609) should be 'freed' of their obligations to the Virginia Company and given 3 acres of arable land each to 'manure and tend'. In return, they would be required to work for the colony for no more than one month in the year, 'which shall neither be in seed time or in harvest', and to contribute two and a half barrels of corn to feed new arrivals, or help pay towards other supplies which had to be imported from England.[29]

The first of the 'ancient planters' to benefit from this scheme was William Spencer. He had arrived with Christopher Newport's first supply mission in 1607, listed as a 'labourer'. He had since acquired a wife called Alice, and possibly a son, also called William. He had served as 'an honest, valiant, and an industrious man' during Smith's presidency, and was made an ensign by Dale.

Spencer's loyalty and persistence was to be rewarded with a plot of his own choosing. Virginia's 'first farmer', as he became known, set about developing his acres by growing Spanish tobacco under Rolfe's guidance.

Spencer thrived, becoming a successful plantation owner. Within a few years he had supplemented his few acres on the banks of the upper James with two plots on Jamestown island (one of 12 acres, the other of 18), and another on the lower side of the river. By 1619, he had become a member of the colonial establishment, being appointed an official tobacco 'taster', with the job of monitoring the quality of tobacco exports on the company's behalf.[30]

For Captain Smith, as for his successors, Spencer's example showed what America could offer. Such men had been denied their proper rewards for venturing into the New World by poor leadership and political chicanery. At last these obstacles were being overcome, allowing even the humblest labourer, through hard work, social responsibility and pious worship, to better himself.

The key seemed to be the private ownership of land – and the chance to plant it with a cash crop. In London, King James himself decried tobacco as containing the 'first seeds of the subversion of all great Monarchies'. In Virginia, it gave men like Spencer their first taste of freedom.[31]

TWENTY

Twelfth Night

ON 3 JUNE 1616, the man-of-war the *Treasurer*, under Captain Samuel Argall's command, entered Plymouth Harbour. As she tied up, Sir Thomas Dale stepped ashore and handed a letter to a messenger, addressed to Ralph Winwood, the King's principal secretary. Having left George Yeardley in command in Virginia, he had brought back to England a ship laden with 'exceeding good tobacco, sassafras, pitch, potashes, sturgeon and caviar', he announced. He also promised to 'give Your Honour great encouragements that this Virginia affords' once he reached London, 'to spur us forward to inhabit there, if His Majesty [wishes] to possess one of the goodliest and richest kingdoms of the world'.[1]

These great encouragements came not just from the ship's cargo. Dale had also brought with him some distinguished passengers. One was Don Diego de Molina, the Spanish captive who had been languishing on Jamestown island for nearly five years. There were now hopes of exchanging him for John Clark, the English pilot captured by the Spanish. His companion, Francis Lembry, had also been aboard, but Dale had ordered him to be hanged from the yardarm as the ship had come within sight of England, after it was discovered that he was not, as he claimed, a Spaniard from Aragon, but an English traitor.

Molina was joined by Pocahontas, or 'Rebecca' as she was now called, on her first visit to England. She was with her husband John Rolfe, and their newborn child Thomas, named after Pocahontas's English patron, Dale. Accompanying her was an Indian delegation sent by Opechancanough, which was led by his priest and chief counsellor Uttamatomakkin, with an entourage of young men and women, including Uttamatomakkin's wife, a daughter of Powhatan called Matachanna.

These passengers would have been welcomed ashore by Sir Lewis Stukeley, Vice Admiral of Devon, a meal at his house providing an opportunity to introduce the Indian delegation to English hospitality, and for the Englishmen to gossip.

The captains had plenty to talk about, not least recent French incursions into the area of North America claimed by the English. Argall, using the ship now at anchor in the harbour, had led an expedition to destroy fledgling French settlements around the Bay of Fundy. In reprisal, some French privateers had recently seized an English ship, which turned out to be carrying none other than Captain John Smith on an expedition attempting to found a new colony in northern Virginia, which he now called 'New England'. Stukeley had just completed an inquiry into the affair.[2]

Having exchanged the latest news and recovered from the voyage, Argall weighed anchor and proceeded to London with the cargo, while the passengers were loaded into carriages to make the 230-mile journey to London.

As the convoy snaked through the West Country, Uttamatomakkin tried to count trees and men, so he could give Opechancanough some estimate of English strength. Marking notches on a tally stick, he continued as they drove through the dense wooded valleys of Devon, and the busy towns and villages of Somerset, across the expanses of Salisbury Plain and passed the ancient monument of Stonehenge, 'till his arithmetic failed'.[3]

Through an interpreter brought by Dale (unnamed, but probably Thomas Savage), he expressed admiration of the large, open fields of fresh green barley and wheat passing by the carriage window. He otherwise reserved his impressions of what must have been a landscape as bewildering to him as Virginia had first been to the English.

London was even more of a shock: noisy, dirty, chaotic, around ten times the number of people crammed into the conurbation's 6 or so square miles as in the 6,000 square miles of the Tsenacomoco.[4]

The carriages clattered through Westminster, past Whitehall Palace and the great noble residences of the Strand, past Cecil's mansion, where plans for the Virginia venture had taken shape, the apothecary shops of Fleet Street advertising their exotic wares with bunches of tobacco leaves and crocodile skins, across the stinking river Fleet, a polluted tributary of the Thames, beneath Lud Gate, supporting statues of Queen Elizabeth and the legendary English King Lud, until they came to a halt outside a tavern, which lay in the shadow of the looming Gothic tower of St Paul's Cathedral: the Belle Savage Inn.

There the company disbanded, members of Pocahontas's entourage being billeted with various members of the Virginia Company, while the rest returned to their homes or took up lodgings. Dale prepared for his onward journey to the Netherlands, where he was to sue his former employer the States General for back pay. Molina was probably allowed to go to the Spanish ambassador's residence at Highgate, to await the planned exchange with the captured English pilot John Clark, who had recently been handed over to the English ambassador in Madrid. In the event, Molina would not reach Spain until the following year, where he was welcomed with a reward of 1,000 ducats from the King, 'in consideration of his long and good service, and the great need in which he finds himself because of his long detention in Virginia'.[5]

The Belle Savage was to be home for Uttamatomakkin, Pocahontas and her husband. The inn was famous in London as one of the first theatres, holding in its galleried inner courtyard regular events including plays, fencing tournaments and other entertainments. It was not considered salubrious in Queen Elizabeth's time, more a place where a man might 'feed his eye with vain delight' and get drunk. Attitudes to theatrical entertainment had eased since then, but there were still qualms. A famous Puritan later claimed that the devil himself put in an appearance at the Belle Savage, during a production of

Christopher Marlowe's infamous play *Dr Faustus*. This notoriety, combined with its name, made it an ideal venue to stage the sensational appearance in the middle of London of an Indian princess.[6]

The Virginia Company was hungry for attention to publicize its ongoing lotteries. These were already regarded with some disdain by the company's critics, who ridiculed those who 'gape at the lottery from morn till even', comparing them to asses. There were also concerns about the 'common sort' being lured into debt.[7]

Nevertheless, the company needed the money, and found Indians one of the most effective ways of keeping up interest. As Shakespeare had quipped in *The Tempest*, Londoners would 'not give a doit [a worthless coin] to relieve a lame beggar' but would 'lay out ten to see a dead Indian' – still more a live one. Eiakintomino, a 'young man from the Virginias', had caused a sensation when he had appeared 'in St James Park or zoo' the year before, displayed alongside a collection of supposedly American birds, animals and plants (they were mostly European). The Virginia Company had included an etching of him, together with an Indian woman, on its lottery advertising.[8]

Now attention was to pass from the zoo to the inn, and to advertise Pocahontas's presence there, the Virginia Company commissioned a portrait. An obscure and therefore relatively cheap Dutch artist called Simon Van de Passe was hired to perform the task. The result was an awkward, probably rushed affair. It showed her dressed in expensively embroidered English dress, her face framed by a huge starched ruff and feathered high hat. Perhaps the most notable feature of an undistinguished work was that she clutched a fan of three white feathers, the insignia of the Prince of Wales, evidently included as a homage to Prince Henry's influence over the venture. The engraving printed up for mass circulation was even less flattering. 'A fine picture of no fair Lady,' sneered one observer, put off less by the engraving's depiction of Pocahontas than the description of her as daughter of the 'mighty prince Powhatan, emperor of Virginia'.[9]

Whatever the quality of the art, the exposure had the desired effect. Sir George Carew, a former member of the Royal Council, wrote to a friend within days of her arrival that the daughter of 'the barbarous prince' Powhatan had arrived. 'The worst of the plantation is past,' he concluded, 'for our men are well victualled by their own industry, but yet no profit is returned.' Around the same date, the newsmonger John Chamberlain sent off a letter to his friend Dudley Carleton in the Hague. Chamberlain was unimpressed with Dale's commodities – 'only some quantity of sassafras, tobacco, pitch and clapboard, things of no great value' – but was scintillated by the prospect of meeting 'some ten or twelve old and young of that country' who had been shipped in with them, 'among whom the most remarkable person is *Poca-huntas*.'[10]

There seems to have been some unease about this use of her. A few years later, a character in Ben Jonson's play *The Staple of News* pointed out that 'a tavern's unfit . . . for a princess', provoking this response:

> I have known a *Princess*, and a great one,
> Come forth of a tavern . . .
> . . . the blessed
> Pocahontas, as the Historian calls her
> And great King's daughter of Virginia . . .[11]

Among the first to come and gawp at her was Samuel Purchas. Purchas was a clergyman from Essex who was close to Richard Hakluyt. Hakluyt was also currently in London, probably suffering from what would prove to be his final illness. He would soon, if he had not already, bequeath his vast archive of travel literature to the 39-year-old Purchas, which Purchas would publish (and often mutilate) under the title *Hakluytus Posthumus*. In preparation of this legacy as England's leading geographer and colonial promoter, Purchas was appointed to lead efforts on behalf of the Virginia Company to scrutinize the visiting Indians.

His first impressions were favourable. He noted Pocahontas's deportment, observing that she 'did not only accustom herself to civility but still carried herself as the daughter of a king, and was accordingly respected not only by the Company, which allowed provision for herself and her son, but of divers particular persons of honour'.[12]

Purchas undertook to interview Uttamatomakkin about religious matters, arranging to do so away from the hubbub of the Belle Savage by having John Rolfe bring him to the nearby house of Dr Theodore Goulston, physician to Edwin Sandys.[13]

In the cramped, chilly chamber of a London house, the old *quiyoughcosuck* and counsellor was a formidable presence, still dressed in his Indian loincloth, mantle and moccasins, a proud, uncompromising ambassador of his people and faith. He also turned out to be a steadfast champion for his religion.

Asked by Purchas about his beliefs, he said that their god was Okeus, who 'doth often appear to them in his house or temple' and who 'by winds or other awful tokens of his presence holds them in a superstitious both fear and confidence'. It was he who had taught the Indians how to plant their corn, and who had prophesied the English invasion of their lands. Indian warriors wore their distinctive long knot of hair 'on the left side hanging down near to the foot' to copy Okeus, Uttamatomakkin explained. Rolfe noted that some of the first English settlers had adopted the fashion.

Purchas asked the priest about the 'Black Boys' ceremony, the ritual first reported by William White while he was living with the Quiyoughcohannock, involving the sacrifice of young boys. Captain Smith had mentioned it in his 1612 work *A Map of Virginia*, describing it as a 'yearly sacrifice of children', fuelling the widespread belief that the Indians, as part of their 'devil-worship', indulged in cannibalism.[14]

The ceremony was called the 'Huskanaw', Uttamatomakkin explained, a local term which meant 'he has a new body'. The aim, as subsequent reports confirmed, was not to sacrifice the boys, but to initiate them into the world of warriors. The boys White had reported seeing 'cast on a heap in a valley as dead' had been given their first dose of an 'Infusion . . . of some Poisonous Intoxicating Roots', which had been administered during a part of the ceremony from which the

English had been excluded. The medicine was made from wisacan, widely used by the Indians as a purge, and jimsonweed (*Datura stramonium*), a hallucinogenic which produces blurred vision, confusion, agitation and combative behaviour (its English name is derived from Jamestownweed, after an episode in 1676, when a group of soldiers was poisoned by a 'boiled salad' containing leaves of the plant, 'the effect of which was a very pleasant comedy'). They were kept in this intoxicated condition for months, held in a wicker cage the shape of a 'sugarloaf' or cone hidden in the woods, watched over by guards who would feed and clean them, as 'they would have wallowed in their own excrements, if they had not been prevented'. At the end of this captivity, they were weaned off the drug, 'and after eleven days, returned to themselves again, not remembering anything that had passed'. They came back to their homes as warriors who had discarded their youth like a snake its skin, and if they so much as recognized their own mothers, they would be forced to undergo the trial again.[15]

John Rolfe confirmed this explanation, recalling that he had met some boys in the 'distracted' state produced by the medicine, who even after returning from the ordeal were 'for a certain time of silent and strange behaviour, and will do anything never so desperate that they shall be bidden'.[16]

Having heard Uttamatomakkin's account of the Indian religion, Purchas now challenged him to abandon it, as Pocahontas had done. It was 'superstition', he said, and he a 'blasphemer' for talking of Okeus as a god, and for wearing his 'devil-lock'. He noted how Indian children were kept in ignorance of their religion, while trickery with tobacco smoke and painted masks was used to keep them in awe of it. Ignorance and trickery: to a Protestant like Purchas, these were the instruments used by the Catholic Church to oppress its people and keep them from a true knowledge of God.

Uttamatomakkin was affronted. He would 'hear no persuasions to the truth', as Purchas put it. The Okeus Purchas termed a mere 'idol' was central to their lives, their very survival. He had 'taught them their husbandry, etc', which was why so many English, facing starvation, had run to Indian towns, and sought the protection of Indian gods. Purchas could try to 'teach the boys and girls which were brought over from thence', but was wasting his time trying to convert the old priest. He was 'too old now to learn'.

Purchas's evangelical efforts were unsuccessful, and left Uttamatomakkin bewildered and furious. He demanded to be returned to his homeland 'in the first ship' available. He was told that this was impossible, as it had 'gone already'.

Purchas ignored this reaction, and over the coming weeks, set about trying to convert the rest of Pocahontas's retinue. His efforts appear to have had some limited success. One boy was apparently christened Abraham at St Andrew Hubbard, the parish church for Sir Thomas Smythe's offices in Philpot Lane. Another, sent to live with the MP George Thorpe, received reading and writing lessons, and ended up both adopting his host's name and becoming his scribe.

Two women also received the English names Mary and Elizabeth, though whether by baptism is unknown. The Virginia Company took responsibility for their upkeep, placing one as a servant with a 'Mercer in Cheapside', the other with a Gough at Blackfriars. Gough's ward became 'very weak of a consumption', and he took 'great care and . . . great pains to comfort her both in soul and body', in recognition of which 'for her recovery' the Company voted to pay 20 shillings a week for two months.[17]

Pocahontas was seen as having a central role in converting her compatriots, and it was probably with this objective in mind that it was arranged for her to meet Dr John King, the Bishop of London, who was also de facto Bishop of Virginia, as the American settlements came under his clerical jurisdiction.

The bishop's palace was within sight of the Belle Savage, squeezed between the giant's buttresses of St Paul's, beneath Lollards' Tower, one of the cathedral's two bell towers. Dr King was one of the foremost Protestant preachers of James's reign. In sermons delivered across London, he had explored a number of profound issues and controversies of the day, from the powers of the King (as befitting a royal appointee, he was an ardent monarchist) to the role of marriage (he believed in a full and equal partnership between husband and wife).

Purchas was on hand to make the introductions, King being his 'honourable and reverend patron'. He noted how the bishop 'entertained her with festival state and pomp, beyond what I have seen in his great hospitality afforded to other ladies'.[18]

No other details concerning the audience are preserved, but soon after, the Virginia Company voted to give Rolfe a special award of £100 if he and Pocahontas undertook 'by their godly and virtuous example' and 'all other good means of persuasions and inducements' to win over the Powhatan people 'to the knowledge of God, and embracing of true religion'.[19]

This now confirmed as her official role, Pocahontas was ready to undertake her most important meeting: with Anne of Denmark, the wife of King James.

The presence of the Indian princess in England had first been brought to Anne's attention by Captain John Smith. Hearing of her arrival at Plymouth, perhaps through the offices of Sir Lewis Stukeley, Smith had sent a letter to the Queen urging her to do Pocahontas 'some honour'. As evidence of her deserving case, he had revealed for the first time the sensational story of how Powhatan's 'most dear and wellbeloved daughter' had saved the captain during his captivity at Werowocomoco. 'At the minute of my execution,' he wrote, she had 'hazarded the beating out of her own brains to save mine'. He also mentioned that, even while her father and the English had been at war, she had become 'the instrument to preserve' Jamestown 'from death, famine, and utter confusion'.

'Though it be from one so unworthy to be the reporter as myself, her husband's estate not being able to make her fit to attend Your Majesty,' Smith urged the Queen to 'take this knowledge' of Pocahontas. 'If she should not be well received,' he added, 'her present love to us and Christianity might turn to such scorn and fury as to divert all this good to the worst of evil.'

Despite the impolitic implication that the Queen would be responsible for the ill effects of ignoring Smith's advice, Anne took it. 'Pocahontas had many Honours done her by the Queen upon Account of Capt. Smith's Story,' according to one account. She was introduced to the Queen by Cecilia, Lord Delaware's wife, and afterwards 'was frequently admitted to wait on her Majesty', who took her to 'many Plays, Balls, and other public Entertainments'. The royal association had an instant effect upon her status, ensuring that she was no longer treated as a spectacle, but 'as a Prince's Daughter . . . and very respectfully receiv'd by all the Ladies about the Court'.[20]

But soon her exposure to such an alien environment began to take its toll, and Pocahontas's health suffered. The Belle Savage was choked in pollution, from coal fires, the foetid river Fleet, and neighbouring Smithfield, a huge cattle market so infamous for the putrid smells of discarded carcasses that one of the streets leading off it and towards Ludgate Hill was known variously as Stinking Lane and Blowbladder Street.[21]

'To be a little out of the smoke of the city,' Rolfe therefore decided to move the family to Brentford, an unremarkable suburb to the west of London. As well as fresh air, Brentford provided access to Thomas Harriot, a mathematician who had visited Raleigh's Roanoke and produced the first English study of the Algonquian language. His cottage, in the grounds of the Earl of Northumberland's residence Syon House, neighboured the village. Still in the employ of the imprisoned Earl, it was there that he was performing a series of astronomical experiments, probably using a telescope that he had taken to Virginia. Harriot's linguistic studies meant that he was one of the few in London capable of conversing with Pocahontas in her native tongue, and she may have enjoyed the opportunity to speak to such a learned and open-minded scholar.[22]

Around this time, Captain John Smith was busily involved in his New England venture. Hitching a lift with a whaling expedition, he had explored the region in 1614 and found 'an excellent good harbour' which would make the ideal location for a settlement. He had named it 'Plymouth', later made famous by the rock on its shoreline, upon which the Pilgrim Fathers supposedly first stepped. Smith had been 'preparing to set sail' there when he heard that Pocahontas was in Brentford, and decided to visit.[23]

His recollection of the event was written seven years later. He remembered arriving and being greeted by John Rolfe and 'diverse of my friends'. Pocahontas then appeared, and Smith offered a 'modest salutation'. She turned away and covered her face. 'And in that humour, her husband with divers others we all left her two or three hours.' When she was ready to speak to him, her first words were furious. She had been expecting him to meet her at Plymouth, or welcome her to London. But he had not been there, and when she had asked to see him, she had been told he was dead. That lie was now compounded by this offhand appearance midway through the visit. Had she not looked after him in Virginia? He had called Powhatan 'father, being in his land a stranger, and by the same reason so must I do you', she said. Smith said he could not be a father to her, 'because she was a King's daughter', but Pocahontas swept aside this spurious appeal to etiquette. With a 'well set countenance', she pointed out that he had not been 'afraid to come into my father's Country', yet 'fear you here I should call you father?' Of course not. Captain John Smith feared nothing and no one. 'I tell you then I will,' she replied, 'and you shall call me child, and so I will be for ever and ever your Countryman.' They parted on these awkward, emotionally charged terms.

Pocahontas's visit ended with an invitation to the royal family's Twelfth Night revels, a series of feasts and entertainments marking the last day of the Christmas festivities. 'The Virginian woman Pocahontas with her father's counsellor have been with the King and graciously used,' noted a fellow reveller. The highlight of the celebrations was a masque by Ben Jonson called *The Vision of Delight*, and she and Uttamatomakkin were guests of honour, being 'well placed' amongst the audience.[24]

By now, Pocahontas's prominence was beginning to rankle.

A catty Chamberlain accused her of getting above herself. 'With her tricking up and high style and titles you might think her and her worshipful husband to be somebody, if you do not know that the poor Company of Virginia out of their poverty are fain to allow her four pound a week for her maintenance.'[25]

These caustic remarks revealed an uncomfortable truth. The company's finances remained dire, and throughout Pocahontas's visit, the treasurer Sir Thomas Smythe had been confined to his offices in Philpot Lane, probing the balance sheet. Smythe was now fabulously rich, thanks largely to his major stake in the organization that had laid the foundations of his remarkable business career: the East India Company. A measure of that company's success is to be found in a meeting of its directors chaired by Smythe, in which there was some discussion about Christmas presents for 'some lords and other [government] officers of whose countenance and favour the Company stand in need'. It was noted that the £1,000 formerly given was not enough to ensure continuing government support for the company's activities. It was therefore decided that between £4,000 and £5,000 would be a more suitable sum – the cost of two supply missions to Jamestown.[26]

The Virginia Company could only dream of handing out such largesse. For the time being, it could not even afford to pay dividends pledged to investors who had bought shares in 1609.

Revealing a true businessman's instinct for self-preservation, Smythe set up an ingenious scheme to ensure at least some measure of protection for the investment he and his closest associates had made in the company. It took the form of yet another company, this one called the 'Society of Particular Adventurers for Traffic with the People of Virginia in Joint Stock'. Smythe was the main investor, and his friend, the grocer and alderman Robert Johnson, became director.

The purpose of the society was to exploit the one asset the Virginia Company could depend upon: its monopoly of trade with Virginia. The society took over responsibility for the 'magazine', the shipping sent on a regular basis to provide the settlers with the supplies they depended upon to survive. These supplies would then be paid for with commodities out of the settlement's common store.

The attraction of this arrangement for Smythe and his little syndicate had become apparent with the arrival of one of the commodities disparaged by Chamberlain: tobacco.

One self-appointed guardian of national morals claimed that 'there is not so base a groom that comes into an alehouse to call for his pot, but he must have his pipe of tobacco. For it is a commodity that is now as vendible in every tavern as either wine, ale, or beer, and for apothecaries' shops, grocers' shops, chandlers' shops they are (almost) never without company, that from morning till night are still taking of tobacco.' He quoted recent figures which showed that there were 'upward of seven thousand houses' in and around London which were selling tobacco. On this basis, he estimated that, if each took half a crown a day (a modest assumption), annual sales would amount to £319,375, 'all spent in smoke'.[27]

Where one man saw a source of moral corruption, Smythe and his associates smelt a commercial opportunity. The tobacco contained in the first few barrels brought by Dale had sold for more than 5 shillings a pound. This was about a quarter of the price of best Spanish tobacco, but that gap was bound to narrow as methods of cultivation and processing improved. Furthermore, they had private reassurances from Rolfe that commercial quantities were already being grown in the colony's plantations, which at the next harvest would yield 10,000 to 20,000 pounds in weight. Such a quantity was worth at least £5,000 sterling, or double the average cost of equipping and shipping a supply.[28]

This meant there was a captive market of settlers wanting supplies, who could be forced to pay premium prices for the goods supplied by the magazine, and had in tobacco a valuable currency with which to pay for them.

Edwin Sandys, one of the most influential figures in the formation of the Virginia Company, and a champion of its business in Parliament, was not part of Smythe's society, probably because of his well-known opposition to commercial monopolies. However, he was now formally admitted as an 'assistant' or director of the Virginia Company itself, and one of his first acts was to move the other major reform that it was hoped would revive the company's fortunes.[29]

At the company's request, John Rolfe had conducted a survey of the lands that were settled. Soon after arriving in London, he had presented the results to Sandys. Six active settlements were identified, most of them on land which the English appear to have purchased from the Indians, though on terms that either the seller or purchaser probably misunderstood, as the Indians and English appear to have had very different concepts of private property and ownership.[30]

The settlements were: Jamestown; Henrico; Bermuda Nether Hundred, accommodating one hundred and nineteen men 10 miles downriver of Henrico, 'which seat containeth a good circuit of ground, the river winding round, so that a pale running across a neck of land from one part of the river to the other maketh it a peninsula'; West and Hurley Hundred, which had twenty-five men working exclusively on growing tobacco; Kecoughtan, at the mouth of the James, where eleven farmers 'maintain themselves'; and Dale's Gift, on the eastern shore, where seventeen men were making salt and catching fish. Rolfe noted that a total of two hundred and five 'officers and labourers', eighty-one farmers and sixty-five women and children were now living in these settlements – 'a small number to advance so great a work' – and if they were to prosper, they needed 'many hundreds more' sent from England to help them. This conclusion was backed up by Dale, who as long ago as 1612 had been urging the sending of another two thousand men.[31]

Sandys evidently agreed with Rolfe's findings, but he saw that a more fundamental change would be needed to assure the colony's future. Until now, Virginia had been seen as a place offering natural resources that might be sold for a profit at home: precious metals, wood, medicinal herbs, furs, anything that might be dug up, cut down, shot, caught, trapped, or bought from the Indians and sold back in England for a profit.

But there was another commodity which Virginia had in abundance, that was in short supply in England, and was a proven source of wealth and power: land. Members of the gentry, for whom Sandys acted as champion in Parliament, were crying out for it. In Virginia, they could apparently take as much of it as they pleased – tens, even hundreds of thousands of acres. Sandys therefore proposed that the company should start selling it off, in return for a modest outlay and commitment to finance the shipping out of the men needed to develop and work it. Furthermore, a tract of good land could be reserved for the company, produce from which would cover government and administration costs.

Such a scheme had already been trialled in Bermuda, and the results in terms of population were spectacular. Despite rats, which had spread 'like one of Pharaoh's plagues' since being brought with a consignment of oatmeal, the archipelago was already sustaining nearly twice as many people as Virginia: six hundred souls, as compared to a total of three hundred and fifty counted by Rolfe.[32]

The first and largest of the new private or 'particular' plantations to be granted in Virginia was 'Smythe's Hundred', named in honour of Sir Thomas, the primary investor, with Sandys and the Earl of Southampton acting as partners. Smythe's Hundred was an enormous plot, over 150 square miles of the land sandwiched between the James and Chickahominy rivers.[33]

The idea of acquiring bits of America as private plots aroused interest outside the company too, particularly among religious dissenters. William Trumbull, the royal agent in Brussels, was warned that a hundred or so Congregationalists 'very near Brownists' had been allowed passage to Virginia, 'and shortly there will be twice as many Puritans'.[34]

Trumbull was referring to the activities of Robert Cushman and John Carver, two prominent members of a congregation of Protestant radicals from Scrooby, Nottinghamshire, who a decade before had emigrated to Leiden, 'a fair and beautiful city', and one of the centres of Dutch religious radicalism. In Leiden, they had found themselves in the midst of another religious dispute with a more liberal Protestant sect known as the Arminians, and decided to look elsewhere for a new home, including 'those unpeopled countries of America, which are fruitful and fit for habitation, being devoid of all civil inhabitants, where there are only savage and brutish men which range up and down, little otherwise than wild beasts of the same'.[35]

It was this business that brought Cushman and Carver to London in 1617. They had come to discuss the possibility of buying one of the new private plots of land in Virginia to found a religious community, and found Edwin Sandys sympathetic to the idea. Though the son of an archbishop, Sandys was a champion of religious tolerance. 'What unity poor Christendom may hope for?' he asked, while religious dogmatism and extremism reigned. For this, his books had been burned in the churchyard of St Paul's, just a few days before the Gunpowder Plot seemed to confirm his point. He was therefore open to discussion with these two Puritans about their plans, particularly as they had come equipped with a list of seven 'articles' designed to reassure the government of their inoffensive intentions, including a pledge not to challenge the authority of the Church of England in America. Sandys promised to consider the matter, and a letter from him following their return to Leiden committed to paper a pledge to use 'all forwardness' to help them. Despite this, it would take another three years for the plan to come to fruition, when a merchant ship called the *Mayflower* was chartered to take these so-called 'Pilgrims' to Virginia, and, diverted by bad weather, instead set them down at the 'excellent good harbour' Captain Smith had called Plymouth.[36]

Meanwhile, other plots of 'virgin' soil were being handed out to all and sundry, including veteran settlers, in lieu of their service to the company. 'In Regard of a great travail, pain and charge undertaken from the beginning of this Christian plantation', Captain John Martin was offered several thousand acres, as well as the right to trade goods privately. Samuel Argall was offered 2,400 acres of land at 'Paspahegh, alias Argall's Town', which it would be his responsibility to develop using labour shipped over at his own expense. In return, he had been given a commission to take over command of the colony from acting governor George Yeardley.[37]

With these arrangements in place, pressure to get new expeditions under way intensified. The first to leave would be a ship called the *George*, which was chartered to take around one hundred and fifty men to settle Smythe's Hundred, plus Pocahontas and her entourage. By mid-January 1617 she was ready to sail, in the hope that she would be in Virginia in good time for the coming planting season.

Uttamatomakkin was by now desperate to return home. He had met Captain Smith soon after the Twelfth Night revels, and complained bitterly about his treatment. During the masque, far from being the centre of attention, he had found himself sidelined. In the hectic celebrations, nobody even bothered to point out the King to him, let alone effect an introduction. 'You gave Powhatan a white Dog, which Powhatan fed as himself,' he pointed out to Smith, 'but your King gave me nothing, and I am better than your white Dog.'[38]

Pocahontas, meanwhile, was preparing to depart 'against her will'. She and her son were now both suffering from an unidentified illness. Waiting for the wind to 'come about to send them away', their condition deteriorated, and John Rolfe began to consider leaving the child behind, fearing it might not survive the voyage.[39]

In the event, they were all aboard the *George* when she cast off sometime around mid-March 1617. They were approaching the wide Thames estuary when Pocahontas's condition became critical, and it was decided to put in at the nearby port of Gravesend. Some comfort to the ailing princess and her infant son would have been sought at a nearby inn, such as the Flushing, run by a Belgian described in a contemporary travel guide as a 'capital fellow'.

All hopes of her recovery now rested either on the purging and blood-letting of English physicians, or the 'extreme howling, shouting, singing, and . . . violent gestures, and Antic actions' performed by priests like Uttamatomakkin as part of their healing rite.[40] Neither was able to save her, and by the end of March, she was dead.

Her body was taken to the local St George's church, where she was buried according to Christian rites. As for her soul, Uttamatomakkin may have feared it would be lost, unless from some tall English oak or elm it might even in this far-off land glimpse the pleasant path that carried her forefathers to their eternal rest. Rolfe believed otherwise. To Edwin Sandys, 'whom I have found a father to me, my wife, and child', he expressed his certain belief that she had found repose in the arms of their Christian god, and 'resteth in eternal happiness'.[41]

The *George* continued round the coast of southern England, heading for Plymouth, where she was to rendezvous with the *Treasurer*. During the journey, baby Thomas Rolfe was probably cared for by Uttamatomakkin's wife Matachanna, but despite her efforts and the 'smooth water' of a calm Channel, the infant remained dangerously ill, and his father fretted about the lack of a wet-nurse to feed him. By the time they arrived at Plymouth, Argall, perhaps more mindful of the distraction to one of his trustiest lieutenants as the danger to the child, insisted that Rolfe leave Thomas in the care of Sir Lewis Stukeley.

Sir Lewis watched the *George* set off in convoy with the *Treasurer* on 10 April, a baby's arm lifted to wave his father off. Thomas remained in Plymouth until an uncle Henry came to take charge of him, and he would not see his mother's native soil for another twenty years.

After a month-long voyage, the *Treasurer* and *George* arrived to find the colony 'in good estate, never enjoying a firmer peace nor more plenty', according to Rolfe. Crops were thriving, cattle were grazing, 'the Indians very loving'. The only disappointment was the state of the 'buildings, fortifications, and of boats', which were 'in great want'.[42]

'Great plenty and peace,' wrote Argall, in a characteristically terse report. 'Silk worms thrive exceedingly,' he added, knowing this news would be welcomed by the King, who had a particular interest in the insects.

The positive assessment was reinforced by activity at Jamestown's dockside. The *George* was prepared for a quick return to London, loaded with the promised 20,000 pounds of tobacco. For the first time in the history of Virginia, an expedition across the Atlantic had actually made some money, without having to resort to piracy.[43]

There were problems, however. Henrico's vicar Alexander Whitaker had drowned, leaving the settlement without a minister. Given the venture's mission to convert Indians, Argall urged the company back in London to arrange the appointment of a successor as quickly as possible.

The 'loving' Indians, meanwhile, showed continuing but subtle signs of realignment, which the English found hard to interpret. Powhatan had withdrawn from active political life, and was now living on the Potomac river with Machumps. Machumps's sister Winganuske had a daughter by the *mamanatowick*, and the child had become 'a great darling of the king's'. Since his loss of Pocahontas, he had decided to spend his declining days with her.

Opitchapam now took Powhatan's place as spiritual leader, but it was Opechancanough who consolidated his place as political leader of the Tsenacomoco Indians. It was to Opechancanough that Uttamatomakkin made his report on his stay in London, and it was damning. He 'rails against England [and the] English people', Argall had heard: the undignified way he was treated, the failure of their *mamanatowick* even to acknowledge his presence, the contempt they showed for their religion, and their ill-disguised design to stamp it out. Asked to estimate the strength of English numbers, Uttamatomakkin told Opechancanough 'to count the Stars in the Sky, the Leaves upon the Trees, and the Sand on the Seashore, for so many People . . . were in England', and so many people would there be in Tsenacomoco, unless they were prevented.

To the Indians, Uttamatomakkin's words must have seemed like the exhalations of Okeus. The trees around them, the beaches where they conducted their daily worship, the very heavens seemed to have become harbingers of terror. Lying upon their cots at night, gazing up through the smoke hole in fearful contemplation, the children of Ahone could see the stars take on the aspect of deadly insects, about to descend upon their heads and destroy them all.[44]

Part Five

TWENTY-ONE

Imbangala

In the summer of 1619, Manuel Mendes de Acunha sailed his ship the *San Juan Bautista* into Luanda, Angola, the newly established Portuguese colony on the west coast of Africa. He carried with him a licence issued in Seville, allowing him to collect a consignment of slaves, due to be shipped over to Mexico.

Portugal was now under the Spanish crown, and its imperial possessions in Africa were providing a useful source of slave labour for mines and plantations in Spanish America. To encourage the trade, Philip III had appointed an *asentista* or special agent called Antonio Fernandes Delvas, a Lisbon businessman. In return for 115,000 pesos paid into the cash-strapped royal exchequer, Delvas was given the right to organize the traffic of Africans to one of two designated ports in central America, Veracruz, Mexico and Cartagena, Colombia. His target was to import five thousand slaves into South America each year, and to help meet it he had issued Acunha with a licence to acquire two hundred, any more than that number to be declared once the mission was complete.[1]

Luanda had been a lucrative source of slaves as a result of a series of wars fought by the Portuguese with the King of Ndongo, whose kingdom, one of the largest in Africa, lay to the east of Luanda, in the highlands between the Kwanza and Lukala rivers. War yielded large quantities of merchandise for slave traders, as vanquished troops could be enslaved en masse. However, in 1589 the Portuguese had suffered a crushing defeat at the hands of the Ndongo along the Lukala river, issuing in a period of uneasy peace that forced traders to buy their slaves from local markets or 'fairs' at higher prices and in fewer numbers.

In an effort to regain control, a new Portuguese governor of Angola was appointed in 1617, Luis Mendes de Vasconcelos. A veteran soldier of the Spanish campaign against the Dutch rebels, he had personally pledged to King Philip that he would drive an army east through the Ndongo kingdom and continue until he reached Portuguese settlements on the opposite side of the continent, in the area of modern Mozambique.

When he arrived in Luanda, Vasconcelos found that his military ambitions far exceeded his resources, which amounted to a motley collection of Portuguese soldiers. It transpired that previous governors had waged military campaigns by recruiting *guerra preta*, 'black armies' from the local chiefs or *sobas* who paid tribute to the Portuguese. However, such troops had proved unreliable, as the *sobas* provided them to play the Portuguese off against dominant local potentates, and as the balance of power shifted from one side to the other, so did loyalties.

Vasconcelos decided he could not rely on the *sobas*, so cast around for alternative sources of military support. This led him to a mysterious, terrifying group of itinerant warriors known as the Imbangala. The earliest known European reference to the Imbangala was by an English sailor called Andrew Battell, whose adventures in Africa while a captive of the Portuguese were published by Samuel Purchas. According to Battell, the Imbangala were marauders, who would 'settle themselves in any country', and live off the locals until their corn and cattle were used up. They would then move on, after looting anything they needed, including adolescent children whom they would use to replenish their numbers, their own offspring being killed at birth.

The Portuguese had already bought slaves from the Imbangala. Battell had witnessed one pillaging campaign during which the Imbangala, with the help of Portuguese weapons, succeeded in capturing sufficient men to fill three slave ships.

At first, Vasconcelos renounced the employment of people who 'are enemies of all living things and thieves of the lands where they enter'. However, the practicalities of attacking the Ndongo without the Imbangala soon persuaded him otherwise, particularly as at that moment, the king of the Ndongo had been left invitingly vulnerable by internal disputes.

And so, as the Portuguese bishop of the region reported to Lisbon, instead of 'leaving off' the Imbangala, Vasconcelos 'embraced them, and he has gone to war with them . . ., killing with them and capturing innumerable innocent people, not only against the law of God but also against the expressed regulations of Your Majesty'.

The result was a stunning military campaign conducted in the autumn of 1618 and spring of 1619, in which Kabasa, the royal capital of Ndongo, was overrun, together with surrounding cities of twenty or thirty thousand people, and 'many prisoners and slaves were carried away as captives'.

Inevitably, the Imbangala renounced their alliance with the Portuguese as soon as was convenient, and rampaged through the rest of the region. However, for the time being, Vasconcelos had a glut of slaves, and at Luanda traders clamoured to buy them. Several hundred were duly lined up on the quayside at Luanda when Acunha's ship the *San Juan Bautista* came alongside and moored.

Acunha passed through the jostling throng, selecting three hundred and fifty 'large and small' men, women and children. Because of the circumstances of their capture, they were likely to have come from a single Ndongo community, probably one of the large cities. They would have been in reasonable physical condition, though, in their leg-irons and neck chains, unable to offer much resistance as they were loaded into the *San Juan Bautista*'s hold. The only formality to be completed before departure was to ensure that they were all christened, a stipulation of Portuguese law.[2]

The *San Juan Bautista* set sail in June or July, and reached the opposite side of the Atlantic by the middle of August. At Campeche Bay, approaching the Mexican destination of Veracruz, it encountered what Acunha described as two 'corsairs' or pirate ships: the *White Lyon* and the *Treasurer*. The ships opened fire on the poorly defended merchantman, forcing Acunha to heave to and allow the captains on board. One, called Daniel Elfrith, claimed to carry credentials or 'letters of marque' awarded to the ship's owner, Robert Rich, Earl of Warwick, by the Duke of Savoy, a French potentate who was in dispute with Spain. This gave Elfrith permission to attack Spanish or Portuguese ships in international waters. The other captain, John Colyn Jope, said his vessel was out of the Dutch port of Flushing, flying under the colours of the United Provinces. As Acunha could tell, both men, and their ships, were very obviously English, a kingdom supposedly at peace with his own.

Accompanied by a chorus of indignant protests at the violation of royal treaties and international law, and threats of Spanish reprisals, Elfrith and Jope went down into the hold and picked out two hundred of the strongest men and women to take with them. Once the slaves had been transferred on to their ships, they sailed off, leaving Acunha 'with only 147, including 24 slave boys' which he was 'forced to sell in Jamaica, where he had to refresh, for he had many sick aboard, and many had already died'.[3]

In April 1619, George Yeardley, who had just received a knighthood from King James, arrived in Virginia. The previous summer, seven years after his untimely return to England, Delaware had finally plucked up courage to return to Virginia. But on the journey over, he had been laid low by one of his ailments, and this time demonstrated beyond doubt its severity by dying. Yeardley had been hurriedly appointed as replacement, to succeed Samuel Argall as governor.

After two years in the job, Argall was only too happy to surrender it. In that time, his relations with the directors in London, and in particular the treasurer Sir Thomas Smythe, had broken down. The bone of contention was Smythe's 'Society of Particular Adventurers', which controlled the magazine and therefore all supplies sent from England.

Since its inception in 1616, the society had held the colony to ransom, pushing up prices by holding back essentials. The cost of tools and clothing, which could not be obtained in America, were shooting up with each new supply mission, and the only commodity with which the settlers were permitted to pay for them was tobacco.[4] The result was a frantic effort to produce more tobacco, and a shortage of necessities such as pickaxes and ploughs. Without tools, the settlers could not develop new land but instead had 'to tend old ground', which was becoming progressively less productive. They were like fish trapped in an evaporating pond.[5]

Argall was not making this up. The same problem was being suffered in Bermuda, the governor, Nathaniel Butler, complaining bitterly of the 'cut-throat prices of the Magazine Ship'. It was 'eating them up'.[6]

However, when such complaints reached the ears of Smythe and Alderman Johnson, they were 'so inflamed with these outrages that they could hardly be contained from running to his Majesty'. In their capacity as owners of the magazine company, they sent a letter to Argall railing at his 'unjust accusations against us and the Magazine, nourishing thereby (instead of pacifying) the malcontented humours of such as seek to bring all to confusion and to overthrow that which is settled upon just and equal Terms'. They pointed out his own sharp practices, accusing him of selling off the company's cattle, employing company labourers and trading with the Indians for his private gain, and, in his self-proclaimed capacity of 'admiral' of the colony, holding back a ship carrying goods belonging to the society.

In a mysterious aside, they also considered Opechancanough's proclamation, reported to them by Argall, that Thomas Rolfe had been declared heir to the Powhatan lands, which the boy was to inherit when he came of age. This Smythe and Johnson dismissed as a 'device' dreamed up by Argall for his own inscrutable purposes.[7]

Nothing more was heard of the Thomas Rolfe issue, but the

charges that Argall had been engaged in piracy and smuggling proved more enduring. It soon became apparent that Kecoughtan, at the mouth of the James, had become a thriving centre of contraband trade. At first, Argall claimed to know nothing about it. Then, the killing of one of the sailors who came ashore to offer some goods for a bale or two of tobacco forced him to act. He immediately circulated a message to the captains at Kecoughtan commanding them 'to send to him of any Ship's arrival', he claimed in a letter back to the London Company.[8]

However, even Argall's friends were aware that he was engaged in some sort of subterfuge. Some of it may have been to circumvent the magazine, but some of his other freelance activities were evidently connected to serving the interests of his patron, the Earl of Warwick, and his privateering policy.

In the autumn of 1618, a ship had arrived at Jamestown, apparently engaged on a fishing expedition. However, witnesses claimed that Argall had manned the ship with 'the ablest men of the Colony', furnished it with supplies for several months at sea, and sent it out to rove the 'Spanish Dominions in the West Indies' in search of prizes. That ship was the Earl of Warwick's man-of-war, the *Treasurer*. On his arrival in Jamestown, Yeardley's first job was to 'give diligent order that the ship be seized upon immediately upon her return, and examination taken of her course and proceeding'.[9]

There was no sign of the *Treasurer* when Yeardley arrived, and any attempt to quiz Argall on the matter was frustrated by the latter's speedy exit from Jamestown. The governor heard too late that his predecessor had departed for England in a swift pinnace sent by Warwick.

Yeardley did not bother giving chase. He had other pressing tasks to perform. He had brought detailed instructions from his patron Edwin Sandys for the wholesale reform of the colony. He was 'to lay a foundation whereon a flourishing State might in process of time by the blessing of Almighty God be raised', and to prepare the way for

'Multitudes who in great Abundance from diverse parts of the Realm were preparing to remove thither'.

His first act was to repeal the draconian 'Laws Divine, Moral and Martial', and implement in their stead 'a laudable form of Government by Magistracy and just Laws for the happy guiding and governing of the people there inhabiting'. Reflecting the hurried manner of Yeardley's appointment, which occurred soon after news of Delaware's death at sea reached London, some aspects of this 'laudable form of Government' had yet to be ascertained, Edwin Sandys awaiting a chance to 'retire himself three or four weeks into the country in which time he would spend his studies in collecting and framing such laws'.[10]

However, the agenda was clear enough for Yeardley to make the first major reform, which was to constitute what came to be called the 'General Assembly'. It was to comprise two bodies: a 'Council of Estate' of six men, which was appointed in London to act as the governor's cabinet, and twenty-two burgesses, two elected to represent each of the eleven settlements now established along the shores of the James.

Elections for the burgesses took place in June and July 1619, and the successful candidates arrived in Jamestown for their first meeting on Friday 30 July 1619. They decided to convene at 'the most convenient place we could find to sit', which was the choir of Jamestown church. It provided an ideal setting, not just because it set a suitably pious tone to the proceedings, but because it allowed the delegates to arrange themselves in the manner of a mini-parliament. Yeardley sat in the pews on one side of the choir, surrounded by his council, the burgesses in the opposite pews. Following the practice of the House of Commons, there was a Speaker (the company secretary, John Pory), who sat somewhere in the middle, and a sergeant-at-arms, who was positioned 'at the bar, to be ready for any service the Assembly should command him'.

The meeting began with a prayer, after which the burgesses were told to take themselves into the church nave. They were then summoned back by name, and 'so every man (none staggering at it) took the oath of Supremacy, and entered the Assembly'. As a final preliminary, the issue of who should be admitted to the assembly was discussed, with the main focus falling upon the representatives of 'Martin Brandon', the private plantation founded by Captain John Martin.[11] Yeardley himself wondered 'whether it were fit that Captain Martin's Burgesses should have any place in the Assembly', as he was exempted from the laws that applied in the colony. It was eventually decided that Captain Martin should be summoned to appear before the assembly in person, 'to treat and confer with us about some matters of especial importance which concerns both us and the whole Colony [and] yourself'.

'These obstacles removed', Pory set about the main business of the meeting, which was to review the 'Great Charter' that Yeardley had brought with him. The very idea that the settlers should get a chance to consider the arrangements that would govern them was entirely novel. One or two querulous voices wondered whether they should even 'presume' to consider a document which the 'Council and Company in England had already resolved to be perfect', but they were ignored. Instead, Pory read out the document.

The charter was to provide the basis of the colony's reformation. Its grand title was suggestive of the Magna Carta, the document signed by King John at Runnymede that was considered the constitutional cornerstone of English liberties. The version of the Great Charter that survives does not have this quality. But crucial sections, in particular concerning the establishment of 'one equal and uniform kind of government over all Virginia', and the powers of the General Assembly, were removed from the copies that have survived, presumably because they were so politically contentious. The remainder dealt with legal and managerial issues, such as the apportionment of land, tenancy terms and finding a way to 'ease all the Inhabitants of Virginia forever of all taxes'.[12]

The charter recommended a complete reorganization of the

colony. 'Ancient planters' – those who had been in Virginia before Dale's departure for England with Pocahontas in April 1616 – would receive a 100-acre 'headright' apiece if they had paid for their own passage to Virginia, or owned shares in the company. Those who had been brought to the colony at the company's expense would have to serve out their term of service (up to seven years) as tenants working on the public land, keeping half of what they earned from their labour, the other half going to the company. After that, they, too, would receive their headright. New planters were offered similar terms, except their headright would be 50 acres, since 'the former difficulties and dangers were in greatest part overcome'.

With so much complex detail to consider, it was decided that the charter should be split up into sections or 'books', and committees of assembly members were appointed to review them.

A hot summer's day beckoned outside, and a chorus of yeas announced a break for lunch. The burgesses dispersed into the fort to dine with friends, or to visit the tobacco fields at the eastern end of the island, where they could check the quality of a ripening crop, or to tour the glasshouse and ironworks beyond the island causeway, to watch manufacturing experiments, or to rest among a copse of trees lining the shore, where they could enjoy the shade and a picnic.

When they reassembled, the committees went off into their huddles elsewhere in the church to discuss the charter, leaving the remainder to consider other business. There followed some discussion on the matter of laws. With the charter missing any specific legal code, burgesses wondered which of 'the instructions given by the Counsel in England to my Lord Delaware, Captain Argall or Sir George Yeardley, might conveniently put on the habit of laws', and whether they could dream up new ones for themselves. Yeardley, perhaps concerned at a growing enthusiasm for self-government, announced that 'for expedition's sake' he would personally peruse the existing instructions, and report back with a suitable selection. However, it was agreed that the 'most severe and cruel' Laws Divine, Moral and Martial implemented by Gates and Dale, which had been 'mercilessly executed, oftentimes without trial or judgment', would not be revived.[13]

After three hours of deliberation, the committees inspecting the charter returned to the church choir, and announced they had reached their conclusions, 'which the whole Assembly, because it was late, deferred to treat of till the next morning'.

The following day, the assembly drew up a series of petitions 'out of the opinions of the said committees', offering the London Company their views on the Great Charter and general course of events. The experience of democratic deliberation, vaguely familiar to any that had belonged to a livery company or city corporation but strange to all as the basis for public government, had fostered a consensual and conciliatory mood, and the petitions were 'humble' and mild. With respect to the allocation of private land, the 'ancient planters' sought reassurances that the land they already possessed should not, 'after so much labour and cost, and so many years habitation, be taken from them'. They asked that male offspring of planters, 'being the only hope of prosperity', should be given shares in the company, that a local treasurer be appointed to see to finances, that more labourers be sent, and that the plantation of Kecoughtan, which under the new arrangements was to become one of the colony's four corporations, should be given an English name.

The rest of the day was taken up with the vexed issue of the price of tobacco. It was decided that this would be the subject of the assembly's first piece of legislation: 'the Establishment of the price of Tobacco, namely, of the best at three shillings and the second [best] at eighteen pence the pound'. Abraham Peirsey, the cape merchant, with local responsibility for the supplies sent by the magazine, was summoned 'to demand of him if he knew of any impediment why it might not be admitted of'. He replied that he took orders from the 'Adventurers', meaning the magazine's investors, not the colonists, a consideration that was swept aside as the assembly's first act was passed.

The following day was a Sunday, so no business was conducted,

Pory merely noting the death of one of the burgesses, a Mr Shelley. On Monday, the assembly, having flexed its legislative muscles, proceeded to exercise them vigorously, passing laws prohibiting attacks on the Indians, 'idleness, Gaming, drunkenness and excess in apparel'. Up to six 'of the better disposed of the Indians' should be allowed to inhabit English settlements at any one time, and their children encouraged to attend education 'in true religion and civil course of life'. Each household should keep a spare barrel of corn for each servant, every planter should plant mulberry trees, grape vines and silk flax, and perform trials with hemp and aniseed.

The final business of that day concerned the magazine. In order that it 'be preserved from wrong and sinister practices', all transactions were to be invoiced, tobacco must be accepted in payment at the agreed price, and if anyone should offer 'any necessary commodity' not available from the magazine, 'it shall and may be lawful for any of the Colony to buy the said necessary commodity of the said party' – a breach of the magazine's monopoly.

In often suffocating 'extreme heat', the burgesses continued in the same vein for the following two days, laws being passed on issues ranging from the registration of births, marriages and deaths to 5-shilling fines for swearing. They also started to exercise their judicial function, hearing cases against such malefactors as the 'lewd and treacherous' Thomas Garnett, a servant of Captain William Powell, who had engaged in 'wantonness with a woman servant of theirs' and accused his master of drunkenness and theft. Garnett was able to bring 'all his fellow servants to testify on his side', but was found guilty, and sentenced by Yeardley to having his ears nailed to the pillory for the duration of the assembly.

On the final day, Wednesday 4 August, Henry Spelman, the young man who had lived over a year with the Indians, was summoned before the assembly. Now in his early thirties, he had been promoted to captain, probably the only one with a working knowledge of the local language. This gave him an important role in dealing with the Indians. It was during a meeting with Opechancanough that an incident took place which was reported by another interpreter, one Robert Poole.

Powhatan had died the previous April, and Opechancanough was now being treated by the English as Virginia's imperial chief. Poole had been with Spelman as part of a delegation that had gone to visit the new *mamanatowick*'s 'court', and, according to Poole, during discussions about changes in the English camp Spelman had spoken 'very unreverently and maliciously against this present Governor' to Opechancanough.

Poole would himself be later charged of telling 'false tales to Opechancanough', and it was perhaps to divert attention away from himself that he accused Spelman. The captain denied the charge, but he did admit to telling Opechancanough that, within a year, 'there would come a Governor greater then this that now is in place'. He was referring to the Earl of Warwick, who was being openly discussed as a possible successor to the deceased Delaware as a new figurehead for the venture.[14]

As Yeardley was well aware, Spelman had expressed a view shared by many. From the time of his appointment, he had aroused snobbish reactions in London, John Chamberlain describing him as a 'mean fellow' whose knighthood, awarded prior to his departure to Virginia, 'hath set him up so high that he flaunts it up and down the streets in extraordinary bravery, and fourteen or fifteen fair liveries [retainers] after him'.

For reasons never fully revealed, Sir Thomas Smythe, the company treasurer in London, had taken particularly against Yeardley, a view that brought him into contention with Edwin Sandys, who had championed his appointment. A group of eight senior colonists and investors, including Sir Thomas Gates and even Yeardley's one-time friend Samuel Argall, had backed this view, arguing that the colony's future depended upon the appointment of someone of greater 'eminence or nobility' than Yeardley.[15]

However, in the governor's presence, no one was willing to endorse Spelman's views openly, and there were many sombre nods of agreement when it was declared that he had brought Yeardley 'in much disteem' with Opechancanough, and as a result made the whole colony vulnerable to his 'slippery designs'. 'Several and sharp punishments were pronounced against him by diverse of the Assembly,' Pory recorded, some inspired by a feeling that Spelman 'had in him more of the Savage then of the Christian'.

But Yeardley, 'hoping he might redeem his fault, being much of childish ignorance', called for clemency, and the defendant was sentenced to being stripped of his captaincy, and forced to serve the colony for seven years as interpreter.

These heated proceedings added to the discomforts of the 'intemperature of the weather', and it was decided to adjourn the assembly and reconvene the following March. Largesse being the principal asset of any political body, 'the last act of the General Assembly was a contribution to gratify their officers', all to receive a share from a levy of a pound of tobacco per adult 'man or manservant'.[16]

A few weeks later, John Rolfe reported the arrival of a 'Dutch man of war' at Point Comfort, under the captaincy of 'Capt Jope'. 'He brought not any thing but 20 and odd Negroes, which the Governor and Cape Marchant bought for victual [food]' Rolfe added. This ship was the *White Lyon*, and the 'negroes' were some of the Angolans captured from the *San Juan Bautista* off Mexico. The *White Lyon* slipped away soon after, and a few days later her consort for that privateering expedition, the *Treasurer*, arrived. She was in a poor state. A correspondent in Bermuda who saw her soon after reported that she was 'so weather beaten and torn as never like to put to sea again, but lay her bones here'.[17]

When Yeardley heard of this at Jamestown, he sent Rolfe to intercept her. She was apparently 'in a very distressed state', but attempts to buy some supplies off the settlers at Kecoughtan were rebuffed. By the time Rolfe arrived, she had already 'set sail and was gone out of the Bay', Eldred's mischievous departing message being a warning that the 'Spaniard would be here the next spring', and that if they had not got 'some Ordnance planted at Point Comfort' by then, 'the Colony would be quite undone'.[18]

The faintest shadow of the fate of those '20 and odd' Angolans sold by Jope for 'victual' can just be detected in the surviving documentation. They had come through a violent and bewildering succession of experiences – bloody war with the fearsome Imbangala, having water and strange incantations cast into their faces by foreign wizards, lying in chains in the black hold of a great ship as it tilted and creaked across impossibly vast and choppy seas, watched friends or family members die in the stifling heat, been dragged from one ship to another mid-ocean, and now deposited on a shore that none of them had ever seen, which might be as far from their homeland as the stars from the earth.

They were handed over to a master, in the case of eleven, Abraham Peirsey. He had claimed an allowance of 200 acres of land in lieu of his work as cape merchant, and needed labour to cultivate and plant it.[19]

They were not Peirsey's slaves. Slavery did not yet officially exist in English law or dominions. It was still associated with the worst atrocities of Spanish conquest, and it was taken as a matter of pride (if not fact) that English imperialism was conducted in a more civilized fashion. Patriots would recall the example of Drake, who had allied with escaped black slaves in South America to defeat their hated Spanish masters (and would forget that it was an attempt to muscle in on the Spanish slave trade that had originally brought Drake into contention with the Spanish). The great Richard Hakluyt had contrasted the 'proud governance of the Spaniards' with the 'gentle government of the English'. Slavery was foreign, freedom English. True Englishmen would rather surrender to death than 'yield our bodies unto such servile halings and tearings as are used in other countries'.[20]

So the Angolans were not officially slaves, but in 'ye service' of Peirsey. This may have amounted to the same thing, since they were employed involuntarily on the basis of a contract or 'indenture' that required them to work for their master for a fixed period, their 'wages' used to cover the costs of buying them off Jope, and their food and lodging. However, in at least some cases, these contracts were honoured, allowing the indentured servant to become a planter, who theoretically enjoyed the liberties of full citizenship.[21]

The reason they had been snapped up so eagerly was a surge in demand for labour, as planters began to cultivate their headrights, the stranglehold of the magazine was loosened, and large private plantations, such as Smythe's Hundred and Martin's Hundred (the latter named after the MP and Sirenaical Richard Martin) were established. Most wanted to plant tobacco, but it turned out to be one of the most labour-intensive commodities to grow, 'for in sowing, planting, weeding, worming, gathering, curing, and making up, it Consumes ten months [a year] at least, if not eleven'.[22]

Hundreds of itinerant workers, orphaned and destitute children, 'Newgatiers' (former inmates of London's notorious Newgate prison) and other felons, unemployed farm workers and apprentices were shipped in from England to meet the demand, most of them on terms similar to the Angolans', indentured to serve a master for a period of time (between three and seven years) in return for their shipping costs (around £6) and maintenance. What drew planters and investors to hire them was a simple calculation that one man could tend two thousand tobacco plants, producing 5 hundredweight of product (560 pounds), which could be worth as much as £80 if of the 'best sort'.[23]

The result was that Jamestown became a prototype American boom town. The skeletal settlers of former years were replaced by a new generation of primped planters. 'We are not the veriest beggars of the world,' John Pory wrote. 'Our cow keeper here of James City on Sundays goes accoutred all in fresh flaming silks, and a wife of one that in England had professed the black art not of a scholar but of a collier of Croydon wears her rough beaver hat with a fair pearl hatband, and a silken suite therto correspondent.'[24]

Jamestown became notorious for the 'enormous excesses of apparel and drinking'. It was seen 'not as a place of habitation but only of a short sojourning'. The old fort burst through its decayed palisades and began to spread across the neighbouring land, the dirt streets lined with rough shacks and teeming with lonely men either flush with tobacco takings or down on their luck, looking for some distraction from hard work and often miserable living conditions.

Stern instructions were sent from London in 1621, ordering the governor and his council 'in particular that you be careful now in the beginning to suppress too much drunkenness, and that all kind of riot both in apparel and otherwise be eschewed'. Settlers even had to be instructed that they were not allowed to 'wear any gold in their clothes', for fear of the civil unrest, such wild displays of extravagance being considered provocative.[25]

Villages and farmsteads were springing up all over the plantations, but as purely temporary structures, so that their owners could profit as much as they could from the surrounding land, before pocketing their cash and returning to England. 'Their houses stand scattered one from another, and are only made of wood,' observed one visitor, 'few or none of them being framed houses but punches [posts] set into the ground, and covered with Boards, so as a firebrand is sufficient to consume them all.'[26]

On the river, it was a free-for-all, with as many as seventeen ships to be seen clustered like flies around the mouth of the James, offering contraband commodities and a variety of 'rotten wines which destroy our bodies and empty our purses' in return for tobacco. Alcoholic products proved particularly lucrative, provoking an outburst of entrepreneurial inventiveness and opportunism. A Mr Russell hawked 'an artificial wine of a [local] vegetable' made to a secret recipe, which he claimed to be an 'excellent preservative against the scurvy and other diseases, and infections'. One smuggler boasted that just four 'butts of wine' were enough to clear the costs of a whole voyage.[27]

Pory found it all impossibly vulgar, and yearned to escape back to Europe. The 'uncouthness of this place' was barely tolerable, he told a friend. He begged for 'pamphlets and relations', the merest scrap of home news to be dropped into a box held 'at the sign of the artichoke', the apothecary shop of George Yeardley's brother Ralph, 'to be sent to me by the first'. Meanwhile, he soldiered on in 'the best and choicest company' – his own – reassuring himself that 'among these Chrystal rivers, & odoriferous woods I do escape much expense, envy, contempt, vanity, and vexation of mind'.[28]

TWENTY-TWO

The Treasurer

On 28 April 1619, a hundred of England's most powerful nobles, knights and merchants, together with a few of London's more successful craftsmen, packed in to the Great Hall of Sir Thomas Smythe's offices at Philpot Lane. They were there for the Easter 'Quarter Court' of the shareholders of the Virginia Company. It was the most important meeting of the year, when the company's directors and officers, from the treasurer down, were elected.

A general hubbub ran round the room as the shareholders awaited the meeting's commencement. For many, it was a highlight of their calendar, a chance to mix with some of the most important members of England's elite. Smythe, the merchant prince, sat resplendent in his treasurer's robes at one end of the hall. Next to him were the earls of Southampton and Warwick, members of the King's Privy Council. Clustered around were assorted knights of the realm, including a substantial number of MPs, the most prominent and controversial being Sir Edwin Sandys.

The meeting was called to order, and an expectant silence fell upon the assembly. Smythe announced the news that many had already heard: having been a founding member of the Virginia Company, and its treasurer for twelve years, he had decided to step down.

There followed a ballot, in which Smythe's merchant friends and court supporters, led by Robert Johnson, were confident of a decisive result in their favour, ensuring the continuation of Smythe's cautious, politically neutral style of management. The nominations were duly announced, and votes cast by each shareholder selecting a ball in a colour representing his favoured candidate, which was dropped into a velvet bag.

When the balls were emptied out, they disclosed a shock result. Not only was Sir Edwin Sandys a clear winner, with fifty-nine balls, but Johnson could boast barely a handful. Hopes of consolation by being voted deputy treasurer were also dashed, as he was defeated by Sandys's close friend Nicholas Ferrar.

Further elections to various committees followed, and with each one it became clearer that the company had undergone putsch, away from Smythe's mercantile clique, to the parliamentarians led by Sandys.

Taking Smythe's place at the table, Sandys offered his predecessor a generous, if doubled-edged tribute, lamenting the 'much sorrow which Sir Thomas Smythe had endured, during the term of Twelve years past from the very infancy of the Plantation unto this present', and congratulating him for having surrendered his post 'at such time as (by the blessing of God) there was hope that the Action might proceed and prosper'.

A grateful company then voted to give Smythe twenty shares in recognition of his service, the generosity of the gesture appearing slightly less than magnanimous, given that the same number were then awarded to a captain who had spent five years in the colony, fulfilling a variety of relatively minor offices.

And on that ambiguous note, the meeting came to an end.[1]

Over the following months, Sandys's new council or board would meet 'daily and weekly' in the 'great hall and other best rooms' at the home of Nicholas Ferrar's father. These now became the company's headquarters, as it still could not afford its own.

To begin with, business appeared to proceed more or less as before. Fears that Sandys would use his position to advance his own political ambitions, and to promote Virginia as an independent 'commonwealth' rather than simply a company asset, appeared to be premature.

But then the cracks began to open. They centred on the magazine, which remained under the control of Smythe's associates. It was encouraging the planting of tobacco to the exclusion of all other crops, which in turn made the settlers ever more dependent upon it for supplies of food as well as tools and clothing. Being the scourge of monopolies, Sandys took it upon himself to wind the operation up, and allow the settlers to buy their supplies from whomever they wished.

Johnson immediately interpreted this as a provocation, and, over subsequent meetings, the antagonism between the two men escalated until 17 July, when they clashed violently, Johnson 'conceiving' that Sandys had 'threatened him in his own person'.[2]

Just as this 'mass of malignity', as Sandys called it, began to threaten the very future of the company, it was given yet more weight by the arrival on the scene of a warmonger: Robert Rich the Younger.[3]

Rich had recently become Earl of Warwick, inheriting the title purchased by his father for £10,000. The new Earl was eager to exploit the influence that came with the promotion, and saw the Virginia Company as providing an opportunity.

Modelling himself on the great Protestant earls of Elizabeth's reign, in particular Leicester and Essex, Rich had decided to challenge James's appeasing policies, and foment war between England and Spain. He felt that the best way to do this was to follow his father's example of sending ships out on the pretext of catching pirates, and, under the colours of foreign dukes in dispute with Spain, attack Spanish shipping.

The result was mayhem across the seven seas. One of his ships, prowling the Red Sea, had given chase to a vessel which turned out to belong to the queen mother of the Great Moghul of India. A passing trading vessel belonging to the East India Company intervened to avert a major international incident, and impounded Rich's ship. This provoked the unabashed Earl to demand reparations from the company's governor, Sir Thomas Smythe.[4]

The legacy of this incident was a deep enmity between the two men, which had encouraged the Earl, through his agent and kinsman Sir Nathaniel Rich, to support Sandys's candidature as treasurer of the Virginia Company. After Sandys's election, Warwick had flaunted his colours by keeping regular attendance at the treasurer's side, standing by him when the argument over the magazine had broken out.

However, their relationship began to change when news came in of the Earl's involvement in more privateering activities in the Spanish Main.

In the autumn of 1619, the crew of a 'Flemish man-o-war of whom 4 were English and 2 Negroes' was found cast away in Somerset.[5] The ship may have been the *White Lyon*, returning from America, though whether the ship itself was lost, or the castaways had escaped from it, is unclear. Two Angolans in the depths of the Somerset countryside must have attracted official attention, and through them and the crewmates, snatches of a story about escapades in the waters off Mexico, and the '20 and odd negroes' landed in Virginia, may have reached Sandys's ears.

Soon after, Sandys received a letter from George Yeardley reporting rumours that the *Treasurer* 'had gone to rob the King of Spain's subjects by seeking pillage in the West Indies, and that this was done by direction from my Lord of Warwick'. Sandys's reaction was cold panic. If the King heard of this incident from the Spanish ambassador, Don Diego Sarmiento, Duke of Gondomar, it could mean the end of the Virginia venture.

Sandys moved quickly to limit the damage, reporting directly to the Privy Council the *Treasurer*'s 'offence against the Spaniard'. He also produced 'public letters' that showed the action had been disavowed by the colonists in Virginia, and 'so likewise the Council and Company of Virginia here joined in the utter disclaiming of the same'. 'It appeared also,' the report continued, that 'by a letter produced' for their Lordships, that the Spanish ambassador Gondomar 'had received satisfaction for the offence aforesaid'.[6]

The intervention seemed to assuage the Privy Council, but by satisfying one side, Sandys merely succeeded in provoking the other.

Rumours began to circulate that he was leaking information about Warwick's other privateering activities to the Spanish ambassador. Sandys denied this, but it was too late. Warwick and his supporters were enraged, and began to agitate for Sandys's dismissal.

When Sandys faced re-election at the Easter Quarter Court of the Virginia Company in May 1620, he knew the Warwick faction was out to get him, but had not anticipated the intervention of an even more formidable enemy: the King. James had sent a messenger to the meeting to announce that, 'out of an especial care & respect he hath to that Plantation', His Royal Highness had decided that only four names were to be nominated for the election of treasurer: Sir Thomas Roe, Alderman Robert Johnson, Maurice Abbott, Sir Thomas Smythe, 'and no other'. Since three were outspoken and passionate supporters of the fourth, and since all other nominations were expressly prohibited, the intent of the King's unprecedented intervention in the company's internal affairs could not be mistaken: it was to put an end not just to Sandys's treasurership, but his entire regime.

The messenger's announcement produced no immediate response. The reflex was to carry on as though nothing had happened. Sandys proceeded to deliver the annual state-of-the-colony address. He reported that there were now 1,261 settlers in the colony, that £9,830 had been raised, mostly through the lotteries, and £10,431 spent, most of it on shipping people and provisions and servicing debts. Taking into account various unsettled items, Sandys defiantly claimed that for the business done that year, 'he had not left the company to his knowledge one penny in debt', though there was between £8,000 and £9,000 still outstanding from Smythe's period as treasurer, whose accounts had yet to be signed off.

He acknowledged several problems, particularly the neglect of the public land at the expense of private plantations, and the excessive cultivation of tobacco and sassafras at the expense of developing other commodities.

After handing over accounts, invoices and a 'catalogue of the names of all persons sent this year' at the company's expense, he offered his thanks to his deputy Nicholas Ferrar and the company 'for their love in choosing him', and so surrendered his seals of office.

'Upon which this great and general Court found themselves upon a deliberate consideration of the matter at an exceeding pinch.' They could not decide what to do: disobey the King and 'incur suspicion of defect in point of duty' or obey him and suffer 'a great breach into their privilege of free election'.

Sandys and the many MPs who were members of the company were familiar with such dilemmas. Parliament had faced them since James's succession, the divide between duty and liberty becoming ever harder to reconcile with each session. And in true parliamentary style, it was agreed that the best way to proceed was to postpone a decision until the next Quarter Court, and appoint a committee, 'to determine of an humble answer unto his Majesty'. Sandys would, in the meantime, continue in his post.[7]

Meanwhile, Sandys privately fumed. A month later, he wrote a long letter to George Villiers, James's new court favourite, recently elevated to Marquis of Buckingham, objecting to the 'late boastings of Sir Thomas Smythe and his partisans' of having set the King against 'my self and my service'. He appealed to Buckingham to intervene on his behalf.[8] But when the King was consulted on the matter, he reportedly took 'great exception' to Sandys 'as principal man that had withstood him [the King James] in Parliament and traduced his government'. 'In a furious passion', he told his advisers he would give 'no other answer but "Choose the Devil if you will, but not Sir Edwin Sandys"'.[9]

In the face of such implacable hostility, Sir Edwin had no choice but to relinquish the post. But he was a wily political operator, and immediately set about reshaping this defeat into an opportunity. If he could influence his succession so that he was replaced by a close ally, then there was hope of assuaging the King while keeping control. Only one man had sufficient authority and status to overcome Smythe's looming presence: Sandys's patron, the flamboyant, idiosyncratic Privy Councillor, Henry Wriothesley, the Earl of Southampton. A few days after being rebuffed by Buckingham, Sandys wrote to Southampton, pleading with him to step into the breach.[10]

At the following Great Quarter Court, held on 28 June, the secret deal Sandys had managed to broker emerged into the open. Sandys surrendered his office once again, and Southampton revealed that the King had 'graciously condescended' to the court holding a free election, withdrawing his nominations and instead merely advising that the shareholders might choose 'such a one as might at all times and occasions have free access unto his royal person'. Southampton was duly elected by general acclamation, before the Smythe faction could organize any sort of counter-offensive.

Sandys's opponents thus succeeded in unseating their enemy, but not his regime. For the remainder of the year, Sir Edwin retained a powerful influence in the company's affairs, even daring to describe the colony publicly as a 'Commonwealth and State', a dangerous challenge to the government's view of it as a possession of the Crown, beyond the reach of Parliamentary scrutiny.[11]

In the early months of 1621, Sandys successfully arranged for Sir Francis Wyatt, a friend and relation by marriage, to replace Yeardley as colonial governor, and started negotiations for a new royal charter to replace that of 1612. No copy of the draft document survives, but his use in November of the term 'commonwealth' hints at its politically charged agenda, as well as an intention to cement the reforms already in place in the colony, such as the presence of elected representatives in the General Assembly, and the institution of a system of civil rather than martial and religious law, enforced through local courts.[12]

Such manoeuvres were undertaken against a background of growing political nervousness. Abroad, there were sinister signs of a new European war brewing, sparked by the decision of James's son-in-law, the German Prince Frederick, to accept the throne of Bohemia (modern Czech Republic), bringing the Protestant German states into direct contention with the Catholic Holy Roman empire. In November 1620, Frederick's forces, deprived of the English support he had asked of his father-in-law, were roundly defeated, forcing him into exile with his wife, James's daughter Elizabeth. Meanwhile, James was entering tentative negotiations to marry his surviving son and heir Prince Charles to the Spanish Infanta. These two developments made his policy of non-alignment with either the Protestant cause or Catholic imperialism increasingly difficult to sustain, and unpopular among his subjects.

At home, farmers were suffering the effects of a freezing winter. Ice floes 'like rocks and mountains' with a 'strange and hideous aspect' blocked the Thames, and before long, food prices began to soar.[13] A series of corruption scandals had also intensified public disgust at the conduct of James's court, and his marital plans for Prince Charles had provoked the publication of violently anti-Spanish pamphlets, many focusing on Ambassador Gondomar's closeness to the King.

War, famine and political unrest only added to the pressure of James's still dire financial position, and he was forced to summon another Parliament, the first for seven years, in the hope of raising taxes. However, after so long in the political wilderness, MPs were not prepared to cooperate until some of their own grievances were dealt with. The catalyst was the issue of free speech. James had in his opening address voiced disapproval of the 'immoderate' talk in the Commons, which was taken as a challenge to the historic right of MPs to speak freely when in the chamber, immune from threats of royal intimidation or legal action. To test the King's intentions, there followed a series of provocative debates on this and other subjects, such as abuses at the royal court and free trade, the latter, thanks to the seventy-nine members of the Virginia Company with seats in the House, provoking particularly lively debate.

Such discussions began to guide the MPs into unknown and dangerous constitutional waters. George Calvert, a member of the Virginia Company since 1609 and now the King's Secretary of State, argued that Parliament had no jurisdiction over colonies as they were the King's direct possessions by right of conquest, and not subject to the laws of England. Sandys countered that, under the royal charters, the land was held privately by the company like any other property in the kingdom, which meant it 'may be bound by the Parliament'.[14]

James rapidly lost patience. He became convinced that, rather than representing the general view, these motions were the work of a core group of conspirators, 'which in number being twelve' was attempting to 'guide the whole House' against him. One of the King's officials reported that he had evidence of this group meeting regularly at the Earl of Southampton's house in Holborn, with Sandys acting as its leader.[15]

It was probably in response to this news that, during a meeting between King James and Spanish ambassador Gondomar, one or the other made the infamous remark that the Virginia Company had become 'a seminary for a seditious Parliament'.[16]

At the end of May, after months of wrangling, George Calvert arrived in the Commons with the news that the MPs had expected, but for which they were not prepared: that James would dismiss them in a week unless they produced what he had asked for. This provoked Sandys into one of the least measured but most effective speeches of his career. 'All things in the country are out of frame,' he lamented. He raged at the government's failure to support French Huguenots displaced by religious strife on the Continent. He bemoaned the corruption of trade by monopolies, how markets and fairs had come to a stand, how farmers were being driven into poverty. 'I had rather speak now than betray my country in silence; if ploughs be rested, cattle unsold, grazing decay, trade perish, what will follow but confusion?'[17]

On 16 June, there was a knock on the door of Sandys's home. It was the Sheriff of London, who had come to take him into custody in the name of the King.[18]

TWENTY-THREE

The 'Viperous Brood'

GEORGE SANDYS, Sir Edwin's younger brother, arrived at Jamestown in October 1621 aboard the *George.* A scholar as well as a gentleman, he had distracted himself from 'the roaring of the seas, the rustling of the shrouds, and clamour of sailors' endured during the voyage by translating two volumes of Ovid's Latin poetry. Sandys had intended to continue with the work after his arrival, but soon found himself overwhelmed by his official duties. He had been appointed to the new post of colonial treasurer, responsible for managing the company's financial affairs in Virginia under the new governor, Sir Francis Wyatt.[1]

Soon after his arrival, Sandys set off to look for a suitable place to establish the 200-acre plot granted to him under his terms of employment, and build a home.

Jamestown was as unimpressive as ever, little more than a collection of wooden hovels, many of which lay beyond the defensive palisade. Across the isthmus connecting the island to the mainland lay the Governor's Land, farmed by a hundred tenants to help pay towards the cost of running the colonial government. A beaten path leading north-west was probably by now Virginia's first permanent road, passing through the blockhouse defending the entrance to the island, across a narrow neck of land between the sandy beach overlooking the James and a swampy creek, and on to the private plantations scattered haphazardly beyond.

The English population had boomed. More than three and a half thousand had been shipped out in forty-two ships since 1619, though it appears many had illegally returned home, offering London-bound skippers promises of rich rewards for their safe delivery. Those who stayed were joined by an increasingly cosmopolitan collection of labourers and tradesmen, Angolans, Armenians, Italians engaged in developing the new glassworks, Frenchmen planting vines for wine production, Germans and Poles manufacturing pitch and tar. George Sandys also saw many Indians wandering freely in and out of the English plantations and towns, 'Eating, Drinking and Sleeping' amongst the settlers.[2]

Sandys staked out his plot somewhere in the vicinity of the Governor's Land, and, with the help of the few indentured servants, proceeded to clear the land and prepare it for crops or grazing. By January 1622, construction of a water-wheel was under way, to complement the windmill built by his neighbour, the former governor Sir George Yeardley.

On 3 March, he wrote a letter home so sanguine in tone, it was quickly reproduced in the Virginia Company's propaganda literature. He was hopeful for the various industries being developed, such as the iron and salt works, and experiments with commodities such as pomegranate, potatoes and silkworms. He also noted that, on the south side of the James, 'there is a fruitful country, blessed with abundance of corn, reaped twice a year'.[3]

Sandys found relations with the Powhatan Indians in good shape, allowing planters to spread themselves and their crops and cattle safely across the land. In July 1621, Yeardley had concluded a peace treaty with the military chief Opechancanough, which promised not only 'a great trade and commerce with [the Indians] hereafter for Corn and other Commodities' but 'good means also for converting them to Christianity and to draw them to live amongst our people'. This agreement came on top of active efforts to establish a college upriver of Jamestown, where Indian children would be educated in the Christian religion, reading and writing, and technical trades. In London, this venture had attracted generous charitable donations of books and land, as well as money, including an anonymous gift of a chest of 'new gold' valued at £550. A central figure in this effort was George Thorpe, who had adopted one of the Indian children brought to London with Pocahontas. He had arrived in Virginia in 1620 to oversee the founding of the college, and provoked murmuring in some quarters for being 'too kind and beneficial' to the Indians, especially Opechancanough, 'to whom he oft resorted, and gave many presents'. In an attempt to demonstrate the material benefits of the English lifestyle, he paid for Opechancanough to move from his 'cottage, or rather a den or hog-sty', as one unsympathetic English commentator described it, 'made with a few poles and sticks, and covered with mats after their wild manner' to a 'fair house according to the English fashion'. When it was completed, Opechancanough had expressed 'such joy, especially in his lock and key, which he so admired, as locking and unlocking his door an hundred times a day'.[4]

For his part, Opechancanough appeared eager to keep friendly relations with the English, if only to secure his own position in the face of continuing threats from the Monacan people beyond the falls. In October of 1619, Yeardley had been approached by Opechancanough's flamboyant lieutenant Nemattanow, known to the English as Jack of the Feather 'by reason that he used to come into the field all covered over with feathers and swan's wings fastened unto his shoulders as though he meant to fly'.[5] In Gates's time, Nemattanow had fought the English as part of Powhatan's efforts to keep them away from the falls, and had earned a reputation for being 'invulnerable, and immortal, because he had been in very many Conflicts, and escaped untouch'd from them all'.[6] But on Opechancanough's behalf, he now asked for 'some 8 or ten English with there Arms to assist him in battle against a people dwelling about a day's Journey beyond the Falls'. The request had been granted.[7]

George Sandys, and the new governor Francis Wyatt, found that the English were still 'in very great amity and confidence with the natives'. Eager to sustain the friendly mood, Wyatt had sent George Thorpe to present the governor's compliments to Opitchapam, the spiritual leader of the Powhatan people, and Opechancanough.

Thorpe had returned with the news that Opechancanough was eager for closer ties, with English families being encouraged to live among the Indians and vice versa. Thorpe had even detected that the great warrior 'had more motions of religion in him' than had been expected, and accepted 'natural principles' over superstitious ones, acknowledging even that the worship of Okeus and such rituals as the Black Boys ceremony might not be 'the right way'. Thorpe also established that, following a tradition that continued to mystify the English, Opechancanough had changed his name to Mangopeesomon.[8]

Soon after this meeting, Nemattanow had appeared at the plantation of one Morgan, possibly a relation of Edward Morgan, a London brewer, who had received shares from Sir Thomas Gates in 1620.[9] Morgan supposedly possessed 'several toys' or ornamental goods for which he would get some 'mighty bargains' if he hawked them among the towns along the Pamunkey river, Nemattanow said. This persuaded Morgan to follow Nemattanow into the woods.

Some days later, Nemattanow returned to the plantation wearing Morgan's cap. There he found two of Morgan's servants, 'sturdy boys' who asked after their master. 'He frankly told them he was dead.' Suspecting 'the villain had killed their master', the boys attempted to arrest him, and in the ensuing struggle, Nemattanow was shot. According to one account, as he was about to expire, he asked the boys not to reveal how he had died, and to bury him among the English, to preserve the mystique of invulnerability among his countrymen.

When Opechancanough heard of the incident, he was reportedly 'much grieved and repined', but reassured the English of his continuing amity 'with the greatest signs he could of love and peace'.[10]

It was in this atmosphere that Richard Pace, a settler living on the riverside opposite Jamestown, welcomed an Indian hunter on 21 March 1622. The hunter had come to see an Indian boy called Chanco, employed by the tobacco planter William Perry, who lived with Pace 'as a son'. The visitor was invited in and, after a supper shared with his hosts, spent the night with Chanco in the pantry or one of the other cramped rooms of Pace's dwelling. As they lay side by side, the visitor whispered to Chanco that a surprise attack against the English was planned for the following day, and he urged Chanco to slay Pace there and then.

The boy got up, armed himself and left the room. But rather than slay his master, he warned Pace of what was about to happen. 'Pace upon this discovery, securing his house, before day rowed over the River to James-City,' a 3-mile journey. He ran into the fort and warned Governor Wyatt.[11]

This was not the only warning that the English received from Indian allies, allowing Jamestown and nearby plantations to be put on alert. Messengers were dispatched to inform the settlements further up the James, in particular Henrico and nearby Charles City, and for those nearer the bay such as Martin's Hundred, but the news came too late.

That morning, Indian workers and traders came into the undefended English settlements, some to share breakfast with their hosts as they discussed deals and fixed prices, others to take up tools and work in the fields with the labourers. More than usual were in evidence – so many that some had been forced to borrow English boats to cross the river – but suspicions were not aroused.

Then, at a prearranged time 'a little before noon', the hostilities began. According to the inevitably one-sided English reports that survive, the Indians drew out hidden weapons or picked up discarded tools, and 'slew most barbarously' any settler they could get their hands on, 'not sparing either age or sex, man, woman or child'. So sudden was this 'execution, that few or none discerned the weapon or blow that brought them to destruction'. Any who tried to run away, were shot down with arrows.

> In which manner they also slew many of our people then at their several works and husbandries in the fields, and without their houses, some in planting Corn and Tobacco, some in gardening, some in making brick, building, sawing, and other kinds of husbandry, they well knowing in what places and

> quarters each of our men were, in regard of their daily familiarity, and resort to us for trading and other negotiations . . .[12]

Several prominent settlers were among the casualties. One was Captain Nathaniel Powell. 'A valiant soldier', Powell had come to Virginia with the first settlers in 1607. He had acted as a councillor, served under Captain John Smith during his exploration of the bay, survived the Starving Time, searched for survivors of the Roanoke venture, been at Smith's side when he had sustained his injuries. The Indians now 'not only slew him and his family, but butcher-like haggled [hacked] their bodies, and cut off his head, to express their uttermost height of cruelty'.

George Thorpe, having recently returned from his meeting with Opechancanough, was tipped off by his servant of the impending attack, but 'out of the conscience of his own good meaning, and fair deserts ever towards them, was so void of all suspicion' that he refused to escape. The 'viperous brood' then fell upon his house, where they 'not only wilfully murdered him, but cruelly and felly [fiercely], out of devilish malice, did so many barbarous despites and foul scorns after to his dead corpse, as are unbefitting to be heard by any civil ear'.[13]

John Rolfe also died around this time, though whether at the hands of the Indians, or of disease, is unclear.

The roll call of less illustrious names was much longer: a Pole called Matthew; the tinker at Captain Spence's house; Henry the Welshman; 'Remember Michel', probably a Huguenot; 'Master Thomas Boise, & Mistress Boise his wife, & a sucking Child'; Francis the Irishman; William Perigo, probably an Italian glassmaker; the maidservants Mary and Elizabeth; 'one old Maid called blind Margaret' who lived at Richard Owen's house; and, most anonymous of all, a man identified only as 'Man', possibly a German.[14]

The assault was carefully coordinated and aimed at extinguishing the entire colony. While groups rampaged through Henrico and Charles City, more than 60 miles away at 'Bennetts Welcome', a plantation established just a month before on the south side of the James, another band set to work. They first descended upon the isolated house of one Baldwin. Baldwin's wife appears to have been outside the house on her own when the Indians arrived, and was so badly wounded, 'she lay for dead'. Baldwin then rushed out of the house with a gun, and 'by his oft discharging of his piece', managed to scare off the assailants.

The Indians moved on to a house half a mile away belonging to George Harrison. It was targeted because Ralph Hamor, one of the colony's most senior captains, was believed to be staying there with his brother Thomas, while his own home was being constructed nearby. The attackers knocked on the door, which was answered by a servant. They said they had come to escort Captain Hamor to Opechancanough. When the servant said the captain was away, they persisted, trying to inveigle their way in 'with many presents and fair persuasions'. When these attempts failed, they set fire to a nearby tobacco barn, luring men out of the house to quench the flames, 'whom the Salvages pursued, shot them full of arrows, then beat out their brains'.

Thomas Hamor had remained in the house, putting the final touches to a letter to be sent back to England on a waiting ship. Disturbed by the hubbub outside, he put down his pen and went to find out what was causing the commotion. As soon as he stepped out of the house, he saw the Indian attackers. An arrow pierced his back as he turned to take refuge in the house. He and a servant desperately barricaded the door, and the Indians set fire to the house to try to flush them out. In panic, a servant let off a pistol 'at random', which caused the Indians to scatter, and provided an opportunity for the injured Thomas Hamor and others to escape to Baldwin's house, where they found the distraught owner tending to his fatally wounded wife.

Meanwhile, the Indians attacked another house, killing all the inhabitants, before retiring back into the woods. There they encountered Ralph Hamor, who was preparing to embark on a hunt. Realizing his life was at risk, the captain fled to his own half-built house, and, with the help of his men armed 'only with spades, axes, and brickbats', launched a counter-attack, which he managed to sustain long enough to repulse the attackers.[15]

> And by this means that fatal Friday morning, there fell under the bloody and barbarous hands of that perfidious and inhumane people, contrary to all laws of God and men, of Nature & Nations, three hundred forty-seven men, women, and children, most by their own weapons; and not being content with taking away life alone, they fell after again upon the dead, making as well as they could, a fresh murder, defacing, dragging, and mangling the dead carcasses into many pieces, and carrying some parts away in derision, with base and brutish triumph.[16]

Thanks to Pace's early warning, Jamestown emerged from the attack unscathed, and it was decided that the rest of the survivors should withdraw from their scattered settlements to the island, along with their cattle. Thus, in a matter of a few hours, Virginia was once again reduced to the state in which it had found itself during the Starving Time, besieged upon the tiny island, unable to tend to the fields that had been cleared, or to graze the cattle they had raised.

The immediate problem was feeding this influx. Being the 'seed time', stores were at their lowest. To make matters worse, the planters, in their efforts to maximize the number of acres devoted to tobacco cultivation, had become almost totally dependent upon the Indians for corn supplies. In 1620, they had even requested that the London Company send beads and other tradeable goods in preference to food, 'to increase and maintain thereby a Christian Commerce and trade with the Savages'.[17] Efforts were under way to plant more fields with wheat and barley, but they had been hampered by a lack of good-quality seed.[18]

Now, with hundreds of men, women and children jostling in the fort, cows crammed into the gardens surrounding the palisade, and ships arriving almost by the day with fresh batches of settlers, survival hopes rested on finding new sources of sustenance. Options were severely limited. One proposal was to launch a counter-attack on the Indians to seize their supplies, even though few were to be had at that time of year. Another was to send a message back to London post-haste, asking the company to organize a speedy supply of food, which would have to be financed out of the company's own coffers, as the settlers had no tobacco to pay for it.

George Sandys took a leading role in both efforts. He drafted a letter to be sent to London describing what had happened and laying out the colony's urgent needs. He also led a foraging expedition to Quiyoughcohannock, later memorialized in a popular ballad:

> Stout Master George Sandys upon a night
> Did bravely venture forth
> And mong'st the Savage murderers
> Did form a deed of worth
> For finding many by a fire
> To death their lives they pay
> Set fire of a town of theirs
> And bravely came away.[19]

Young Richard Frethorne had the misfortune to be among the new settlers shipped into the grisly aftermath of these events. Little is known about his background, though informed guesswork suggests that he came from the county of Gloucestershire, in the west of England. His father was possibly a local merchant or tradesman, or even a member of the minor gentry, fallen upon hard times following a famine in 1621 or the decay of the local manor. The family was well enough connected to be on correspondence terms with Robert Bateman, a successful London merchant, MP and prominent member of the Virginia Company in London.[20]

Finding young Richard unsettled at home, his father had accepted, possibly through Bateman, an undisclosed sum to put Richard into the service of William Harwood, a member of the Council of Estate in Virginia who was also governor of Martin's Hundred.[21] Harwood undertook to ship Richard out to Virginia, where he would work for one of the planters in Martin's Hundred.

Frethorne had arrived some time in late 1622 or early 1623 to find the plantation at its lowest ebb, barely recovered from the assault of the previous March. Martin's Hundred had been among the worst hit. Of the one hundred and twenty settlers living there at the beginning of 1622, just twenty-two had survived, crowded into the two houses that remained intact, 'and a piece of a Church'.

The colony was by now engaged in bitter reprisals, with soldiers destroying crops before the Indians got the chance to harvest them, leaving local granaries empty.[22]

Vengeance may have been sweet, but it was not nourishing, and by the autumn of 1622, Jamestown was overcome by a dreadful famine from which it would not recover for months. George Sandys gave a picture of the situation in a letter he sent home soon after Frethorne's arrival. The previous October, yet another consignment of settlers had been shipped in 'with not above a fortnight's provision'. Just eleven of perhaps fifty survived through to the following year. Sandys managed to place five with plantation owners, leaving four on his own hands, all of whom tried to run away to the Indians. Only two were successful, the others succumbing to an infectious sickness now rampaging through the settlements.

To make matters worse, there was hardly any tobacco to pay for imported supplies, 60 hundredweight being 'the most that this year's crop hath produced' – a third less than the quantity of 'good' tobacco shipped back in 1619, when prices were much higher and commercial planting only just beginning in earnest.[23]

It was in these conditions that Richard sat down early in March 1623, and wrote a letter to his patron Robert Bateman setting out how he was 'in a most miserable and pitiful case both for want of meat and want of clothes'. He explained that he was awaiting emergency provisions that were due to be sent aboard the *Sea Flower*. Furthermore, the colony was in a critical condition, as 'at every Plantation all of them for the most part were slain and their houses and goods burnt', survivors being taken away by the Indians and 'kept alive'.

Governor Harwood estimated that it would cost more than £3,000 to repair the damage done to Martin's Hundred by the Indian attack, but that year they could plant 'but a little tobacco' to pay for it. Instead they must endure the condition, and plant 'corn for bread' to make up for the shortfall in supplies, 'and when we have done, if the Rogues come and cut it from us as they have sent all the plantations word that they will . . . then we shall quite be starved'.[24]

A month passed, and still the *Sea Flower* failed to appear on the horizon, so Frethorne picked up his quill again, and wrote to his 'loving and kind Father and Mother':

> This is to let you understand that I your child am in a most heavy case by reason of the country, [which] is such that it causeth much sickness, [such] as the scurvy and the bloody flux and diverse other diseases, which maketh the body very poor and weak.

There was nothing there to comfort him, no food but peas and 'loblolly', ship's gruel. 'As for deer or venison I never saw any since I came into this land. There is indeed some fowl, but we are not allowed to go and get it,' presumably because security prevented hunting parties from leaving the plantations.

Instead, he was forced to 'work hard both early and late for a mess of water gruel and a mouthful of bread and beef'. Day and night, people cry out 'Oh! That they were in England without their limbs' rather than have to endure such misery.

They lived in fear of the Indians 'every hour'. 'We are in danger, for our plantation is very weak by reason of the death and sickness of our company.' They had lost eighty souls in the Indian attack, half of the twenty merchants who had accompanied Frethorne on his voyage over had since died of disease, and two more were expected to succumb any moment.

'And I have nothing to comfort me, nor is there nothing to be gotten here but sickness and death.' Those with private means were better off. They could buy butter and beef from the smugglers at Kecoughtan, at least that was what Frethorne's 'fellows' told him. As for himself, 'I have not a penny, nor a penny worth, to help me to either spice or sugar or strong waters.'

All that had kept him alive was the help of 'Goodman Jackson', the gunsmith at Jamestown, and his wife, 'he like a father and she like a loving mother'. One of Frethorne's tasks was to help deliver consignments of tobacco and other commodities grown on the plantation by barge to Jamestown, 10 miles upriver of Martin's Hundred. This would involve an overnight stay, sometimes in the boat, and 'if it rained or blowed never so hard, we must lie in the boat on the water and have nothing but a little bread'. One day, Frethorne had taken refuge in the fort, where he met Goodman Jackson. The gunsmith took pity on the lad, and made him 'a cabin to lie in always when I [would] come up, and he would give me some poor jacks [fish] [to take] home with me, which comforted me more than peas or water gruel'.

In their conversations, Jackson would marvel that Frethorne's parents would 'send me a servant to the Company. He saith I had been better knocked on the head.'

And so the catalogue of adolescent self-pity continued, until he came to the nub of the matter: 'If you love me,' he wrote, 'you will redeem me suddenly, for which I do entreat and beg.' In other words, he wanted his parents to buy out his indenture so that he could return home. 'And if you cannot get the merchants to redeem me for some little money, then for God's sake get a gathering or entreat some good folks to lay out some little sum of money in meal and cheese and butter and beef.'

Realizing the latter plea to be more realistic, he gave detailed instructions on what should be sent and how it should be packaged.

Oil and vinegar were 'very good', but liable to leak out of containers on the journey over. If they sent cheese, 'it must be very old cheese', the sort sold at the cheesemongers for twopence farthing or a halfpenny a pound. It was to be packed into a barrel with 'cooper's chips', wooden slats, between each cheese 'or else the heat of the hold will rot them'. Finally, the barrel must be clearly addressed to 'Goodman Jackson, at Jamestown, a gunsmith', as 'there be more of his name there'.

If he was to die before the cheese came, Frethorne added plaintively, 'I have entreated Goodman Jackson to send you the worth of it, who hath promised he will.'

With pledges of love to his friends and kindred, his brothers and sisters, he signed off with a final plea to his father to reply, as 'the answer of this letter will be life or death to me'.

In the copy of the letter that has survived, he appended a list of 'the names that be dead of the company came over with us to serve under our Lieutenants', identifying thirteen men by name, and 'a little Dutchman, one woman, one maid, one child'.[25]

Nothing more is heard of Frethorne. He is not mentioned in any other correspondence or official papers, and his name does not appear in a census taken the following year. Perhaps his parents stumped up the money to bring him home, or perhaps he succumbed to the epidemic of disease afflicting the colony.

His letter came into the hands of John Pory, at whose suggestion the roll call of casualties was added at the end. Pory would have been expected to show it to Governor Wyatt, as all letters were supposed to be vetted before being sent home. Perhaps he did so, and fearing it might be censored, made a copy for himself, which he packed up with some of his own correspondence to be sent on the next ship to his patrons Sir Nathaniel Rich and the Earl of Warwick. It was thanks to this small act of subterfuge that Frethorne's sorry story, unlike its author, survived, and that the Virginia Company was brought a step closer to its dissolution.[26]

TWENTY-FOUR

The Unmasked Face

THE VIRGINIA COMPANY'S 'Great and General' Easter Quarter Court of 1622 took place two months after the Indian attack on the colony, on 22 May. Despite this, no one in London yet knew of the catastrophe.

The first morning session of the court was sparsely attended. The current treasurer, the Earl of Southampton, was absent. Otherwise, the names of just twenty-one members of the company were recorded in the official minutes, and heading the list was Sir Edwin Sandys.

Following his arrest the previous June, Sandys had disappeared from public view for several months. He was still nowhere to be seen when a somewhat nervous Parliament had reconvened in November 1621, prompting his fellow MPs to start asking questions. Eventually, on 21 November, the Speaker had produced a letter from Sandys, in which the MP said he was detained by sickness and asked for the house's leave to remain away until he had recovered.[1]

This sparked off an open debate about whether or not Sandys had been detained after the last session, 'for parliament business' which would be an infringement of parliamentary privilege. No, the King's Secretary of State Calvert answered emphatically, and it was suggested that the house should stop talking about Sandys and revert its attention to the matters in hand. However, some of Sandys's supporters persisted, claiming to be 'pinioned', as they did not know if an open discussion of the delicate issues they were expected to discuss would lead to them facing the same treatment as their former and absent colleague. Sir Dudley Digges opened a sensitive debate on foreign affairs by saying that he hoped his speech could navigate the dangerous straits between the 'sands of the one side, and the rocks of the other' – clearly a pun on Sir Edwin's name.

A few days later, Sir William Spencer, the Earl of Southampton's son-in-law, could restrain himself no longer. Announcing that Calvert's assurances could not be relied upon, he insisted that the MPs needed to know directly from Sandys whether or not he had been arrested for what he had said in Parliament. Spencer was backed by another MP, who described Sandys's detention as 'notorious throughout the kingdom', and a 'breach of our privilege'. If the MPs did not defend themselves from such breaches, 'we deserve to be hanged'. A delegation of three MPs was duly appointed to go and interview Sandys directly on the matter.

The delegation quickly tracked Sandys down to his London home at Holborn, to where he had just returned from his country home in Kent. 'By the fire side, at ten of clock at night, the servants being put out', they discussed Sandys's months of confinement.

He must have struck them as a somewhat dejected figure. He felt he was 'frowned upon by so great persons, and I would that to frown on me were the worst they meant me', as he had put it in a letter to Nicholas Ferrar's brother John.[2]

Following his arrest the previous June, he had spent five weeks in detention, and had been interrogated about a number of issues, most of them to do with parliamentary business. He had been asked whether he had agitated against the King's command for Parliament to be dismissed, and whether he had encouraged others to thwart the King's desire to see 'some few bills to be passed'. His interrogators also wanted to know 'what conference he had' with particular members of the House of Lords, probing for evidence that he was conspiring with the Earl of Southampton.

While he was answering these questions, his home had been searched for incriminating papers, provoking more questions. One concerned a theological paper 'found in his closet, begun to be penned by him' about the powers of 'kings of the earth . . . and of their power and right' which he 'was willed to explain'.

They also found a letter from 'one Mr Brewer, living in Amsterdam'. This was Thomas Brewer, who had been involved in organizing the *Mayflower* expedition along with Sandys's correspondents Robert Cushman and John Carver. His interrogators were anxious to know when he had received it, what answer he had made, and whether or not he was involved in any other correspondence with the 'Brownist' exiles in Holland.

In the event, the interrogators had been unable to establish anything incriminating, and Sandys was released on orders from the Privy Council to remain 'confined to his said house [at Northborne, Kent] and within five miles compass of the same until further order be given by his Majesty'. In November, 1621 in an attempt to lower the political temperature prior to the new parliamentary session, James released a number of high-profile captives, including Sandys and the long-suffering Henry Percy, Earl of Northumberland, who had remained in the Tower since his arrest for complicity in the Gunpowder Plot.[3]

Sandys had decided to stay in Kent until the agitations of the House of Commons drew him back to the capital, and within weeks of his return he found himself thrust back into the fray not just of parliamentary politics, but the Virginia business.

The Quarter Court of 22 May 1622 was to provide Sandys with an opportunity to test the intensity of royal displeasure towards the company since his return to active membership. The morning's business was taken up with the company accounts, which, like the company, were in a grim state. One of the consequences of the government's antagonism towards Sandys was that, just before his arrest, the Privy Council had issued an order banning the lotteries, on the grounds that they had been subject to fraud and abuse. This had deprived the company of its main source of income. Sandys might have hoped that he would be able to reinstate the lotteries in a new charter, but that effort had also foundered with his arrest.

With each setback, another obstacle appeared. The lack of a new charter meant there was no chance of reinstating the company's right to trade tobacco free of duty, which had expired in 1619. Now, all imports were subject to swingeing customs duties, justified on the grounds of royal disapproval of smoking.

Meanwhile, the Virginia Company had to compete with a revived New England project, led by Sir Ferdinando Gorges. In 1620, the King had awarded Gorges the patent for settling lands between 40 and 48 degrees of latitude, roughly from modern Philadelphia to the northern tip of Maine. To the alarm of the Virginia Company directors, this meant his territory overlapped theirs, including the site of the fledging Plymouth Colony founded by the 'Brownists' or Puritans Sandys had been interrogated about. This new patent, in contrast to Virginia's, was aimed at settling the land according to feudal principles, with patents granted to nobles giving them not only ownership of the land in the King's name, but the right to govern it like a principality. There were suspicions in the Virginia Company that Gorges's New England venture 'had greater compliance with the King and Court interest, the more to divide the strength, and weaken the power of the Council and Company of Virginia'.[4]

Against this background, it was hard to see how the company's already stretched finances could bear the strain. According to the latest accounts, the income for the year just ended was £6,756 2s 3d. Together with the sums needed to pay off the debts arising from Sir Thomas Smythe's time, this meant there was a shortfall of £1,400. A further £800 was due to pay for the latest supply sent to Virginia. The only hope of making up the deficit was through the sale of tobacco, but, with tobacco profits under pressure, it was thought that current stocks were 'not likely to discharge but a little of the said Debts'.

The Quarter Court's afternoon session was much more crowded than the morning, as proceedings approached 'after their accustomed manner to the election of [the] Treasurer, Deputy and other Officers'. Also in the now accustomed manner, this prompted a royal intervention. A Mr Hamersley, alderman, stood up amidst the throng, and announced that he had come with a message from the Secretary of State George Calvert. Hamersley was not initially recognized, and he first had to explain his 'seldom coming to Court' which he blamed on the company's 'Officers negligent warning of him'.

He then proceeded to deliver his message, which he said he had received the night before from Calvert in person. The King, he announced, while not wanting 'to infringe the liberty of their free choice', urged the company's shareholders to consider one of five nominations for the post of treasurer. This time, Sir Thomas Smythe and his immediate circle did not feature, but a partisan flavour was still detectable in the list, with the King favouring merchants who had proved helpful with the royal finances.

Hamersley's message was taken as a sign of the King's benevolence 'unto the Plantation and of his gracious meaning not to infringe the privilege of the Company and liberty of their free election', and the shareholders held a vote of thanks. They then reinstated the Earl of Southampton, with Nicholas Ferrar continuing as his deputy.[5]

Southampton formally received his seals of office at a meeting held two weeks later. The mood was buoyant, with Southampton announcing that there were 'now greater hopes then ever of a flourishing State and Common Wealth in Virginia'. Lord Cavendish then revealed a more troubling item of news. He had been commanded to communicate to the company the King's response to Southampton's reappointment. It seemed His Majesty was affronted that, 'out of so large a number' of nominations 'by him recommended', the company had chosen not one. He wanted a merchant in charge of the company, 'instancing' as a good example Sir Thomas Smythe, 'in whose times many Staple Commodities were set up which were now laid down' in favour of tobacco. The King had a particular interest in the production of silk and wine, and was disappointed to find that, despite his interest and support (which included the donation of mulberry trees to feed the silkworms), nothing had been achieved.

This expression of royal displeasure was but the opening salvo in an apparently co-ordinated barrage of petitions and proposals that threatened to undermine the company's independence. Captain John Martin caused consternation by petitioning the King directly for an area of land encompassing 80 miles of woodland to become a royal forest, which Martin would manage on James's behalf. It transpired that the area concerned encompassed not only several plantations, but Jamestown itself, reducing this core area of the colony to a royal 'demesne'. The company thus found itself in the embarrassing position of having to deny the King his forest, which it did by accusing Martin of having 'ruined' his own land in the colony by making it a 'receptacle of vagabonds and bankrupts and other disorderly persons'.

But an even bigger threat was tobacco. With no other source of income, the company's addiction to it was becoming lethal. At the beginning of June, the King's Lord Treasurer Cranfield craftily proposed that the Virginia and Bermuda companies should come together and apply for a monopoly over imports of all tobacco, whether from Virginia or Spanish America. Despite having protested in Parliament about the abuse of royal monopolies, for the sake of the company Sandys and his fellow directors had no choice but to accept, which initiated protracted negotiations over the terms.[6]

In the midst of these delicate discussions and heated debates, a ship called the *Sea Venture*, named after the vessel wrecked at Bermuda, appeared at a London dockside from Virginia, its decks packed with returnees full of grim stories of the Indian attack. News spread quickly through the capital of what had happened, reviving the perennial speculation that the Virginia venture was once again on the threshold of extinction.

Any hope of a cool, pragmatic or even sympathetic response from the company was lost beneath panic and fury. A letter was sent back to Jamestown on 1 August 1622, castigating the colonists for having left themselves so vulnerable to attack. As for their urgent 'want of corn', it was a request that 'doth much perplex us' as the year before they had been forewarned that the company's funds were 'utterly exhausted', and that they would have to rely on their own resources. The company would send a ship sometime over the next few months, but instead of carrying corn, it would be loaded with some rusting arms together with 'four hundred young men . . . to repair with advantage the number that is lost'.[7]

In the late sixteenth and seventeenth centuries, the word 'massacre' had a special meaning for Protestants, holding the same political potency as 'holocaust' would for Jews in the twentieth. Until the 1560s, it had simply been the French word for a butcher's block. Following the brutal religious wars that broke out in France during that period, it acquired its modern meaning. It was used to describe the Spanish attack on Huguenots in Florida in 1562, but the crystallizing event was the infamous St Bartholomew's Day Massacre of 1572. The Catholic French monarchy was accused of conniving with Catholic extremists to slaughter thousands of Huguenots lured to Paris on the pretext of attending the wedding of their leader Henry of Navarre to the King's Catholic sister.[8] It was probably to show solidarity with the victims of that event that Captain Gosnold had been christened Bartholomew, and it would remain in the minds of many Englishmen as European Protestantism's defining event. Now, here was another massacre, resonant of those in Florida and Paris, with hapless Protestants being lured into peaceful co-existence, only to be slaughtered for their trust. Perhaps the fact that they had survived this massacre as they had the last was decisive evidence that the empire of Protestant civilization was now too powerful to be wiped off the map, and would one day cover the globe. This, as the great Huguenot geopolitical thinker Du Plessis had put it, was the 'very genealogy of the world'.[9]

However unjust it may now seem to call the Indians' attempt to reclaim their homeland a massacre, it was the word the Virginia Company and the settlers used to provide them with a readymade justification for their past behaviour and future actions. These were set out in the company's official response to the attack: *A declaration of the state of the colony and affairs in Virginia, With a relation of the barbarous massacre in the time of peace and league, treacherously executed by the native infidels upon the English*, by Edward Waterhouse. The book also provided a list of casualties, so 'that their lawful heirs, by this notice given, may take order for the inheriting of their lands and estates in Virginia', a clear indication that, as far as the company was concerned, this did not mark the end of the venture, but a new beginning.

The central argument of Waterhouse's work was that the 'massacre' was good for the colony, just as a blood-letting will 'make the body more healthful'. 'Our hands which before were tied with gentleness and fair usage, are now set at liberty by the treacherous violence of the Savages,' he argued. They were no longer constrained by the company policy of appeasing the Indians, they were no longer required to buy land from the Indians 'at a valuable consideration to their own contentment'. Instead the colonists 'may now by right of war, and law of nations, invade the country, and destroy them who sought to destroy us . . . Now their cleared grounds in all their villages (which are situate in the fruitfullest places of the land) shall be inhabited by us, whereas heretofore the grubbing of woods was the greatest labour.'

The English, in other words, felt they no longer had any responsibility to civilize these 'rude, barbarous, and naked people', but must conquer them, 'by force, by surprise, by famine in burning their corn, by destroying and burning their boats, canoes, and houses, by breaking their fishing weirs, by assailing them in their huntings, whereby they get the greatest part of their sustenance in Winter, by pursuing and chasing them with our horses, and bloodhounds to draw after them, and mastiffs to tear them, which take these naked, tanned, deformed savages, for no other than wild beasts'.[10]

As Waterhouse anticipated, the rush of blood seemed to revive the venture's weakening pulse. John Donne delivered a famous sermon proclaiming himself 'an adventurer, if not to Virginia, then for Virginia'.[11] Michael Drayton, the author of the stirring 'Ode to the Virginian Voyage', was roused again, reminding readers in his monumental 'Poly-Olbion' of Virginia's Elizabethan roots. It was in the battles 'against the Iberian rule' and in Holland's 'sure defence', that Elizabeth had sent her ships

> . . . to that shore so green,
> Virginia which we call, of her a Virgin Queen.[12]

Such efforts, combined with tough economic conditions at home, kept up recruitment, even after news of the 'Massacre of March 22' had broken. Not all, or even most of those who signed up were willing participants. John Hagthorpe, a friend of Captain John Smith's, reflected the circumstances of many when he wrote that 'suits of law . . . compel me to transport myself and my family into Virginia'.[13] Nevertheless, when the company had written to Wyatt and his council in Virginia that there were 'many hundreds of people' ready to leave for America, it was not exaggerating.[14]

While the company's public face was thus sanguine, the body was full of bile. Negotiations over the tobacco contract, begun in a mood of cooperation, became paralysed by delays and prevarications. By December 1622, Sandys had been promised a final draft of the contract, but when he went to enquire about its progress, he discovered that 'the Lord Treasurer's going to Newmarket to the King, had caused that to be forgotten'.[15]

Many shareholders took this as a sign that it was now every man to himself, and Sandys's once sturdy body of support among the company's membership began to disintegrate, each meeting becoming more quarrelsome than the last.

The following April 1623 the company discovered that a report had been presented to the King with the threatening title 'The Unmasked face of our Colony in Virginia'. Its author was Nathaniel Butler, the Earl of Warwick's close associate, who had been governor of Bermuda. He had visited Virginia that winter, at the time of Frethorne's arrival, when the colony was at its lowest ebb following the Indian attack. The 'Unmasked face' was his withering assessment of the colony's condition. He had found the plantations to be 'seated upon mere Salt Marshes full of infectious Bogs and muddy Creeks and Lakes'. The settlers' houses were worse than 'the meanest Cottages in England'. There was not 'the least piece of fortification', Henrico and Charles City had been 'left to the spoil of the Indians'.

Perhaps the most devastating charge was that, of the 'not fewer than ten thousand souls transported thither . . . there are not through the aforenamed abuses and neglects above two thousand of them at the present to be found alive'. The former figure was an exaggeration, the latter a bone of contention, but the evidence for mass mortality was nevertheless compelling. Furthermore, Butler argued that if the problems faced by the colony were not 'redressed with speed by some divine and supreme hand, it instead of a Plantation will shortly get the name of a slaughter house, and so justly become both odious to ourselves & contemptible to all the world'.[16]

These ominous words turned out to be the first salvo in a concerted attack upon the Sandys administration launched by the implacable Warwick. It also marked the beginning of Virginia's erasure from popular history as the birthplace of English America. Since being reported by Sandys to the Privy Council over the *Treasurer*'s piratical activities, the Earl of Warwick had been quietly preparing his retaliation. The reports of the Indian attack and its aftermath now provided him with the opportunity to unleash its full force. Besides publicizing Butler's report, he released a portfolio of letters and papers documenting the company's troubles, which had been assiduously gathered by the Earl's agent Sir Nathaniel Rich. Exhibit A was Richard Frethorne's moving and plaintive letter to his parents. In addition, Sir Nathaniel had managed to get hold of the private correspondence of Sir Edwin's brother George, including embarrassing letters to John Ferrar lambasting the 'very weak council' in the colony, and to Sir Thomas Wroth, Sir Nathaniel's brother-in-law, which revealed that 'the living have been hardly able to bury the dead through [the London Company's] imbecility'.[17]

On 7 May 1623 at a 'very full' company meeting, the consequences of Warwick's efforts were revealed. The government had decided upon the 'dissolution of the [tobacco] contract', ending all hopes of staving off the company's financial crisis through the import of tobacco. The company was also told that Smythe's sidekick Alderman Johnson '& those others that had opposed the Contract' had 'delivered unto his Majesty a very bitter & grievous petition against ye government & carriage of the Company these four Last years'.[18]

A royal commission was set to investigate the charges. To aid its inquiries, the company's directors were ordered to surrender 'all Charters Books (and by name the blurred Book or Books), Letters, Petitions, Lists of Names and Provisions, Invoices of Goods, and all other writing whatsoever, and Transcripts of them, belonging to them'. These were duly delivered and, under mysterious circumstances, all those relating to the Smythe era disappeared soon after.[19] The commission also sought written testimony from various witnesses, including the company's most prominent critics.

Only one submission survives: that of Captain John Smith. The captain's relations with the company had deteriorated since his return from Virginia. In 1621, short of money and desperate to raise finance for a rival colony in New England, he had approached the company's directors, in the hope of being compensated for his services in the early years. At the Virginia Company's Great Quarter Court of 2 May, 1621 he had presented a petition, claiming that during his time in Virginia he had 'discovered the Country and relieved the colony' and asked in return that the company would 'reward him either out of the Treasury here, or out of the profit of the generality in Virginia'. His request was referred to a committee, and appears to have been denied if it was not simply ignored.[20]

In the light of this experience, the captain had little reason to be generous, and this was reflected in his blunt testimony to the commissioners. The first question he was asked, according to his own account of the proceedings, was why the plantation 'hath prospered no better since you left it in so good a forwardness'. Because of 'idleness and carelessness', he replied, which had 'brought all I did in three years . . . to nothing'. In his view, the root of the company's problem was its open and public structure. Having begun with 'but six patentees', it was now owned by 'more than a thousand' shareholders. This made for a 'multiplicity of opinions' and officials, preventing decisive action. He proposed that the company be dissolved, and taken over by the King. His Majesty should then sack all those officers, and send over a squad of soldiers together with some labourers to sort out the mess. Such measures could be paid for by a tax – 'a Penny upon every Poll, called a head-penny; two pence upon every Chimney, or some such collection' – levied in England.[21]

The commission appears to have been heavily swayed by the captain's testimony, and by July it was actively investigating ways of recalling the company's patent.

This prompted all-out war between the Sandys and Warwick factions. Despite an order from the Privy Council warning them to avoid 'all bitterness and sharpness', their antagonism grew 'so violent' that 'they seldom meet upon the exchange or in the streets but they brabble and quarrel'. The strength of feeling goaded Sandys into penning a vicious attack on Warwick's conduct, which led to him and his associates being 'restrained of their Liberty, and confined to their several Lodgings or Houses, as persons guilty of Contempt against the directions and commands of [the Privy Council], where they are to remain until His Majesty or this Board shall give further order'.[22]

Sandys was released soon after, but the disputes continued, and climaxed at a meeting of Bermuda investors when Sandys 'fell foul of the Earl of Warwick'. Lord Cavendish sprang to Sandys's defence, prompting Warwick to challenge Cavendish to a duel. To avoid the King's intervention, the two men arranged to meet in Holland, and on 17 July made their way for the Sussex coast. Cavendish fell ill, and was detained by the county sheriff, but Warwick, disguised as a fisherman, made it as far as Ghent, where a royal command summoned him back to England.

In the view of the newsmonger John Chamberlain, the feuding had become so intense 'that if that society be not dissolved the sooner, or cast in a new mould, worse effects may follow than the whole business is worth'.[23] This turned out to be the view of the government, and in November it acted. The King exercised one of his most powerful royal prerogatives, a *quo warranto* (Latin for 'By what authority?'). This gave the Privy Council virtually limitless powers to investigate the company's affairs, interrogate its officers, and, if it so decided, rescind its charter.[24]

In a last desperate attempt to save the situation, Sandys drew up a petition appealing to Parliament for help. Proclaiming the venture to be 'not simply matter of trade, but of a higher nature', the petition called upon Parliament to protect 'this Child of the Kingdom, exposed as in the Wilderness to extreme danger and as it were fainting and labouring for life'. The nature of that protection was too sensitive a matter to set out explicitly, beyond a vague invitation that MPs should hear a 'full relation of those oppressions and grievances' the company had suffered.[25]

Sandys's petition was presented to the House of Commons on 26 April 1624, and was referred for consideration by a committee. On the morning of 29 April, the Commons received a letter from the King, forbidding in the most respectful but threatening terms the House from considering Sandys's petition. 'By general Resolution', it was withdrawn. With so much else at stake, there was no stomach to make Virginia the issue over which the MPs would confront the Crown.

A month later, the Virginia Company was dissolved, and King James formally took it over, along with the tobacco trade, which he made a royal monopoly. He delegated the company's management to a commission or council of royal appointees, which included Sir Thomas Smythe and Alderman Johnson. It was to meet every Thursday afternoon 'at Sir Thomas Smythe's in Philpot Lane, where all men whom it should concern may repair'.[26]

And so, in London, Virginia's brief political effervescence, the hope of turning it into a new 'Commonwealth and State' under Parliamentary control, seemed to dissolve away into Smythe's quiet, amoral, pragmatic corporatism. But in America, such hopes were not so easily quashed.

On 27 May 1623, the *John & Francis*, the veteran privateering vessel that had brought Virginia's first supply in 1607, arrived at Jamestown. One of those who expectantly gathered at the quayside for the ship's arrival was Robert Bennett, with his sixteen-year-old son or nephew Richard.

Robert was one of three brothers from a Puritan family of London merchants.[27] Around 1620, his elder brother Edward, a prominent grocer, author of a pamphlet in support of the Virginian tobacco trade, and the Virginia Company's 'largest adventurer', had been awarded a patent to set up a private plantation at Warraskoyack, on the south bank of the James river (modern Burwells Bay).[28] Two years later, Edward had sent out one hundred and twenty settlers, including several members of his family, under the command of Captain Ralph Hamor, to found the new plantation, named 'Bennetts Welcome' in honour of its benefactor.

Just a month later came the Indian attack. A total of fifty-three of the planters were killed that day, forcing the plantation to be abandoned. Nine months later, the situation was even worse. In a letter home, one survivor wrote that just ten of the 'men and boys' who had come with him from England were still alive.[29]

In the spring of 1623, Edward had commissioned the ships *John & Francis* (100 tons) and *Godsgift* (80 tons) to sail to Virginia with twenty-two fresh recruits. The *Godsgift* had been sent first, carrying brother Robert and young Richard to start work on the reconstruction effort.[30]

The *John & Francis* had followed, and when Robert Bennett was handed the ship's manifest, he was delighted to discover that it was loaded with supplies: 19 butts of 'excellent good wines', 750 jars of oil, 16 barrels of raisins, 18 barrels of rice, 2 half-hogsheads of almonds, 3 half-hogsheads of wheat, 18 hogsheads of olives, 5 firkins of butter and a cheese. As well as food, he also found a chest and two barrels packed with candles, three packs of linen carrying brother Edward's 'mark' and 'two dryfats of Mr King's' cloth.

The delivery put Robert in optimistic mood, reflected in the letter he sent by return to his brother, confirming delivery of the goods 'safe and well conditioned'. 'The Fort is abuilding apace,' he reported. Food was still in short supply, but 'our men stand well in their health, God be thanked', he added.

Robert had already conducted a survey of Bennetts Welcome, and was convinced the territory covered by his brother's patent to be 'the best estate in all the land'. Despite official warnings against returning to outlying territories, a crop of tobacco and corn had already been planted in the fields surrounding the remnants of the settlement. Robert promised Edward that, by the next ship, his agent in London 'shall receive a good parcel of Tobacco from me with good profit'. He was also confident that by the following year he would have 'set out all this land, and houses' at Bennetts Welcome, making it a plantation capable of sustaining two or three hundred men.[31]

Not everyone was as sanguine as Robert. Dephebus Canne sent a letter a month later in which he reported that the settlers were still 'destitute of food, and they pray for relief'. 'Would to God that the apparel and frieze [woollen cloth]' recently imported 'were turned into meal, oatmeal and peas'.

However, even Canne was forced to admit that 'the weather has been good and seasonable'. Ships were expected daily with supplies of fish, among them the *John & Francis*, which after unloading her cargo of goods for the Bennett planters had been sent on a fishing expedition off the coasts of Newfoundland and Canada. There were also 'great hopes' of a good harvest of both corn and tobacco that autumn.[32]

This feeling of recovery and opportunity was joined by a sense of outrage when news reached the Chesapeake of Nathaniel Butler's and Alderman Johnson's attacks on the state of the colony and on Sandys's regime. The colony's Council of Estate was made up of Sir Edwin Sandys's appointees, including his brother George, and was bound to leap to its patron's defence. However, the entire General Assembly, including the elected burgesses, seemed united in a feeling of collective indignation when it met in April 1623 to discuss the issue. Suspicions about the King's interference in the matter became so intense that a few months later, the Privy Council was forced to send a message from London reassuring the colonists that the 'King has no other intention in reforming and changing the present government of Virginia, than the remedying of bad effects that tend to endanger the whole population' and that 'every man's estate shall be fully preserved'. Despite such reassurances, the colonists continued to press for guarantees that the status quo be maintained, including the parliamentary form of government instituted by Sandys. The 'slavery' they had suffered during Smythe's time 'has been converted into freedom', they boldly informed the Privy Council, and they begged that they 'may retain the liberty of their General Assembly. Nothing can more conduce to our satisfaction or the public utility.'[33]

The General Assembly was equally forthright when it came to formulating an official response to its critics in London. Johnson was dismissed as Smythe's stooge, 'inseparably chained' to the 'offences and infamies' of Smythe's regime, during whose rule the colony 'for the most part remained in great want and misery, under most severe and cruel laws'.[34]

Nathaniel Butler's 'Unmasked face' provoked similarly dismissive remarks. His depiction of the colony on the threshold of extinction was clearly biased, and distorted by his failure to take into account recent events:

> The time that this informer came over, was in the winter, after the massacre; when those wounds were green, and the earth

> deprived of her beauty. His ears were open to nothing but detraction, and he only enquired after the factious, of which there were none among us, and how he might gather accusations against those in the government, being, as it should seem, sent over for that purpose.

But if the beauty of prose is any measure of the feelings that gave rise to them, it was his remarks about the territory itself that caused the most consternation, and in particular his reference to the colonists living on 'Salt Marshes full of infectious Bogs and muddy Creeks and Lakes'.

> In this he traduceth one of the goodliest rivers in the habitable world, which runs for many miles together within upright banks, till at length, enlarged with the receipt of others, it beats on a sandy shore, and imitates the sea in greatness and majesty.[35]

This was the reaction of a people who felt they were no longer visitors but inhabitants. Whatever their mistakes, whatever problems they faced, whatever the rights and wrongs of their occupation of Indian territory, this land was their land. It was this feeling which seemed to inspire Robert Bennett's optimism, and it must have been infectious, as young Richard Bennett quickly took to his new home. Within five years he had become a member of the colony's council, aged just twenty years old. In 1652, he became Virginia's first Puritan governor, steering the colony through a tumultuous period of English history, when the antagonisms wrestled with by Sir Edwin Sandys and his supporters culminated in Parliament rising up and executing James's son and successor King Charles I. Charles's replacement was a new republican regime of a sort never before seen in Europe, and the term it adopted to describe itself was that used for the colony and since by the state of Virginia: a commonwealth.[36]

The defining moment of the colonists' act of possession was the 'Indian massacre'. Through their response to it, they crafted and honed their American identity. Robert Bennett became aware of this soon after his arrival, when a group led by Captain William Tucker took twelve men up the Potomac to collect some English held by the Indians, 'and withal in colour to conclude a peace with the great King' Opechancanough. Terms were agreed, and concluded upon the banks of the river with speeches at which Tucker proposed a toast to peace. Several of the assembled Indians drank the toast, and were killed by a poison specially prepared by the colony's physician, a Dr John Potts. A battle ensued, during which the English claimed to shoot forty or fifty Indians, including, Tucker mistakenly believed, Opechancanough himself.[37]

When news of this action reached England, there was disapproval of the underhand manner of the attack, and calls for Dr Potts to be disciplined for supplying the poison. In Virginia, however, such qualms were swept aside. 'Whereas we are advised by you to observe rules of justice with these barbarous, perfidious enemies, we hold nothing unjust that may tend to their ruin,' Governor Wyatt told the Virginia Company in London.[38]

There followed a series of coordinated assaults aimed at clearing the Indians off the land surrounding that 'goodliest river' the James. The campaign began in the autumn, following the Indians' harvest, with Bennett, under Tucker's command, attacking Warraskoyack, the territory surrounding Bennetts Welcome. He took the Indians' corn and destroyed their houses. By the following spring, he had re-established the plantation, and settled thirty-three men there.[39]

The following spring, the council in Virginia sent a report on its progress to the Virginia Company, which continued to meet at Nicholas Ferrar's house, even as it was in its final death throes. 'We have to our uttermost abilities revenged ourselves upon the Savages having upon this river, cut down their Corn in all places which was planted in great abundance upon hope of a fraudulent peace, with intent to provide themselves for a future war,' the colonists boasted, 'burning down the houses they had re-edified [rebuilt], and with the slaughter of many enforced them to abandon their plantations.'[40]

And so Virginia was taken. Her 'lovely cheeks' had been 'lately blushed with Virginian-English blood', noted Samuel Purchas, in a sermon commemorating the conquest.

> And thy blush being turned to indignation, thou shalt wash, hast washed thy feet in the blood of those native unnatural Traitors, and now becomest a pure English Virgin; a new other Britain, in that new other World: and let all English say and pray, GOD BLESS VIRGINIA.[41]

Among those whose blood had been shed in this brutal act of purification was Henry Spelman. As part of the campaign waged against the Indians that autumn of 1623, he had taken a small bark carrying twenty-six men up the Potomac. Having reached deep into the interior, near a place called Nacotchtanck, he went ashore. As he disappeared into the trees, some of the local Pawtuxunt people came alongside the bark in their canoes. They started climbing aboard, causing such alarm among the English soldiers that one of them let off his gun. The Indians leapt overboard, and were 'so distracted with fear, they left their canoes and swam ashore'.

The Indians climbed up the river bank, and followed Spelman into the trees. From the boat, the English soldiers heard a commotion, which ended with Spelman's severed head being hurled from the trees, and landing on the bank.[42] In panic, the soldiers heaved up the anchor and sped off towards the Chesapeake Bay as fast as their oars and sails would take them.

Spelman's head had come to rest in a 'low pleasant valley from whence distill innumerable sweet and pleasant springs'. Beyond stretched a place of 'mighty Rocks', tinctured by a 'spangled scurf'.[43]

According to the Indian tradition, Henry Spelman's spirit would have risen from his remains, and climbed to the top of a nearby tree. From there could be seen 'a fair plain, broad pathway' stretching east-wards to the afterworld, 'on both sides whereof doth grow all manner of pleasant fruits, as mulberries, strawberries [and] plums'. The pathway was crowded with future generations of Indians, walking away from their ancestral lands, 'toward the rising of the sun, where the godlike Hare's house is', to join their forefathers, living 'in great pleasure in a goodly field, where they do nothing but dance and sing and feed on delicious fruits with that Great Hare, who is their great god'.[44]

Displacing them was an influx of traders and planters, Catholics and Puritans, adventurers and refugees, indentured servants and African slaves, wave after wave coming upon the land, clearing it, manuring it, fencing it in, building upon it, fighting over it. To the south, along the Piscataway, the Diggeses and Calverts, whose ancestors had battled in Parliament over the colony's future, established their vast plantations of tobacco. Along the opposite shore of the Potomac lay the Little Hunting Creek, which became Mount Vernon, home of the Washington family. And to the north, where the river branched, a great diamond-shaped plot of marshy land was cleared of its trees, scattering wildlife into the surrounding hills. There, a complex of white buildings, colonnades and boulevards suddenly arose, a new Rome. It was the 'Federal City' of a new republic, Washington DC.

The words echoing in Spelman's ears as he left England were from the Book of Genesis, quoted by the Reverend Symonds in a sermon to send the fleet on its way: 'I will make thee a great nation'. And so, around the site where he had fallen in 1623, after the shedding of much blood and the ruination of many lives, it came to be.

NOTES

The primary sources relating to the early years of the Virginia venture are widely scattered and often hard to find, so where possible the notes refer to recent and accessible editions, notably Edward Wright Haile's excellent *Jamestown Narratives*, which includes all of the main texts with modernized spellings, and a useful Who's Who of the main participants.

BL	British Library, London.
Bodleian	Bodleian Library, Oxford.
Cecil	Manuscripts of Robert Cecil, First Earl of Salisbury, at Hatfield House, Hertfordshire.
CSP	*Calendar of State Papers*, HMSO.
DNB	*Dictionary of National Biography*, Oxford University Press, 1885–1900.
HMC	Historical Manuscripts Commission.
LOC	Library of Congress, Washington D.C.
New DNB	*Oxford Dictionary of National Biography*, Oxford University Press, 2004.
PRO	Public Records Office, Kew, since 2003 officially renamed the National Archives.
RO	Records Office.
SP	State Papers (held at the PRO, see above).

PROLOGUE: THE GREAT WHITE FLEET

1 Paul Embler's diary is held in the Manuscripts Department of the Library of the University of North Carolina at Chapel Hill, ref 5122-z. I am grateful to Paul A. Embler for granting permission to quote from the diary, and to Ron Branson of the County History Preservation Society (www.countyhistory.com) for his help in acquiring the transcript.

2 Still, 1956, p. 260.

3 *New York Times*, April 27, 1907, pp. 1–2.

4 http://www.npl.lib.va.us/sgm/oldlobby/archives/james.html.

5 Morris, E., 2002, p. 502; Zimmermann, 1998.
6 *New York Times*, Dec. 16, 1907.
7 Tonnage figures taken from the US Naval Historical Center, http://www.history.navy.mil/. Morris (2002), p. 501, puts the tonnage as 348,000, perhaps because he included auxiliary ships in the estimate.
8 'Indian' is used in this book to refer to native Americans. It is the term chosen by representatives of the Native Americans of Virginia to refer to themselves and their tradition. See King (2000), p. 222, for a recent academic discussion of this issue.

PART ONE

1 *A Feast of Flowers and Blood*

1 Quinn, 1979, vol. 2, pp. 356, 372.
2 Davenport, 1917, pp. 60–61.
3 Quinn, 1979, vol. 2, p. 372.
4 Quinn, 1979, vol. 2, pp. 384–9.
5 Quinn, 1979, vol. 2, p. 396.
6 Quinn, 1979, vol. 2, p. 374.
7 Quinn, 1979, vol. 2, p. 374.
8 Quinn, 1979, vol. 2, p. 448.
9 Quinn, 1979, vol. 2, p. 400.
10 Quinn, 1979, vol. 2, p. 462.

2 *Machiavelli*

1 Aubrey, 1962, p. 254.
2 Handover, 1959, pp. 275–6; Nelson, 2003, p. 36; Husselby, 2002.
3 Handover, 1959, pp. 230–1.
4 Jardine, 1999, p. 326.
5 Quinn, 1955, pp. 613–4.
6 Cecil MS 119, folio 6; Quinn, 1974, pp. 241*ff*; Quinn, 1979, vol. 5, doc. 785.
7 Elliott, 2002, p. 286.
8 *Eastward Hoe*, Act III, scene iii, ll. 61–2, 140–4; Chapman, 1914, pp. 835–6; Drummond, 1842, p. 20.
9 HMC Salisbury, vol. 18, p. 84 ; Quinn, 1979, vol. 5, doc. 784.
10 Hakluyt, 1599, vol. 3, pp.1, 306, 258, 349.
11 Handover, 1959, pp. 275–6.

3 *The Adventurers*

1 Quinn, 1979, vol. 5, pp. 192–3.
2 Konig, 1982, pp. 357–8.
3 Barbour, 1969, doc. 1; HMC Salisbury vol. 9, p. 269; vol. 18, pp. 133–4.
4 Kingsbury, 1906, vol. 1, pp. 107*ff*.
5 Barbour, 1969, p. 34. The idea that Smythe summoned a meeting on the day the 'Instructions for Government' received the royal seal is supposition.
6 Barbour, 1969, pp. 37–8.
7 Haile, 1998, p. 868; Barbour, 1969, p. 47.
8 Smith, 1986, p. 213.
9 Wingfield, 1993, especially pp. 7–76. There are numerous references to Wingfields in Fox's *Actes and Monuments*, e.g. see Fox, 1576, book 9, p. 1304; book 8, p. 1202
10 Stow, 1631, pp. 1,017–8; Gookin, 1949, p. 185.
11 Purchas, 1625, p. 1,649.
12 Smith, 1986, vol. 1, p. 203.
13 Brown, 1890, doc. LXXXV, 'Extract from Fishmongers' Records. April 24, 1609'. This lists 'those persons of this Company that have before this day adventured to Virginia' and mentions a William Day as 'Adventurer with Capt. Gosnell, Capt. Archer & Timothy Lodg'. Gosnold had left for Virginia by December 1606, and did not return, so Day must have negotiated the investment with Gosnold and Archer before then. Timothy Lodge is unidentified.
14 The papers relating to the Isle of Wight incident only mention a 'Captain Gosnold' (Cosnoll, Gosnall), omitting his first name. In the Calendars of State Papers (*CSP Dom – James I* (Edward VI and Mary, vol. 9, pp. 1611–18)), this captain is assumed to be Bartholomew's uncle Robert. A study of all the documentation to which the calendars refer (Hatfield MSS 190/96–7, PRO SP 14/9/56 etc.),

and other papers, suggests otherwise. Robert was unlikely to have had the title 'captain' before 1609, when he was granted the office of Keeper and Captain of St Andrew's Castle, Hampshire, in reversion after William Christmas (*CSP Dom – James I*, p. 496). A 'known companion' of this Captain Gosnold that evening was Bowyer Worsley. The Worsleys were a prominent local family with interests in Virginia. A Richard Worsley, probably Bowyer's younger brother, was among those who signed up to go to Virginia under the first patent (Smith, 1986, vol. 1, pp. 222–3). Bowyer applied for a patent to set up a private plantation in Virginia in 1622 (Kingsbury, 1906, doc. CCXXXVI). Isle of Wight RO, MSS Oglander, OG/TT/10; Hampshire RO MSS Daly (Southwick and Norman Court Estates) 5M50/338; Brown, 1890, p. 1,060.

15 Bartholomew's aunt Elizabeth was the granddaughter of the Wingfield patriarch Sir Anthony, and several of Sir Anthony's other grandchildren had estates neighbouring those of the Gosnolds. A Henry, possibly Bartholomew's uncle, left to posterity a reminder of the link between the two families by scratching his name on a window pane at Otley Hall, the Gosnold country seat. (Gookin, 1963, p. 18.)

16 LOC Irene Wright MSS Box 7, 'Christopher Newport and William Parker'.

17 Bovill, 1968, p. 148; Read, 1960, p. 439.

18 Smith, 1986, vol. 1, p. 85.

19 Purchas, 1614, p. 830; *New DNB*, 'Charles Leigh'.

20 *Henry IV Part 1*, Act 2, scene iv, ll. 326–7; Act 5, scene iv, ll. 122–3.

21 Nicholls, 1992, p. 312.

22 Batho, 2000, p. 43; *New DNB*, 'George Percy'; Barbour, 1971, pp. 7–9.

23 Smith, 1986, vol. 2, p. 420.

24 Smith, 1986, vol. 3, pp. 153, 271; Beckwith, 1976.

25 Smith, 1986, vol. 3, pp. 153–4. Barbour argues in his biography of Smith that he had meant to write 'His parent's dying', i.e. 'With the death of a parent', meaning his father (Barbour, 1964, p. 8).

26 Barbour, 1964, p. 58.

27 Smith, 1986, vol. 3, pp. 186–7.

28 Purchas, 1614, p. 830; *New DNB* 'Leigh, Charles'. Smith may also have been introduced to the Virginia venture via a Bristol connection. Thomas Packer, the main beneficiary of Smith's will, married Joan, daughter of Richard Smyth, a Bristol alderman who had extensive links with that city's colonial adventures. A Richard Smyth, eldest son of John Smyth of Epping, migrated to Bristol, while a brother called Nicholas went to Lincolnshire, to live near where John Smith was brought up. A Robert Smith (the spelling is incidental), brother of Nicholas, was headmaster of the Free Grammar School of King Edward VI in Louth, and Barbour speculates that John Smith's family was related to that of Robert and Nicholas. This suggests that John might have kept in touch with his Bristol relatives, to the extent that he befriended their acquaintance Packer. Given Richard's links to Bristol's colonial activities, he could have recommended Smith to Hakluyt (who was based in Bristol around 1600) or one of the grandees associated with the Virginia venture. (Smith, 1986, pp. *lvi–lvii*, doc. *iv*; Bristol RO ref AC/WH/3/1, dated 25 January 1609/10.) Smith's time with Robert Bertie in London is mentioned in a letter written by Bertie's wife Elizabeth to him from Lincolnshire, in which she refers to missing 'Mr Smith' (Lincolnshire RO, MS 10ANC/* Lot 340/1). Peregrine the Elder's link to the Roanoke

venture is mentioned in a letter from John Stubbe to Lord Willoughby while he was in Denmark, dated 15 November 1585 (Lincolnshire RO 8ANC1/38; Quinn, 1955, p. 222).

4 *Departure*

1 Smith, 1986, vol. 3, p. 64.
2 Barbour, 1969, p. 59.
3 The idea of a navigable route through North America to the Pacific was not as far-fetched as it may seem. In Russia, a river system had recently been discovered by the English navigator Anthony Jenkinson that connected Archangel, Russia's most northerly port, to Persia, via the Dvina and Volga. (Quinn, 1955, p. 264; O'Mara, 1982, p. 5.)
4 Barbour, 1969, doc. 4.
5 Reports of the numbers taken to Virginia in the first expedition vary from one hundred to one hundred and eight. The precise figure is probably one hundred and five, of whom one died en route, though this cannot be verified. (Bernhard, 1992, p. 601.)
6 Rabb, 1998, p. 13.
7 *The eyght tragedie of Seneca*, John Studley, London, 1566; *Treason pretended against the King of Scots by certaine lordes and gentlemen*, Christopher Studley, London, 1585.
8 BL MS Lansdowne 88 folio 51.
9 Smith, 1986, vol. 1, pp. 208–9. Smith gives a slightly different list of names in *Generall Historie* (Smith, 1986, vol. 2, pp. 140–42).
10 The departure date was 19 December, according to Smith (Smith, 1986, vol. 2, p. 137), Saturday 20 December according to Percy (Haile, 1998, p. 85). The dates are Old Style.
11 Haile, 1998, p. 85.
12 There has been controversy over the size of the ships and the numbers that they carried. Percy wrote that the fleet carried eight score in total, i.e. one hundred and sixty (Haile, 1998, p. 88). Purchas specifies a more precise one hundred and forty-four, a total of seventy-one on the *Susan Constant*, fifty-two on the *Godspeed* and twenty-one on the *Discovery* (Bernhard, 1992, pp. 599–601; Camfield, 1994; Bernhard, 1994). No plans or descriptions have survived for any of these ships, so estimates of their dimensions vary. Those given are taken from a rule of thumb used by Thames shipwrights in the mid-seventeenth century to calculate the tonnages of ships: $t=((k*b)*(b/2)/94)$, where t is the tonnage, k the length of the keel, and b the beam or width of the ship at its widest point. The total length of the ship (excluding the bowsprit) would typically be 1.6 times the length of the keel. (Bushnell, 1664, pp. 8, 59, 62.); Thompson (1959).
13 Haile, 1998, p. 200.
14 Miller, 1948, pp. 495–6; Smith, 1986, vol. 3, p. 296.
15 Smith, 1986, vol. 1, p. 212.
16 Renshaw, 1906, p. 54; Barbour, 1969, doc. 6.
17 Smith, 1986, vol. 2, p. 137. Percy's brother Henry was known as the 'Wizard Earl' for his interest in unorthodox ideas and association through Harriot with Raleigh's 'school of atheism'. Wingfield later faced accusations of being an atheist, as it was subsequently discovered that he had not brought a bible with him to America. It is possible, though less likely, that it was Percy who had learned of Hunt's misdemeanours, as Heathfield was in the diocese of Chichester, which was near the Percy country home at Petworth.
18 Smith later described Martin as 'very honest'. (Smith, 1986, vol. 1, p. 33.)
19 Wingfield refers to the accusation that he named Smith as being part of the 'intended and confessed mutiny of Galthropp' (Haile, 1998, p. 200). This may refer to a later incident. See

also Smith, 1986, vol. 2, p. 139 and *n* for a discussion of this episode.

20 Smith, 1986, vol. 2, p. 139 and *n*; Haile, 1998, p. 85*n*.

21 Haile, 1998, p. 86.

22 Casas, 1583, ¶2[r].

23 Taylor, 1935, p. 309. Purchas picked up this theme. In a marginal note accompanying his edition of Smith's *Proceedings*, he claimed Spanish actions had produced 'desperate depopulations', though also thought English 'gentleness' had made the Indians 'proud'. (Smith, 1986, vol. 1, p. 237.)

24 The account of the outward voyage and arrival in America is based on Percy's account (Haile, 1998, doc. 1), unless otherwise specified.

25 Smith, 1986, vol. 3, p. 236.

26 Smith, 1986, vol. 1, p. 29.

27 Quinn, 1979, docs 473–4.

28 Brown, 1890, doc. 4.

29 Barbour, 1969, doc. 8.

30 Brown, 1890, docs 14–15.

31 Barbour, 1969, doc. 9.

32 HMC Salisbury, Part 18, p. 452.

PART TWO

5 *Tsenacomoco*

1 Sidney, 1595, B3[v].

2 Jefferson, 1905, vol. 2, p. 132; Smith, 2002, pp. 87*ff*.

3 The following is a retelling of Indian history based upon various ethno-historical, archaeological and anthropological sources, principally Strachey (Haile, 1998, doc. 41, mainly 'caput 7', pp. 651–61), Spelman (Haile, 1998, doc. 29), Uttamatomakkin (Haile, 1998, doc. 55) and Smith's *Map of Virginia* (Smith, 1986, vol. 1, pp. 160*ff*). The history draws upon Gallivan (Gallivan, 2003, especially chap. 8) who develops the theory of a change in social dynamics towards the development of powerful chiefdoms in the Late Precontact period. Rountree provides a full list of other sources on Powhatan Indians (Rountree, 1989, pp. 3–6), and summarizes what they contain. Besides Rountree, the main secondary sources on Indian culture are: Gleach, 1997; Hantman, 1990; Kupperman, 2000, especially pp. 115*ff*.

4 Haile, 1998, pp. 569–661.

5 Strachey was the first to mention Tsenacomoco as the Indian term for the land of the Virginian Indians (Haile, 1998, doc. 41, p. 598). Its exact translation is disputed. According to the philologist Professor James A. Geary in Quinn's *Roanoke Voyages* (Quinn, 1955, vol. 2, p. 854), the word, which some have translated as 'long house people', probably comes from the terms for 'close together' and 'land dwelt upon'. This has been accepted as the most likely meaning, e.g. in Gallivan (Gallivan, 2003, p. 11).

6 On the derivation of 'Chesapeake' see Quinn, 1955, vol. 2, pp. 854–5.

7 Gallivan discusses the likely impact of climate change in the 'Late Woodland' (*c.* AD 900–1500) period (Gallivan, 2003, pp. 17–19; Kupperman, 1982; Blanton, 2000).

8 Haile, 1998, p. 881; Rountree, 1989, pp. 69–70.

9 Smith, 1986, vol. 2, p. 121. Okeus is particularly well documented in contemporary English ethno-historical accounts, see Strachey (Haile, 1998, pp. 645–6), Uttamatomakkin (Haile, 1998, pp. 881–2), Smith in his *Map of Virginia* (Smith, 1986, vol. 1, p. 169) and *Generall Historie* (Smith, 1986, vol. 2, pp. 125, 144) and Beverley (Beverley, 1947, p. 198).

10 Haile, 1998, pp. 595–6.

11 *Manitu* is discussed by Gleach (Gleach, 1997, p. 40). Gallivan (Gallivan, 2003, pp. 160*ff*) writes of the changes in Powhatan society just

before the 'contact period' (when Europeans first arrived), observing through a close study of archaeological data a transition from the 'peer polities' of Powhatan villages, which emphasized group activity, to chiefdoms, with a high degree of social inequality, which occurred around the time of, but was not necessarily caused by, the arrival of Spanish and English explorers and settlers.

12 Hantman, 1990, p. 680; Gallivan, 2003, p. 155; Smith, 1986, vol. 2, p. 184, vol. 1 p. 165.

13 The location of Powhatan is near modern Richmond, Virginia. Gleach's (Gleach, 1997, p. 33) preferred translation of Powhatan is 'one who dreams', but most others have adopted the earlier translation by Trumbull (Trumbull, 1870, p. 10).

14 Gerard, 1904, p. 314; Lederer, 1672, p. 4. Smith refers to two words (or, more likely, different pronunciations of one) which the Powhatans apparently used to refer to the English: Uttasantasough (Smith, 1986, vol. 2, p. 130) and Tassantessus (Smith, 1986, vol. 2, p. 246), which he assumed to mean 'stranger'. The latter is used by Strachey (Haile, 1998, p. 616).

15 Strachey (Haile, 1998, p. 614); Gallivan, 2003, p. 163.

16 Hantman, 1990.

17 Quinn, 1955, vol. 1, pp. 259–60.

18 Strachey (Haile, 1998, p. 652); Smith (Smith, 1986, vol. 1, p. 170).

19 Strachey (Haile, 1998, p. 662)

6 *Soundings*

1 Kuin, 1998, p. 561.

2 Haile, 1998, pp. 86–90; Smith, 1986, vol. 2, pp. 137*ff*; vol. 1, pp. 27*ff*.

3 Barbour, 1969, doc. 3.

4 Bemiss, 1957.

5 Smith, 1986, vol. 1, p. 234; vol. 2, p. 106.

6 This first landing may have been in the vicinity of modern Buckroe Beach, Hampton, Virginia.

7 Rountree, 1989, p. 60. The size of the village, Kecoughtan, is described by Strachey as 'sometimes of a thousand Indians and three hundred Indian houses' (Haile, 1998, p. 626); Smith calculated the number of houses to be eighteen, surrounded by 3 acres of cultivated land (Smith, 1986, vol. 1, p. 37). Smith had visited the village a number of times, and the discrepancy may arise because Strachey was writing of its historical size, before it was taken by the English in 1610.

8 Quinn, 1955, pp. 259–60*n*.

9 Haile, 1998, p. 123.

10 Rountree, 1989, p. 56; Quinn, 1955, p. 886; Smith, 1986, vol. 2, pp. 12, 77; Haile, 1998, p. 93; Smith, 1986, vol. 1, p. 34. The name Percy recorded for the town has caused some confusion, as 'Rappahannock' was later used interchangeably with 'Tappahannock' to refer to another town, located near the modern city of that name on the Rappahannock river. Smith's map labels the town visited by Newport that day 'Quiyoughcohannock'.

11 Haile, 1998, p. 198.

12 Smith, 1986, vol. 3, p. 295.

13 Barbour, 1969, pp. 51–2, 25; Smith, 1986, vol. 2, p. 138.

14 Haile, 1998, pp. 94–6.

15 Hakluyt, 1599, p. 339; O'Mara, 1982, p. 5.

16 Gleach argues that the name Powhatan was derived from the 'proto-Algonquian root paw, dream'. Other ethno-historians, notably Rountree, continue with the conventional interpretation that the name means 'falls on a rapid stream', and that the *mamanatowick* was named after the place of his birth. (Gleach, 1997, p. 32, Rountree, 1989, p. 11.)

17 Haile, 1998, p. 595.

18 Haile, 1998, p. 97.

19 Smith, 1986, vol. 2, p. 116; vol. 1, p. 162; Haile, 1998, pp. 630–1, 671–3; Rountree, 1989, pp. 62, 88–9; Kupperman, 2000, pp. 147*ff.*
20 Smith, 1986, vol. 1, p. 160; Rountree, 1989, pp. 46*ff.*
21 Haile, 1998, p. 622.
22 Beverley, 1947, pp. 61–2; Gleach, 1997, pp. 34–5. For the translation of Opechancanough, see Gerard, 1904, p. 314.
23 Smith, 1986, vol. 1, pp. 31–2; vol. 2, p. 139. Archer put the number injured in the incident at ten (plus one fatality); Smith put the number at 'thirteen or fourteen' (*True Relation*) and seventeen (*Generall Historie*). Smith also doubled Archer's estimate of the size of the Indian army, claiming it was four hundred strong.
24 *Tamburlaine the Great*, Part 2, Act III, scene ii, ll. 62–90; Marlowe, 1998, p. 8; Fourquevaux, 1589, pp. 4*ff*; Batho, 1960, p. 249; Quinn, 1955, p. 132; Kelso, 1995–2001, vol. 3, pp. 31–6; Kelso, 2004, pp. 77–9.
25 Smith, 1986, vol. 2, p. 325; vol. 3, p. 26; Kelso, 1995–2001, vol. 3, p. 38.
26 Barbour, 1969, doc. 18; Beverley, 1722, p. 17. Cope describes to Cecil how the soil sample came from under 'Two Turffes of earth' dug up during 'all ther fortyfycations'. Given the tone of the rest of the letter, it is possible he was not being literal.
27 Smith, 1986, vol. 3, p. 276.
28 Smith, 1986, vol. 2, pp. 420–1.
29 Smith, 1986, vol. 3, p. 273; Barbour, 1969, p. 25; Haile, 1998, pp. 194, 198; Smith, 1986, vol. 3, p. 272.
30 Smith, 1986, vol. 2, pp. 139–40.
31 Smith, 1986, vol. 2, p. 143.
32 Brewster had been 'keeper of recusants' at Framlingham Castle, Suffolk, where in 1601 he had summoned Robert Gosnold, Bartholomew's uncle, to help capture three escaped Catholic priests. In April 1603 he had complained to Cecil that, because his prisoners were so poor, he was having to subsidize their incarceration, and would be 'utterly undone if you relieve me not'. Relief duly came, in the form of his recruitment to the Virginia venture. (*Acts of the Privy Council*, vol. 32, p. 255; HMC Salisbury, Part 10, p. 202; Part 15, Brewster to Cecil, 20 April 1603; N., 1868; Haile, 1998, pp. 43–4.) There are difficulties in separating the Jamestown William Brewster from the better-known Pilgrim Father (1566/7–1644). (Barbour, 1969, doc. 17.) On Brewster's connection to the Pilgrim Father, see PRO PROB 11/65, will of William Brewster of Hedingham Castle, Essex; *Notes & Queries*, 4th series, 2 : 32 (8 August 1868), pp. 125–6.
33 Haile, 1998, p. 98.

7 *The Spanish Ambassador*

1 Barbour, 1969, doc. 18; Quinn, 1979, doc. 463; Quinn, 1974, pp. 386*ff.*
2 Barbour, 1969, doc. 10.
3 Brown, 1890, doc. 18.
4 Barbour, 1969, doc. 9.
5 In fact, the expedition was a failure, due to under-investment. (Hunter, 1899, vol. 1, p. 286.)
6 Barbour, 1969, doc. 19.
7 Barbour, 1969, doc. 20.
8 Barbour, 1969, docs 21, 4; Williams, 1995, p. 484.
9 'I am credibly informed,' Cope wrote to Cecil, that Hazell had 'gotten away Captain Weymouth'. They were 'as far as Deal Castle outwards in their way towards Spain. I pray god they may be stayed, lest we repent their going later.' (Barbour, 1969, p. 109.) On 27 October 1607, two months after being stopped at Deal, Weymouth received a pension from the King 'till further advancement', i.e. employment. This was clearly in compensation for his role in the mission to Spain. (*CSP Dom – James I*, vol. 28.)

10 Barbour, 1969, doc. 11.
11 *New DNB*, 'Croft, Sir Herbert'.
12 Barbour, 1969, doc. 24.
13 Brown, 1890, doc. 16; Barbour, 1969, doc. 23.
14 Barbour, 1969, doc. 25.
15 Barbour, 1969, p. 112.
16 Barbour, 1969, docs 20, 22.
17 Smith, 1986, vol. 1, pp. 222–4, 165.
18 Smith, 1986, vol. 1, pp. 161–3; Haile, 1998, pp. 132–3.

8 *Bloody Flux*

1 E.g. Francis Perkins Sr and Jr (Smith, 1986, vol. 2, p. 161).
2 Barbour, 1969, p. 145. Thomas Dekker uses the term in his satirical pamphlet 'The Blacke and White Rod', referring to a benefit accidentally acquired (Dekker, 1630, A2v). See also Christopher Marlowe, *The Tragical History of Dr Faustus*, Act IV, scene ii, l. 35.
3 The lack of data on the number of runaways is one of the factors that makes demographic information about Jamestown so unreliable, and a subject of considerable debate. See Bernhard, 1992; Camfield, 1994; Bernhard, 1994.
4 Virginia Company, 1610, p. 14.
5 At the time, the English believed the village to be called Rappahannock or Tappahannock. Quiyoughcohannock was established as the name in later English narratives. (Whitaker, 1613, p. 40; Smith, 1986, vol. 1, p. 34; Blick, 2000.)
6 Haile, 1998, p. 622.
7 Barbour, 1969, pp. 145–6; Blick, 2000.
8 Haile, 1998, p. 671; Rountree, 1989, p. 109.
9 Rountree, 1989, pp. 39–40; Stahle, 1998.
10 Wingfield writes of presenting the chief of the 'Tapahanah' with a red waistcoat on 7 July 1607. (Haile, 1998, p. 185.)
11 The term comes from an Algonquian term meaning 'he who goes about glowing', and seems to refer to deities taking on a human form. (Quinn, 1955, pp. 373, 888.)
12 Haile, 1998, p. 100.
13 Unless specified, the account that follows comes from Smith's *Generall Historie*, book 3, chap. 2 (Smith, 1986, vol. 2, pp. 142*ff*), Percy's *Observations* (Haile, 1998, doc. 1) and Wingfield's *Discourse* (Haile, 1998, doc. 13).
14 Philip Barbour promoted this view, drawing on an anachronistic conception of the English class system in his attempt to promote John Smith as champion, even progenitor, of a more meritocratic model that is assumed to have developed in America (Barbour, 1964, p. 141). Smith, for example, is cited as criticizing those 'that never did know what a day's work was', but he was referring to the labourers, who turned out to be 'for the most part footmen' or servants of 'Adventurers', meaning the richer gentlemen.
15 Smith, 1986, vol. 2, p. 153; vol. 3, pp. 176, 17.
16 Hammer, 1998, p. 61.
17 Smith, 1986, vol. 2, p. 143.
18 Bernhard, 1992, p. 602.
19 Corn (maize) provided a good source of calories, with the advantage over wheat or barley of requiring little processing. Assuming a man required 2,500 calories (though ongoing experiments in 'living archaeology' at the Jamestown Settlement exhibition near the site of the original fort suggest 4,000 to 8,000 calories were needed to fuel the settlers' physically demanding work, see *The Free Lance-Star*, Fredericksburg, Virginia, 5 June 2004), and that corn yields nearly 400 calories per pound (probably less in the seventeenth century), the settlement would need about 0.8 tons a week to feed forty men, the number who survived the summer. The annual corn consumption of the Indians is based on studies in New

England (Thomas, 1976, p. 12; Bennett, 1955, pp. 392*ff*).

20 Kelso, 1995–2001, vol. 4, pp. 6*ff*.

21 Drew (or Dru) Pickhouse (or Piggase, Pigesse, Piggays, Piggayes, Pickers, Pickesse, Pickas, Pycas, Pykas – even by seventeenth-century standards, a wide variety of spellings) was MP for East Grinstead, Sussex, in 1586 and in 1589 inherited from his father the rustic Manor of Brambletye, an estate subject to a long-running dispute with the local Lord Buckhurst. (Hills Wallace, 1906, pp. 112–3; Smith, 2001; Leppard, 2001.) I am grateful to Jack Russell for his help on the local history of East Grinstead.

22 Barbour, 1962; Haile, 1998, p. 99.

23 The link between Ratcliffe and Cecil rests on his letter to Cecil from Jamestown (Haile, 1998, doc. 19), and his will (PRO prob/11/117, dated 1 June 1609), which names 'Richard Percivall, Esquire my lovinge friende' as his executor. (*DNB* 'Percevall, Richard'.) I am grateful to Patrick Martin for making me aware of the link between Percival and Cecil, and other help in connection with researching Ratcliffe. It was thought that he was a Captain Ratcliffe who served in the Netherlands and was captured at the battle of Mulheim in 1605, but this is unlikely, since this Ratcliffe was back in Dutch pay by 1607. (Markham, 1888, p378; *New DNB*, 'Ratcliffe John'. For other Ratcliffes mentioned in the state papers, see SP 14/20/52; 14/7/86; Cecil MSS 190/35; 160; Dodd, 1938.

24 Haile, 1998, p. 646.

25 A different account appears in *True Relation*, which does not refer to the appearance of Okeus or the troop of Kecoughtan in war colours. Smith, 1986, vol. 1, p. 35, 37.

26 Smith, 1986, vol. 1, p. 35.

27 SP 14/18/66.

28 Haile, 1998, p. 453; Barbour, 1962, p. 311.

29 In *Generall Historie*, Smith conflated Kendall's and Wingfield's escape attempts, but other accounts show that they were distinct (Haile, 1998, pp. 194, 453; Smith, 1986, vol. 2, p. 145).

30 Smith, 1986, vol. 1, p. 47; Haile, 1998, p. 622.

31 Smith, 1986, vol. 1, p. 161.

32 Haile, 1998, p. 617.

33 Purchas recorded that White witnessed Casson's execution (Purchas, 1614, p. 767). The participation of the Quiyoughcohannock in the hunting party that captured Smith is based on the assumption that Casson would not have been brought 30 miles to be interrogated at their village (Smith, 1986, vol. 2, p. 146). Smith records his visit to 'Topohanack', Quiyoughcohannock, in his *True Relation*, in which he claimed that, having no commission to 'spoil', he left the food stores untouched.

34 Accounts of Casson's execution, all based on White's testimony, are provided by Purchas (Barbour, 1969, p. 145), Smith (Smith, 1986, vol. 2, p. 127) and Strachey (Haile, 1998, p. 617).

35 There are no further references to White in the historical record, unless he was the William White buried in Virginia on 12 September 1624. (LOC Peter Force Collection, Series 7, item 53.5, 'Muster of Inhabitants . . .'.) On Indian executions, see Haile, 1998, pp. 491–2; Smith, 1986, vol. 1, p. 175.

36 Ackroyd, 2001, p. 426.

37 Lamb, 1995, p. 239; Haile, 1998, p. 133.

38 Smith, 1986, vol. 2, p. 149.

39 Strictly, the date is 2 January 1607 Old Style; Wingfield puts the date of Smith's return as 8 January (Haile, 1998, p. 195), but seems to be mistaken. *True Relation* puts the number of Smith's escorts at four (Smith, 1986, vol. 1, p. 57). By the time he wrote *Generall Historie*, the number had grown to twelve.

40 Smith, 1986, vol. 1, p. 61; vol. 2, p. 152.

41 There is no chapter of Leviticus relevant to the death of Smith's soldiers, and in any case Levitical law, which sets out the code of sacrificial, ceremonial and criminal law under which the Israelites were supposed to live, was under English jurisprudence superseded by common law. (Pont, 1599, n.p. 'The second sermon'.) Both Wingfield (Haile, 1998, p. 196) and Smith (Smith, 1986, vol. 2, p. 152) mention the use of Levitical law. This use of Biblical law has received little scholarly attention. Konig merely notes that it happened, suggesting that it may reflect a growing uncertainty about the role of common law in the administration of the colony (Konig, 1982, p. 12). When trying to annul his first marriage, to his older brother's widow Catherine of Aragon, Henry VIII had argued it violated Levitical law. George Elton points out that, in the 1530s, MPs considering an expansion of the treason law designed to regularize Henry VIII's marriage to Anne Boleyn wanted the 'opinion of the Levitical law' left out of the legislation, because such matters were already dealt with in other acts. (Elton, 1968, pp. 221–2; Barbour, 1969, pp. 37–8.)

9 *True Relations*

1 Haile, 1998, p. 133. Brian Dietz (Dietz, 1991, pp. 14–20) lists two ships that might be the *John & Francis* sent to Virginia, one of London built 1598, the other of Plymouth 1607. The tonnage of the former was 250, of the latter 246. He lists two ships named *Phoenix*, both of London, one built in 1595 of 180 tons, the other in 1597 of 250 tons.

2 Smith, 1986, vol. 1, pp. 215.

3 Haile, 1998, pp. 133–4; Smith, 1986, vol. 1, pp. 217, 79.

4 The story of Smith's capture is based on his own accounts contained in the *True Relation* (Smith, 1986, vol. 1, pp. 43–57), *Proceedings* (Smith, 1986, vol. 1, pp. 212–3), and *Generall Historie* (Smith, 1986, vol. 2, pp. 11, 146–52).

5 The number of warriors mentioned in *True Relation* is two hundred (Smith, vol. 1, p. 47), in *Generall Historie*, three hundred (Smith, vol. 2, p. 146).

6 From his description, in particular the reference to the spheres, Smith appears to have followed the orthodoxy of the time, which placed the earth at the centre of the universe. The Copernican model, putting the sun at the centre, was still acknowledged by only a few scholars in England, notably John Dee and Thomas Harriot.

7 The location and timing of the following episodes is unclear from Smith's accounts, but internal evidence suggests he was taken to Uttamussack. Menapucunt and Cinquoteck were respectively homes of Powhatan's two brothers Kekataugh (Smith, 1986, vol. 1, p. 51) and Opitchapam (Smith, 1986, vol. 1, p. 77). The description of Uttamussack comes from Strachey (Haile, 1998, p. 652).

8 Itoyatan (spelled Itopatin in some accounts) was also known as Opitchapam. He lived in Cinquoteck, according to a reference elsewhere in Smith's *True Relation* (Smith, 1986, vol. 1, p. 77), near Opechancanough.

9 In *True Relation*, the initiation ritual experienced by Smith is described as though it appeared after his meeting with Powhatan. However, according to *Generall Historie*, the ritual was enacted while he was still in Opechancanough's care (Smith, 1986, vol. 1, p. 59; vol. 2, pp. 149–150). Barbour reorders the corresponding

sections in the recension introducing *True Relation* (Smith, 1986, vol. 1, pp. 10–15). Many attempts have been made to make sense of the rituals described by Smith. Gleach (Gleach, 1997, pp. 112*ff*) offers the most persuasive interpretation.

10 The bay is modern Purtan Bay, about 20 miles from the mouth of the York river.

11 Smith, 1986, vol. 1, p. 173.

12 *True Relation* gives the name as Anchanachuk (Smith, 1986, vol. 1, p. 55). 'They inhabit the river of Cannida,' Smith, 1986, vol. 1, p. 232.

13 A 1588 translation of a famous romantic work by the French author Claude Colet, entitled *The Famous, Pleasant, and Variable Historie of Palladine of England, Discoursing of Honourable Aduentures, of Knightly Deedes of Armes and Chiualrie* features a Lady Nonpareila who was 'imprinted' on the heart of the book's hero. This is the sort of literature that probably inspired Smith's choice of description of Pocahontas (Colet, 1588, p. 51). Her age was guessed by Smith to be between ten and twelve years old – not much younger than the 13-year-old Juliet whom Shakespeare imagined in his late 1590s play *Romeo and Juliet.*

14 'Demi-God', *Proceedings* (Smith, 1986, vol. 1, p. 213); 'one of their owne Quiyouckosucks', *Generall Historie* (Smith, 1986, vol. 2, p. 147).

15 Smith, 1986, vol. 2, p. 152.

16 Powhatan said he would release Smith four days after his arrival at Werowocomoco, and having dispatched him a day early, apparently ordered the guides sent with him to delay their arrival at Jamestown by a day (Smith, 1986, vol. 1, pp. 53, 61, 215*n*).

17 Haile, 1998, p. 630.

18 Smith, 1986, vol. 2, p. 154. Smith provides the only account of Newport's visit to Powhatan: *True Relation* (Smith, 1986, vol. 1, pp. 61–77); *Proceedings* (Smith, 1986, vol. 1, pp. 214–9); and *Generall Historie*, Book 3, chap. 3 (Smith, 1986, vol. 2, pp. 153–8).

19 Cheshire and Chester Archives, Cholmondeley of Cholmondeley collection, ref DCH/F/151.

20 Tyndall is not mentioned in Smith's account, but the naming of the Tyndall's Point near the mouth of the river (modern Gloucester Point, opposite Yorktown) suggests he had returned with Newport on the *John & Francis* and had been brought along for this expedition (Mook, 1943, p. 386).

21 Smith, 1986, vol. 1, p. 55, Barbour, 1969, p. 267.

22 In the British Library copy of *True Relation*, a marginal note reads, 'This author I find in many errors, which they do impute to his not well understanding the language'. In 1630, Smith admits to some of the earlier inadequacies of his understanding of the Algonquian language spoken by the Powhatan people. Kupperman, 2000, p. 32.

23 Barbour, 1969, vol. 1, pp. 63*ff.*

24 Barbour, 1969, doc. 40; p. 274. The only evidence for Hunt returning to England is the listing of a 'Master Hunt', the title by which he is referred to by Smith (e.g. Smith, 1986, vol. 2, pp. 204, 207, 214, 217) in the list of settlers brought with the Second Supply (p. 241). Barbour speculates that the settler may have been the Thomas Hunt listed as an adventurer in the 1609 charter.

25 Smith, 1986, vol. 2, pp. 187–8.

26 Smith, 1986, vol. 2, p.160; vol. 1, p. 210.

10 *The Virginian Sea*

1 Haile, 1998, p. 199.

2 Moryson, 1617, Book 3, Part 3, p. 156.

3 Bottigheimer, 1979, pp. 46*ff.*

4 Churchyard, 1579, Diii^{v}: '[Gilbert's] maner was that the heddes of all those (of what sort soeuer thei were) whiche were killed in the daie, should bee cutte of from their bodies, and brought to the place where he incamped at night: and should there bee laied on the ground, by eche side of the waie leadyng into his owne Tente: so that none could come into his Tente for any cause, but commonly he muste passe through a lane of heddes, whiche he vsed *ad terrorem*, the dedde feelyng nothyng the more paines thereby: and yet did it bryng greate terrour to the people, when thei sawe the heddes of their dedde fathers, brothers, children, kinsfolke, and freendes, lye on the grounde before their faces, as thei came to speake with the saied Collonell. Whiche course of gouernemente maie by some bee thought to cruell, in excuse whereof it is to bee aunswered. That he did but then beginne that order with theim, whiche thei had in effecte euer tofore vsed toward the Englishe.'
5 Quinn, 1940, p. 490.
6 Quinn, 1940, p. 15, doc. 13, appendix 1.
7 Wingfield, 1993, p. 36.
8 Campion, 1633, 'Edmund Spenser's View of the state of Ireland', p. 9; Haile, 1998, p. 588.
9 Spenser, 1590, vol. 1 : 12–13.
10 Canny, 1973, pp. 576*ff*.
11 Barbour is dismissive of Wingfield's remark, considering it 'an idle comment, recorded in scorn by an offended highborn gentleman' (Barbour, 1964, pp. 83–4), but he acknowledges that the matter 'remains open' (Smith, 1986, vol. 3, p. 134).
12 Smith, 1986, vol. 3, p. 280.
13 Kupperman, 2000, pp. 31–2.
14 Smith's account of his discoveries of Chesapeake Bay is contained in his *Proceedings* and *Generall Historie*, chaps 5–6 (Smith, 1986, vol. 1, pp. 224*ff*; vol. 2, pp. 162*ff*).
15 Smith, 1986, vol. 3, pp. 158–9; Barbour, 1964, p. 21.
16 Smith, 1986, vol. 2, p. 119.
17 Smith calculated 'thirty leagues', 90 miles, but appears to have exaggerated the distance.
18 On prices of medicines, Guybert, 1639, G3^{r}–H2^{r}.
19 Quinn, 1955, pp. 267–8. Lane records that the company had agreed to eat two bull mastiffs, whereas Smith mentioned only a single dog. Otherwise his account is remarkably accurate to the original.
20 Rountree, 1989, p. 70.
21 Smith refers obliquely to Ratcliffe wanting to 'divide the country' in his letter to the Virginia Council in London (Smith, 1986, vol. 2, p. 188), and may be here referring to that plan.
22 Smith, 1986, vol. 3, p. 438. Smith quotes Dr John Dee's *General and Rare Memorials*.
23 Smith is confused as to the dates. In his *Proceedings* he gives 20 July as the departure date for the second expedition, even though he writes that the first ended on 21 July. In *Generall Historie*, he changes the date to 24 July.
24 BL Add MSS 36774/21 folio 14, 'Circular letters of safe-conduct for . . . Thomas Webbe, going on a mission to Germany', 10 June 1591; SP 12/239/20, 'Pardon for Thomas Webbe, convicted of coining and uttering Elizabeth shillings'; Wilding, 1999, p. 15; Dee, 1998, pp. 253, 263, 264. HMC Salisbury vol. 11, p. 117; Bruce, 1953, p. 93.
25 LOC, Peter Force Collection, Series 7, Item 53.5; Dietz, 1991, pp. 14–20.
26 Northumberland RO, ref QSI/1, folio 36r: 23 April 2 James I; PRO SP 14/20/45.
27 Hammer, 1998; Lewkenor, 1595, E1^{v}–E2^{r}; Quinn, 1940, p. 74; HMC Salisbury vol. 14, p. 173, *CSP Dom* –

James I, 1603–1610, item 81; PRO SP 94/13/2.

28 Barbour, 1964.

29 Barbour, 1969, pp. 163–4, HMC Salisbury, Part 6, pp. 229*ff.*

30 Cronin, 2001, p. 45. Powhatan's coronation has been consistently interpreted by historians as a gesture of subordination, making Powhatan a 'vassal' of King James (Gallivan, 2003, p. 168), or an 'obedient sub-king to the Majesty of James I' (Barbour, 1964, p. 237), even though the status of a 'sub-king' was unknown to Early Modern English politics. Even if it were, a coronation would not be the form of rite deployed to establish such status.

31 An alternative interpretation of the coronation is supplied in a 1919 article 'The Coronation of Chief Powhatan Retold' by the Yankton Sioux writer Zitkala-Sa. In her view, the awkwardness of the occasion arose because Powhatan wanted nothing to do with white men's symbols of power. 'To the liberty loving soul of Powhatan, this royal camouflage was no comparison to the gorgeous array of Autumn in that primeval forest where he roamed at will.' Totten, 2005, p. 110.

32 Haile, 1998, pp. 90, 192; Smith, 1986, vol. 1, p. 151.

33 Smith, 1986, vol. 1, p. 162.

34 Haile, 1998, p. 809; Blanton, 2000; Stahle, 1998.

35 Haile, 1998, doc. 15; Huntington, Ellesmere 1683, 'To the Honourable Knight, Sir John Egerton, at York House,' 26 November 1608.

PART THREE

11 *El Dorado*

1 Trevelyan, 2002, pp. 143–4.

2 James I, 1604, B[r].

3 HMC Report 6, 'MS of Duke of Northumberland at Syon House', pp. 228*ff*; Batho, 1957, p. 449.

4 Brown, 1890, vol. 1, p. 99.

5 Lorimer, 1979, pp. 131–2. The account that follows is based on the research contained in this paper.

6 E.g. in December 1607, a statement was issued to the Lord Chief Baron of the Exchequer 'of the cause between the Farmers of Tobacco and John Eldred, who refuses to pay the imposts on tobacco which he has imported'. *CSP Dom – James I, 1603–1610, item 143.*

7 Trevelyan, 2002, p. 419.

8 Cromwell, 1655, p. 9.

9 Hariot, 1588, p. 16; Quinn, 1955, p. 898.

10 Arents, 1939.

12 *The Mermaid*

1 Barbour, 1969, doc. 37.

2 Johnson, 1609, p. 5.

3 Brown, 1890, doc. XVII; Barbour, 1969, p. 163.

4 Rabb, 1964, pp. 140*ff*; *New DNB*, 'Cecil, Robert'.

5 Barbour, 1969, p. 163.

6 Inigo Jones, Richard Connock and Sir Robert Phelips were known to have connections to Henry's court (Shapiro, 1950, p. 10). Coyrate (who was absent at this time) and Jonson dedicated works to the Prince. William Hakewill's brother George was one of Henry's chaplains. Several others wrote elegies to Henry, including Henry Goodere and Hugh Holland.

7 Strong, 2000, pp. 4, 8, 12, 16, 166.

8 Brown, 1890, doc. CXLVII.

9 *CSP Venetian*, vol. 11, p. 237.

10 Smith, 1986, vol. 1, pp. 7–8, 116–7.

11 Smith, 1986, vol. 2, pp. 45–6. Healey probably went to Virginia in 1609, a journey which Thorpe alludes to in the dedication to the Earl of Pembroke in Healey's magnum opus, *City of God*, Augustine, 1610,

(Augustine, 1610, $A3^{v}$).

12 *New DNB*, 'Thorpe, Thomas'; Martin, 2003.

13 Coryate, 1616, pp. 37*ff*; Shapiro, 1950; Simpson, 1951; Strachan, 1967; Pritchard, 2004. The link between the Sirenaicals and Virginia has not been previously noted. At least eighteen of the twenty-two names identified by Coryate, the self-proclaimed 'beadle' of the group, had links to the Virginia venture.

14 Smith, 1986, vol. 1, pp. 29, 31.

15 Price, 1609, $F2^{r}$–$F3^{v}$.

16 *New DNB*, 'Symonds, William'; Smith, 1986, vol. 1, p. *lv*.

17 Symonds, 1609, pp. 1, 26.

18 Benson, 1609, p. 92.

19 Symonds, 1609, pp. 10–12; Crakanthorpe, 1609 $D2^{v}$; Crashaw, 1610, $C3^{r}$; Gray, 1609, $C3^{v}$.

20 Gray, 1609, $B2^{v}$, $B3^{v}$.

21 Johnson, 1609, C^{v}–$C2^{r}$, $E3^{r}$.

22 Hakluyt, 1609; Lescarbot, 1609, ¶¶^{v}.

23 Hall, 1613, pp. '66' [96], 215.

24 Greenblatt, 2005, p. 235.

25 Foster, 1987; Rabb, 1966, p. 224; Martin, 2003, p. 11; *New DNB*, 'Thorpe, Thomas', 'Hakewill, William'. Hakewill is not listed among the Second Charter adventurers, but is among the 'List of Subscribers. November, 1610, to February, 1611', where he is listed alongside the likes of Edwin Sandys (Brown, 1890, doc CLXII). A. L. Rowse suggests that W. H. could be Sir William Harvey, Southampton's stepfather. Harvey has no known connection with the Virginia or any other trading company at that time.

26 Shapiro draws up a list of twenty-two names based on several sources, including a letter by the travel writer Thomas Coryate addressed to Laurence Whitaker, 'the High Seneschal of the right Worshipful Fraternity of Sirenaical Gentlemen, that meet the first Friday of every Month, at the sign of the Mermaid in Bread Street in London' (Shapiro, 1950). They were John Bond, Christopher Brooke, Robert Bing, Richard Connock, Thomas Coryate, Robert Cotton, Lionel Cranfield, John Donne, George Garrat, Henry Goodere, William Hakewill, Hugh Holland, John Hoskyns, Arthur Ingram, Inigo Jones, Ben Jonson, Richard Martyn, Doctor Mocket, Henry Neville, Robert Phelips, John West and Laurence Whitaker. In late 1608, Donne had applied to become secretary to the Virginia Company, probably through Sir George More, a member of the Royal Council. Donne had secretly married More's daughter Anne in 1601, his friend and fellow-Sirenaical the barrister Christopher Brooke being a witness at the wedding. More had been furious, and was in a position to have Donne thrown into the Fleet Prison. However, father- and son-in-law had become reconciled, and this might have raised Donne's hopes of preferment in the venture. Around the same time, Christopher Brook was also involved with Virginia, helping to draw up important documents for the London Company. (Barbour, 1969, p. 247; Kingsbury, 1906, vol. 3, doc. 6; *New DNB*.) Jonson had no formal connection with the Virginia venture, but had already risked his career writing about it in *Eastward Hoe*, and mentioned it in several other plays. Inigo Jones would go on to design the sets of the 'memorable masque' of 1613, written by Jonson's friend George Chapman, which featured richly attired 'Virginian' priests and princes. The same production involved Richard Martin (see below) and Christopher Brookes (Herford, 1986, p. 339; Chapman, 1613; Shapiro, 1950, p. 13). The most senior Sirenaicals (Sir Robert Cotton, Sir Henry Goodyear, Sir Arthur Ingram, Sir Henry Neville,

Sir Robert Philips, Sir Thomas Roe, and Lionel Cranfield, son of a merchant, who would become the first Earl of Middlesex) all invested heavily in Virginia in the hope of building up estates to match their new status. There are references to these men scattered throughout Virginia records and State Papers, e.g. Cotton: Brown, 1890, doc. CCCLXI; Goodyear: Brown, 1890, doc. CLXII; Ingram: Kingsbury, 1906, vol. 2, 5 June 1622; Neville: Bodleian MS Ashmole 1147 (Second Virginia Charter), folio 159; Phillips: Kingsbury, 1906, vol. 1, 13 November 1620; Roe: Kingsbury, 1906, vol. 1, 16 February 1619; Cranfield: HMC Report 4, MS Earl de la Warr, p. 283. George Garratt, William Hakewell, John Hoskyns and Richard Martin, respectively a merchant, a grocer, an MP and a lawyer, were all more or less involved in the Virginia Company, Martin (no relation of Captain John Martin) being the company's lawyer. (Garrett: Kingsbury, 1906, vol. 2, 4 February [pm], 1622/3; Hakewell: Smith, 1986, vol. 2, p. 278; Hoskyns: Brown, 1890, doc. CLXII; Martin: *New DNB*.) A Robert Byng, described by a contemporary as a 'mere good-fellow, a man of no estate, who, for saucy conduct before the Council table, and offensive behaviour to Lord Southampton, had been committed to the Marshalsea', is likely to have been the man involved in later investigations into the Virginia Company's affairs. (*CSP Colonial* 1574–1660, p. 65; Kingsbury, 1906, vol. 1, 'List of Records'.)

27 Sandys, 1629, pp. 188–9, 195; Rabb, 1966, p. 40; Rabb, 1998, pp. 34–5.

28 Barbour, 1969, doc. 44.

29 Barbour, 1969, doc. 47.

30 *CSP Venetian*, vol. 11, p. 237.

31 Rabb, 1966, pp. 92*ff*.

32 Bodleian MS Ashmole 1147, folios 155–71 (a contemporary copy of the charter); Quinn, 1979, doc. 801.

33 Barbour, 1969, doc. 36; Brown, 1890, p. 317.

34 Brown, 1890, docs LXXII, LXXIII.

35 O'Brien, 1960, p. 139; Brown, 1890, doc. LXXIV.

36 Brown, 1890, docs LXXVII, LXXVIII.

37 Thomas Harriot may also have been consulted. Taylor, 1935, p. 501.

38 Lorimer, 1979, pp. 133, 137*n*.

39 O'Brien, 1960, pp. 140–4.

40 Brown, 1890, doc. LXX. For income of carpenters and builders, see Boulton, 1996, pp. 275*ff*, fig. 1. Wage rates are difficult to calculate, partly because wages made up a smaller proportion of an individual's income in pre-industrial England (Woodward, 1981, pp. 29*ff*). Some craftsmen could hope to earn several pounds a day (e.g. Seaver, 1985, p. 122, though this relates to the mid-1600s).

41 Strong, 2000, pp. 35–6.

42 Brown, 1890, pp. 252–3.

43 Bernhard, 1992, p. 608; Barbour, 1969, p. 277. Barbour comments: 'There is no other account of the time which mentions such quantities of livestock, and it is highly doubtful that any considerable amount was actually sent. All evidence militates against it.' However, there are later reports of livestock in Virginia which could have been introduced in 1609. The pilot John Clark taken by the Spanish in 1611 told his captors that the English had 'brought over 100 cows, 200 pigs, 100 goats, and 17 mares and horses' (Haile, 1998, p. 545). Sir Thomas Dale, who arrived in 1611, found 'cattle, cows, goats, swine, poultry, etc. to be well and carefully on all hands preserved, and all in good plight and liking' (Haile, 1998, p. 523).

44 Brown, 1890, doc. CXLIV. The dating of this letter is uncertain.

45 Essex RO Q/SR 51/17–22, Sessions Rolls, Michaelmas 1574.

46 Culliford, 1965, pp. 48–53.

47 Barbour, 1969, p. 267; Smith, 1986, vol. 2, p. 189.
48 Fuller, 1662, 'Dorcetshire', p. 283; Pope, 1911.
49 Haile, 1998, p. 350; Barbour, 1969, doc. 49.
50 Brown, 1890, p. 325, a letter from Dudley Carleton to John Chamberlain. The letter is dated 18 August 1609, but appears to be referring to events in May. Kingsbury, 1906, vol. 1, doc. VI.
51 BL Add MS 34599 folio 24.
52 Peterson, 1988 , p. 44. The ship's draft weight is given by Strachey (Haile, 1998, p. 415).
53 Barbour, 1969, p. 275; Brown, 1890, p. 343. The evidence that this mission was secret lies in Zuñiga's description of Argall's vessel as a 'fisherman's ship' (Barbour, 1969, doc. 56).
54 PRO 11/117; *New DNB*, 'Ratcliffe, John'.

13 *Promised Land*

1 Melville, *Moby Dick*, chap. 35.
2 The basis of this account is Strachey's 'True Reportory of the wrack and redemption of Sir Thomas Gates' (Haile, 1998, doc. 22) and Gabriel Archer's letter from Jamestown, 31 August 1609 (Haile, 1998, doc. 18).
3 Horace, *Carmina*, 3.27.23–24.
4 There seems to have been some confusion in the reporting of the rendezvous location. According to the company's *True and Sincere Declaration*, in the event of ships going astray, 'they should steer away for the West Indies and make for the Baruada, an island to the north of Dominico' (Virginia, 1610, p. 13). Hakluyt uses the name 'Baruada' to refer to modern Barbuda, lying at a latitude of 17 degrees north of the Equator. Barbuda, however, is not north of the island of Dominica (at a latitude of 14 degrees, according to Hakluyt). It is possible Baruada was a misprint for the channel of Bahama, at 27½ degrees. (Hakluyt, 1599, p. 624.)
5 Wright, 1920, p. 464.
6 Smith's account comes from his *Proceedings* (Smith, 1986, chaps 9–12).
7 Haile, 1998, p. 354.
8 Haile, 1998, pp. 482*ff*.
9 Rountree, 1989, p. 76.
10 Haile, 1998, p. 502.
11 Haile, 1998, pp. 503*ff*.
12 Haile, 1998, p. 559; Shirley, 1949, pp. 237–8.
13 Smith, 1986, vol. 2, p. 126; Gerard, 1904, p. 315.
14 Percy mentions that Ratcliffe went with 'Powhatan's son and daughter aboard his pinnace', but does not mention their names (Haile, 1998, p. 504).
15 Smith, 1986, vol. 1, p. 244. Michael was apparently unrelated to John Ratcliffe, alias Sicklemore.
16 According to Spelman he took fourteen or fifteen, though this can be taken to mean the number who made the final leg of the journey with him. The account of Ratcliffe's capture and death comes from Spelman.
17 Smith, 1986, vol. 2, p. 232.

14 *The Astrologer*

1 Emerson, 1984, p. 206.
2 Cook, 2001, pp. 167, 184.
3 Traister, 2001, pp. 173–5.
4 Roberts, 1990, p. 156. Forman identifies him only as 'Mr Staper', but given Richard Staper's involvement in a number of trading companies (the East India, Levant as well as a number of other ventures), the identification seems likely. If Mr Staper was not Richard, he may have been Hewett, another merchant also involved in trading ventures (Rabb, 1966, p. 382).
5 Barbour, 1969, doc. 56.
6 Haile, 1998, p. 352.
7 Johnson, 1612, pp. 10–11.
8 Haile, 1998, p. 359.

9 Virginia, 1610.
10 Cook, 2001, p. 184.

15 *Devil's Island*

1 The following account is taken from Strachey (Haile, 1998, doc. 22) and Jourdain (Jourdain, 1613).
2 For the armaments of the *Sea Venture*, see Peterson, 1988, p. 42.
3 Jourdain, 1613, p. 11.
4 MacCulloch, 2004, pp. 376*ff*.
5 Craven, 1937, p. 191; Lefroy, 1882, pp. 14–5.
6 Haile, 1998, p. 429; *Aeneid*, I: 263; Culliford, 1965, p. 123; Hulme, 2000, p. 35. Virgil writes of '*Italia populosque ferocis*', and '*ferocis*' (*ferox*) is translated in some instances as 'proud', but the context suggests 'wild' or 'savage'.
7 The account that follows is based on Strachey (Haile, 1998, pp. 416*ff*), Jourdain (Jourdain, 1613, pp. 14*ff*), Smith (Smith, 1986, vol. 2, pp. 231*ff*, 350).
8 LOC, MS Peter Force Collection, series 7, item 53.5, 'Muster of inhabitants of James Citty'.
9 The description of the Starving Time is taken mostly from Percy (Haile, 1998, pp. 505–6), with additions from the Brief Declaration by the 'Ancient Planters of Virginia' (Haile, 1998, pp. 895–6).
10 The account of Collins's crime became notorious, being repeated in the Virginia Company's official account of the Starving Time (Haile, 1998, pp. 473–4).
11 Smith, 1986, vol. 2, p. 233; Bernhard, 1992; Camfield, 1994; Bernhard, 1994. Bernhard calculates a total of four hundred and seventeen losses from the arrival of the first three ships in May 1607 to the beginning of the Starving Time in September 1609, out of about eight hundred who had been sent during that period.
12 Wright, 1920, p. 464. The 'many women and children' were observed by an Indian sent to spy on the area by a chief based in Florida who was interviewed by the Spanish. Smith noted that the Powhatan Indians 'seldome make warre for lands or goods, but for women and children' (Smith, 1986, vol. 1, p. 163); Rountree, 1989, p. 121.
13 Brown, 1890, p. 418.
14 Smith mentions that the *Sea Venture* carried one hundred and fifty 'in all', meaning the crew as well as the passengers. Around one hundred may have been passengers, with a few losses (such as Ravens and his men) along the way (Smith, 1986, vol. 2, p. 348).

16 *Deliverance*

1 Smith, 1986, vol. 2, p. 234.
2 Crashaw, 1610, D^{v}, L1^{v-r}.
3 Fitzmaurice, 1999, pp. 38–9.
4 Smith, 1986, vol. 2, pp. 234–5. A version of the latter part of the passage was first published in the Virginia Company's own account of the Starving Time (Haile, 1998, pp. 474–5).
5 Bodleian MS Ashmole 1147, folios 191–2.
6 Haile, 1998, pp. 458–9.
7 De la Warr, 1611, [A2^{r}]; HMC Report 6, MS of Duke of Northumberland at Syon House, p. 229b; Bohun's name was variously spelled Bownde, Bohune, Bounaeus, Bowne, Bownre. 'BOWNDE, Laurence/ Laurentius', *Physicians and Irregular Medical Practitioners in London 1550–1640: Database* (2004). URL: http://www.british-history.ac.uk/report.asp?compid=17261&strquery=Bownde. Date accessed: 13 October 2005.
8 Haile, 1998, pp. 459–60.
9 Brown, 1890, p. 751.
10 Haile, 1998, p. 434.
11 Haile, 1998, pp. 509, 436.

12 Haile, 1998, pp. 475, 897–8.
13 Culliford, 1965, p. 188.
14 Haile, 1998, p. 467.
15 This event, reported by Percy, may be the same as that mentioned by Strachey (Haile, 1998, pp. 624–5) regarding the taking of 'Tackonekintaco, an old *weroance* of Warrascoyack, whom Captain Newport brought prisoner with his son Tangoit about 1610'.
16 Smith, 1986, vol. 2, pp. 210, 212, 214.
17 Haile, 1998, pp. 509–11.
18 Cf Shakespeare, *Henry V*, Act IV, scene vii, ll. 1–2.
19 Quinn, 1979, doc. 820.

PART FOUR

17 *A Pallid Anonymous Creature*

1 Elliott, 2002, p. 300.
2 Morris, 1998, diagram 17.4. The exchange rate is approximated using a report, by Robert Thomson, who visited Mexico in 1555. He recorded one peso as being equivalent to 4*s* 8*d*. Subsequent inflation will have changed that figure (Hakluyt, 1599, p. 452).
3 Elliott, 2002, p. 301; Casey, 1999, p. 154.
4 Kamen, 2003, p. 160.
5 Elliott, 2002, p. 290.
6 Wright, 1920, p. 450.
7 Loomie, 1963, pp. 195*ff*.
8 Wright, 1920, p. 464.
9 PRO SP 14/58/157.
10 Barbour, 1969, p. 157*n*.
11 Haile, 1998, doc. 24.
12 Wright, 1920, p. 454.
13 LOC, David Quinn Collection, MSS 97, folder 2 ('Letter from the minor chaplain of Your Majesty, Friar Balthasa López, from the monastry of St Francis in city and garrison of San Augustín, 12th December 1599' and 'An Account of the Population of the Coast of Florida, and obstacles that were encountered for its fortification and defence.' [n.d.]); 96, folder 6 ('AGI Est. 147 Caj. 5, Leg. 16').
14 Haile, 1998, pp. 537–540; 542–546.
15 Brown, 1890, p. 522.
16 Wright, 1920, p. 457.

18 *Strange Fish*

1 Haile, 1998, doc. 33; HMC Report 75, vol. 3, p. 85.
2 Brown, 1890, doc. CLXXI; HMC Report 75, vol. 3, p.106.
3 Mayes, 1957, p. 22.
4 Brown, 1890, docs CLXXXII, CLXXXIV, CLXXXV, CLXXXIX.
5 Brown, 1890, doc. CCVII.
6 Honan, 1999, p. 298.
7 Whether or not *The Tempest* can be counted as Shakespeare's last play is hotly disputed, but it was certainly a late work, and the plays that may have followed it are of considerably less interest. (Shakespeare, 1987, p. 64; Greenblatt, 2004, p. 373.)
8 Roberts, 1962, p. 125; Shakespeare, 1999, p. 9.
9 Shakespeare, 1987, pp. 30*ff*; *The Tempest*, Act II, scene i, ll. 124–140.
10 Fowler, 2000.
11 {Dee, 1604 #835}. Simon Foreman records meeting Dee at 'Mr Staper's' on July 2, 1604, indicating he was in London around this time ({Roberts, 1990 #571}, p. 156).
12 {Vaughan, 2000 #836}, p. 49.
13 *Tempest*, I, ii, 218–221, I, ii, 229, II, ii, 24–32, II, ii, 56–57, II, ii, 164–165, II, ii, 176, II, i, 144–165, II, i, 172, V, i, 181=184; {Smith, 1986 # 453}, VI, p. 93.
14 Gayley, 1917, pp. 45*ff*; Shakespeare, 1987, p. 32.
15 Following Strachey's biographer S. G. Culliford (Culliford, 1965, pp. 151–4), the preferred candidate for the 'excellent lady' is Dame Sarah Smythe, the daughter of William Blount, and Sir Thomas Smythe's third wife. The Reportory was not an authorized account, and if Strachey

had wanted it to survive intact, with a view to future publication, he might not have wanted it to reach Smythe's hands. Also, his reference to her as a 'noble' lady strongly suggests she was an aristocrat, a category that did not encompass Dame Sarah.

16 Virginia, 1610, pp. 23–4.
17 Brown, 1890, doc. CLXI.
18 The Muscovy and other corporations had this stipulation. Kingsbury, 1906, vol. 1, Introduction, §4.
19 Kingsbury, 1906, vol. 3, doc. XVI; Brown, 1890, doc. CXCVIII.
20 HMC Report 75, vol. 4, p. 215; Kingsbury, 1906, doc. XVI.
21 W., 1612.
22 Brown, 1890, docs CCXX-CCXXII.
23 Johnson, 1612, pp. 5, 13–14, 21.
24 Brown, 1890, docs CCIX, CCXI, CCVI, CCXXI, CCXXII.
25 Croft, 1991, p. 794.
26 Walne, 1962, pp. 33, 307.
27 'Our plantations go on, the one doubtfully, the other desperately, Ireland with all our money and pains not yet settled in any fashion to assure us either profit or safety,' Calvert added. The Irish reference was to the Ulster Plantation, where Scottish settlers were being shipped over in their thousands following the Flight of the Earls. HMC Report 75, vol. 3, p. 344.
28 Brown, 1890, doc. CCXIX.
29 Cornwallis, 1641, pp. 24–75.
30 Strong, 2000, pp. 166–7; Haile, 1998, p. 841.

19 *The Good Husband*

1 *New DNB*, 'Rich, Robert, first Earl of Warwick (1559?–1619)'.
2 Haile, 1998, p. 801.
3 Haile, 1998, p. 868; Strachey, 1612; Greene, 1991, p. 58.
4 Haile, 1998, pp. 899–901.
5 Haile, 1998, p. 779.
6 Haile, 1998, pp. 868–9.
7 Haile, 1998, p. 754.
8 Smith, 1986, vol. 2, p. 243; vol. 3, p. 187.
9 The account of the taking of Pocahontas comes from Hamor (Haile, 1998, pp. 802*ff*), Smith (Smith, 1986, vol. 2, pp. 243–4) and Argall (Haile, 1998, pp. 754*ff*).
10 Haile, 1998, p. 618.
11 Smith, 1986, vol. 1, p. 247.
12 Haile, 1998, p. 662.
13 Haile, 1998, pp. 600, 680. Smith, 1986, vol. 2, p. 262.
14 Haile, 1998, p. 742.
15 Raleigh, 1596, pp. 51–2; Stymeist, 2002.
16 *Eastward Hoe*, Act III, scene iii, ll. 17–18
17 Beverley, 1947, p. 38.
18 Brown, 1890, p. 290; Deuteronomy, 7.
19 Haile, 1998, pp. 850–56.
20 The following account of Dale's expedition to the Pamunkey river comes from Hamor and Dale (Haile, 1998, pp. 806*ff*, 843*ff*). Smith's account in the fourth book of his *Generall Historie* (Smith, 1986, vol. 2, pp. 244*ff*) is based on Hamor's.
21 Hamor calls the town 'Matchcot', which Smith identifies as a town neighbouring Werowocomoco.
22 Beverley, 1947, p. 45; Gleach, 1997, p. 140; Purchas, 1617, p. 956.
23 Dale's account, based on the published version of a letter to a 'D. M.' (Haile, 1998, pp. 844–45), is garbled and confused, whereas Hamor's (Haile, 1998, pp. 808–9) is detailed and consistent.
24 Genesis, 25: 23–27.
25 Smith, 1986, vol. 2, pp. 246–7.
26 Haile, 1998, pp. 830*ff*.
27 Smith, 1986, vol. 2, p. 350; White, 1630, pp. 53–4.
28 'Some, seizing upon what was little more than an experiment with commissary problems, have even suggested that the early history of Jamestown was tinged with the colour of communism. There was,

in fact, about as much communism involved as may be found in a modern chain gang. Indeed, this specious and discredited view would merit not even the notice of a dismissal were it not for the fact that it remains to the present day for many Americans the most telling argument against the theories of Karl Marx.' (Craven, 1937, p. 319.) This view persisted into 2005. A columnist in the *Washington Times* argued that the 'real problem' faced by Jamestown 'was political'. 'The company that sponsored Jamestown provided for settlers to be fed from a common store. There was no incentive to be productive. But communism did not work. Gentlemen settlers spent their time hunting for gold – they found none. John Smith later instituted a new rule: Those who do not work shall not eat. That provided an incentive to produce food.' ('Seasonal smells of success', Edward Hudgins, *Washington Times*, 24 November 2005. http://www.washingtontimes.com/commentary/20051123-100556-7034r.htm.)

29 Haile, 1998, pp. 814–5. Hamor wrote that Dale allotted the land to 'every man in the colony' but other evidence (Haile, 1998, p. 902) suggests that it was only the 'ancient planters'.

30 LOC, Peter Force Collection, series 7, item 53.5; McCartney, 2003, figs 4–5, p. 46; Records, vol. 1, doc. LXXXVII; Smith, 1986, vol. 2, p. 247.

31 James I, 1604, A3^{v}.

20 *Twelfth Night*

1 Haile, 1998, doc. 54.

2 Smith, 1986, vol. 2, p. 430.

3 Haile, 1998, doc. 54. The evidence concerning this journey is ambiguous. Dale wrote in his letter to Winwood,'I shall with the greatest speed the wind will suffer me present myself unto you.' Purchas, who provides the account of Uttamatomakkin's visit, mentioned that he 'landed in the west parts', meaning Plymouth. It is possible that Dale, whose letter was written upon arrival, changed his mind about going to London by ship because of a change in weather conditions.

4 Rountree (Rountree, 1989, p. 15) discusses population estimates for Virginia at the time of contact with the Europeans, with the minimum number agreed to be about fourteen thousand. However, like figures for London, they are necessarily very approximate.

5 Wright, 1920, pp. 456–7.

6 Berry, 1989, p. 134; Salkeld, 2004; Prynne, 1633, p. 556.

7 London, 1616; Davies, 1617, epigram 129; Heywood, 1612, a2^{v}; Walne, 1962.

8 Vaughan, 2005, pp. 57–8.

9 An oil painting which may be de Passe's original is currently in the National Portrait Gallery, Washington DC, NPG.65.61. Kupperman, 2000, p. 200.

10 Brown, 1890, docs CCCLVII and CCCLVIII.

11 *Staple of News*, Act 2, scene v.

12 The accounts of Uttamatomakkin's interview are in Purchas, 1617, pp. 954–5 and Purchas, 1625, p. 1,774. Haile, 1998, doc. 55.

13 Rabb, 1998, p. 340; *Records*, vol. 1, doc. CXLIII.

14 Smith, 1986, pp. 171–2. Smith added a claim made by the Quiyough-cohannock chief that not all the children died, but that some were 'kept in the wildernesse by the yong men till nine moneths were expired'.

15 Beverley, 1947, pp. 207, 139; Gerard, 1907, p. 94; http//www.nlm.nih.gov/medlineplus/ency/article/002881.htm.

16 Purchas indicates in a marginal note that Rolfe contributes to the discussion.

17 Vaughan, 2005, p. 59; Kingsbury, 1906, vol. 1, 12 May 1620.

18 Haile, 1998, pp. 883–4.
19 Vaughan, 2005, p. 63.
20 Smith, 1986, vol. 2, pp. 258–60; Beverley, 1947, pp. 43–4.
21 Weinreb, 1987, p. 432. Stinking Street has since been renamed King Edward Street.
22 Beverley, 1947, pp. 42–3.
23 Barbour, 1964, p. 312; Smith, 1986, vol. 1, pp. 319, 340.
24 Chamberlain, 1939, vol. 2, p. 50.
25 Chamberlain, 1939, vol. 2, pp. 56–7.
26 Stephenson, 1913, p. 94.
27 Rich, 1615, pp. 25–6.
28 Brown, 1898, p. 268. The *George* returned in 1617 with 20,000 pounds of tobacco.
29 A lack of official company documentation for the period 1616–19 means that Sandys's role in the company during this period can only be guessed at. (Rabb, 1998, pp. 332–5.)
30 Rountree, 1989, p. 114. The English record only in passing the terms by which they acquired Indian territory, e.g. Kingsbury, 1906, vol. 3. doc. CXVI. 'That the lands of this country were taken from them by conquest, is not so general a truth as is supposed,' wrote Thomas Jefferson (Jefferson, 1905, vol. 2, p. 221). 'I find in our historians and records, repeated proofs of purchase.' For the early colonial period, such proofs are scarce and never detailed.
31 Haile, 1998, doc. 53; p. 554.
32 Brown, 1890, p. 634; Smith, 1986, vol. 2, p. 356; Craven, 1937, p. 189, *New DNB*, 'Elfrith, Daniel (*fl.* 1607–1640)'.
33 The policy is fully set out in the company's instructions to George Yeardley of 18 November 1618. (Kingsbury, 1906, vol. 1, doc. XLVII.)
34 HMC, Report 75, 'MSS of the Marquess of Downshire', vol. 5, p. 487.
35 Bradford, 1981, pp. 17, 21, 26.
36 Sandys, 1605, S4^r; Kingsbury, 1906, vol. 1, doc. XCIII.
37 University of Virginia, MS8963-A, 'Copy of agreement "Found in Old Map Room during cleaning Dec. 1968"'; Kingsbury, 1906, vol. 1, doc. LXV.
38 Smith, 1986, vol. 2, p. 261.
39 Chamberlain to Carleton, 18 January 1616–17, in *CSP Dom – James I*, xc, No. 25.
40 Mossiker, 1996, p. 278; Rountree, 1989, p. 130; Smith, 1986, vol. 1, p. 59.
41 Haile, 1998, p. 889.
42 Haile, 1998, p. 888.
43 Brown, 1898, p. 268.
44 Haile, 1998, p. 620; Purchas, 1617, pp. 956–7; Kingsbury, 1906, vol. 1, doc. XXV; Beverley, 1947, p. 43.

PART FIVE

21 *Imbangala*

1 Sluiter, 1997.
2 Thornton, 1998; Purchas, 1625, pp. 974*ff*; Thornton, 1999, pp. 100*ff*.
3 Sluiter, 1997, p. 397.
4 Payment could also be made in sassafras, but this was not considered an agricultural commodity.
5 Kingsbury, 1906, vol. 3, doc. XXXVIII; vol. 2, 7 May 1623.
6 Craven, 1930, p. 460; PRO Manchester Papers MS 278.
7 Kingsbury, 1906, vol. 2, 19 June 1622.
8 Kingsbury, 1906, vol. 3, docs XLI, XLIII.
9 Kingsbury, 1906, vol. 2, 7 May 1623; vol. 3, docs CLII, LIX.
10 Kingsbury, 1906, vol. 1, 3 April 1620.
11 Martin Brandon was a plantation distinct from Martin's Hundred, which was named after Richard Martin, the MP and Sirenaical.
12 Kingsbury, 1906, vol. 3, doc. XLVII. The identification of this document, entitled 'Instructions to George Yeardley', with the Great Charter is revealed by a short section quoted in Pory's report of the General Assembly's meeting (Kingsbury, 1906,

vol. 3, doc. LXV). Wesley Craven argued that, despite the name 'Great Charter', it was a 'misinterpretation' to see these instructions as some sort of 'Magna Carta', because it 'deals with the question of government and political rights only in the most general manner'. This is true of the copy that Kingsbury transcribed, which, as he puts it, is 'slightly mutilated' (Craven, 1964, pp. 52–3). But one of the missing sections clearly concerns the colony's government. Its introductory sentence happens to be quoted in Pory's report of the General Assembly: 'And forasmuche as our intente is to establish one equall and uniforme kinde of government over all Virginia &c.' What remains in the surviving copy is 'And forasmuch as our intent is to Establish one Equal . . .' and after a 'blank of several lines', as Kingsbury puts it, continues '. . . Plantations, whereof we shall speak afterwards, be reduced into four Cities or Burroughs'. This clearly shows that a key part of the document is missing. When or why it was removed is unknown, but it means that Craven's assertion that the document lacks political significance is unproven.

13 Haile, 1998, p. 912.
14 Kingsbury, 1906, vol. 3, doc. XCVI.
15 Chamberlain, 1939, vol. 1, p. 188; Craven, 1964, p. 81; Kingsbury, 1906, vol. 3, docs XCVI, LXXXIX.
16 Kingsbury, 1906, vol. 3, doc. LXV.
17 McCartney, 2003, p. 28.
18 PRO Manchester Papers 279; Kingsbury, 1906, vol. 3, doc. XCIV.
19 Ransome, 1992, reel 1, 159; Hotten, 1962; Thorndale, 1995; McCartney, 2003; Kingsbury, 1906, vol. 1, 17 November 1619.
20 Morgan, 1975, pp. 7*ff*; Taylor, 1935, pp. 142–3; Guasco, 2005, p. 240.
21 In the surviving documents of the period, slavery is only ever referred to as a punishment, and usually in connection with the ill-treatment arising from Dale's laws, such as when the Ancient Planters collectively complained of the 'extreme slavery' they had suffered during that time (Haile, 1998, p. 900).
22 Kingsbury, 1906, vol. 3, doc. XCVIII.
23 Morgan, 1971, pp.176–8; Nathaniel Butler makes several references to 'Newgatiers' being shipped to Bermuda (Manchester Papers, MS 284). An example of a surviving indenture shows a Robert Coopy, 'husbandman', of Gloucestershire being offered transport and living costs in return for promising 'faythfully to serve . . . Sr Willm [Throckmorton], Richard George and Iohn [Smyth of Nibley] for three yeares from datye here of his landinge in the land of Virginia'. After that time, he would be a 'free man of the said Cuntry theirby to enioy all the liberties freedomes and priviledges of a freeman there' (Kingsbury, 1906, vol. 3, doc. LXXVI).
24 Kingsbury, 1906, vol. 3, doc. LXXXII.
25 Kingsbury, 1906, vol. 3, doc. CLXXXII.
26 Kingsbury, 1906, vol. 3, doc. CCLII; Morgan, 1971, pp. 179–80.
27 Kingsbury, 1906, vol. 1, p. 303; Morgan, 1971, pp. 181–2.
28 Kingsbury, 1906, vol. 3, doc. LXXXII.

22 *The Treasurer*

1 Kingsbury, 1906, vol. 1, 28 April 1619.
2 Kingsbury, 1906, vol. 1, 17 July 1619.
3 Kingsbury, 1906, vol. 3, doc. LXXXI.
4 Craven, 1930, pp. 461*ff*.
5 PRO Manchester Papers, 252.
6 *Acts of the Privy Council*, 1619–21, p. 142.
7 Kingsbury, 1906, vol. 1, 17 May 1620.
8 Kingsbury, 1906, vol. 3, doc. CXIV.
9 Chamberlain, 1939, vol. 2, p. 305; Rabb, 1998, p. 349.
10 Sandys sent Ferrar a 'discourse' which appears to have related to this matter, as well as suspicions about

John Pory secretly acting on behalf of the Warwick faction. Sandys asked the deputy to forward the discourse to Southampton (Kingsbury, 1906, vol. 3, doc. CXVII).

11 Kingsbury, 1906, vol. 1, 15 November 1620.

12 Beverley, 1947, p. 50. 'Anno, 1622, Inferior Courts were first appointed by the General Assembly, under the Name of County Courts, for Tryal of Minute Causes; the Governour and Council still remaining judges of the Supream Court of the Colony.'

13 Jardine, 1999, p. 442.

14 Rabb, 1998, pp. 234–5. For a discussion on the issue of tenure in Virginia, and the concept of 'free and common socage' used in the Royal Charters, see McPherson, 1998.

15 Rabb, 1998, p. 215.

16 Woodnoth, 1651, p. 4. Craven (Craven, 1964, pp. 10–11) considers this remark to be of 'doubtful historical value', mainly because it challenges his argument that James's view of the Virginia Company was shaped by commercial rather than political considerations. It first appeared in Woodnoth, who was writing in about 1644, and was later repeated by Ferrar (Peckard, 1790, pp. 115–6), who claimed to have heard it from Southampton, who in turn had heard it from a fellow lord at court. This roundabout route may demote it to hearsay (though it should be pointed out that most unofficial royal quotes arise from court gossip and memoirs), but reveals that it was at least current in the highest reaches of the court, and reflected contemporary beliefs.

17 Rabb, 1998, p. 254.

18 Thompson, 1997, p. 781.

23 *The 'Viperous Brood'*

1 Davis, 1955, chap. VI.

2 Beverley commented on the intermingling of the Indians and the English at this time. Beverley, 1947, p. 50.

3 Copland, 1622, pp. 12–13. The original letter has not survived.

4 Kingsbury, 1906, vol. 1, 10 July 1621, 22 February 1619/20; vol. 3, doc. XLIX; Waterhouse, 1622, pp. 15–16.

5 Haile, 1998, p. 517.

6 Beverley, 1947, p. 52.

7 Kingsbury, 1906, vol. 3, doc. LXXXVI.

8 Kingsbury, 1906, vol. 3, doc. CCXII; Waterhouse, 1622, p. 16.

9 Kingsbury, 1906, vol. 1, 31 May 1620.

10 Smith, 1986, vol. 2, p. 293.

11 Waterhouse, 1622, pp. 20–21; Kingsbury, 1906, vol. 3, doc. CCCXIX. Waterhouse only identifies Pace and Perry by their surnames. Richard Pace was a tobacco planter in Virginia at this time (Kingsbury, 1906, vol. 3, doc. CLXXXVI), and, in 1624, another planter, William Pace, returned to England, where he applied for help with the maintenance costs of an 'Indian Boy'. Pace's request was granted, perhaps because he was the Indian who warned of the impending attack (Kingsbury, 1906, vol. 2, 26 April 1624). Kingsbury transcribed Chanco as 'Chauco'.

12 Waterhouse, 1622, p. 14.

13 Waterhouse, 1622, p. 17.

14 Waterhouse, 1622, pp. 35*ff*.

15 Smith, 1986, vol. 2, pp. 294–5.

16 Waterhouse, 1622, p. 14.

17 Kingsbury, 1906, vol. 1, 15 November 1620.

18 Kingsbury, 1906, vol. 3, doc. CCXII.

19 Davis, 1955, p. 129.

20 Rose, 2005, p. 103.

21 Kingsbury, 1906, vol. 3, doc. CLXXXVI.

22 Kingsbury, 1906, vol. 4, doc. CCCLXVIII.

23 Kingsbury, 1906, vol. 4, doc. CCLXXXV; vol. 3, p. 258; Review, 1922, p. 526. Pory mentions '10m' of the 'best tobacco' being shipped that

year. This is taken to mean £10,000. The quantities of tobacco produced in Virginia around this time are hard to assess, as much of it was being sold in Holland to circumnavigate the royal impost.

24 Manchester Papers 317; Kingsbury, 1906, vol. 4, doc. CCXCIII.

25 Manchester Papers 325; Kingsbury, 1906, vol. 4, doc. CCXCVIII.

26 Rose, 2005, pp. 106–7.

24 *The Unmasked Face*

1 Thompson, 1997, p. 782.

2 Rabb, 1998, pp. 267–8.

3 Rabb, 1998, pp. 260–5; Bradford, 1981, p. 44.

4 Woodnoth, 1651, p. 4.

5 Kingsbury, 1906, vol. 2, 22 May 1622.

6 Kingsbury, 1906, vol. 2, 5 June 1622.

7 Kingsbury, 1906, vol. 3, doc. CCLII.

8 Greengrass, 1999, p. 69. The Church of England issued a pamphlet in 1606 entitled *Prayers and thankesgiuing to be vsed by all the Kings Maiesties louing subiects, for the happy deliuerance of his Maiestie, the Queene, Prince, and states of Parliament, from the most traiterous and bloody intended massacre by gunpowder, the 5 of Nouember 1605*. A 1607 work by William Russell reported on the 'bloudie and terrible massacre in the citty of Mosco with the fearefull and tragicall end of Demetrius the last Duke'. Otherwise, before 1622, the word massacre had generally referred to events that took place during the French Wars of Religion, in particular the St Bartholomew's Day Massacre in Paris, 1572.

9 Kuin, 1998, pp. 568–9; Mornay, 1587, pp. 117–8.

10 Waterhouse, 1622, pp. 22–4.

11 Donne, 1623, A3[r].

12 Drayton, 1622, p. 265. A series of economic, legal and educational tracts also appeared in 1622 and 1623, promoting the venture: Malynes, 1622; Brinsley, 1622.

13 Hagthorpe, 1622, A3[v]; Smith, 1986, vol. 3, p. 51.

14 Kingsbury, 1906, vol. 3, p. 669.

15 Kingsbury, 1906, vol. 2, 4 December 1622.

16 Kingsbury, 1906, vol. 2, pp. 374–7.

17 Kingsbury, 1906, vol. 4, doc. CCCXXXIX. Rich added a note to a summary list of the evidence, drawn up in May or June 1623, that he had provided an 'abstract' of Frethorne's letter, but evidently did not have it to hand, as it 'must be added out of the copy at large'. Rose writes, 'If Rich did not have Frethorne's letter at hand, how did he know the indentured servant would see things his way? One explanation is that he was not waiting to get the letter from Frethorne's parents in England, but already knew what the tenor of its contents would be.' It is, of course, possible that he did not have it simply because, having perused it, he was still circulating it among Warwick's partisans. However, as Rose points out, there is evidence that the Warwick faction was not above forging, or perhaps creatively editing, the evidence it used to support its case (Rose, 2005, pp. 106–7).

18 Kingsbury, 1906, vol. 2, 7 May 1623.

19 Kingsbury, 1906, vol. 1, pp. 108–9.

20 Kingsbury, 1906, vol. 1, 2 May 1621.

21 Smith, 1986, vol. 2, pp. 327–32.

22 Craven, 1964, p. 308.

23 *Acts of the Privy Council*, 1621–3, pp. 490–1; Craven, 1964, pp. 307–10.

24 Craven, 1964, p. 315.

25 Kingsbury, 1906, vol. 2, pp. 526–8.

26 Craven, 1964, pp. 320–1, 326–7; Smith, 1986, vol. 2, p. 332.

27 Kingsbury, 1906, vol. 2, pp. 104–5; Bennett, 1936; Boddie, 1936; Wight, 1899, p. 206. Richard Bennett may have gone later to Virginia (*New DNB*, 'Bennett, Richard

(1608–1675)'). Another Robert Bennett was in Virginia at this time, resident of Elizabeth City. LOC Peter Force MS, Series 7, Item 53.5, muster of inhabitants 23 February 1624/5.

28 Bennett, 1620.

29 *CSP Colonial* vol. 1, p. 36.

30 Kingsbury, 1906, vol. 2, 9 November 1623.

31 Kingsbury, 1906, vol. 4, doc. CCCLXVIII.

32 *CSP Colonia*l, vol. 1, p. 48.

33 *CSP Colonial*, vol. 1, pp. 53, 58.

34 Haile, 1998, p. 912.

35 Davis, 1955, p. 180.

36 Tyler, 1893.

37 Kingsbury, 1906, vol. 4, doc. CCCLXVIII; *CSP Colonial*, vol. 1, p. 48.

38 Kingsbury, 1906, vol. 4, p. 451.

39 Boddie, 1933, p. 118.

40 Kingsbury, 1906, vol. 4, doc. CDXXIX.

41 Purchas, 1623, p. 100. Purchas quotes, '*O quam te memorem virgo?*', *Aeneid*, 1. 327.

42 Smith, 1986, vol. 2, pp. 320–1. According to the reports reaching Smith, it was not certain the head was Spelman's, but he was never seen again and was reported dead a few months later.

43 Smith, 1986, vol. 1, p. 148; vol. 2, p. 167.

44 Haile, 1998, p. 661.

BIBLIOGRAPHY

Ackroyd, P. (2001), *London: The Biography* (London, Vintage).

Arents, G. (1939), 'The Seed from which Virginia Grew', *William and Mary College Quarterly,* vol. 19: 2, pp. 123–129.

Aubrey, J. (1962), *Aubrey's Brief Lives* (Ann Arbor, Michigan, University of Michigan Press).

Augustine, S., Bishop of Hippo, J. Healey, et al. (1610), *St. Augustine, Of the citie of God vvith the learned comments of Io. Lod. Viues. Englished by I.H.* (London: Printed by George Eld).

Barbour, P. L. (1962), 'Captain George Kendall: Mutineer or Intelligencer?', *Virginia Magazine of History and Biography,* vol. 70.

Barbour, P. L. (1964), 'The Identity of the First Poles in America', *William and Mary College Quarterly,* vol. 21, pp. 77–92.

Barbour, P. L. (1964), *The Three Worlds of Captain John Smith* (London, Macmillan).

Barbour, P. L. (1971), 'The Honorable George Percy, Premier Chronicler of the First Virginia Voyage', *Early American Literature,* vol. 6, pp. 7–17.

Barbour, P. L. and S. Hakluyt (1969), *The Jamestown Voyages Under the First Charter, 1606–1609: documents relating to the foundation of Jamestown and the history of the Jamestown colony up to the departure of Captain John Smith, last president of the council in Virginia under the first charter, early in October 1609* (Cambridge, published for the Hakluyt Society at the University Press).

Batho, G. R. (1957), 'The Finances of an Elizabethan Nobleman: Henry Percy, Ninth Earl of Northumberland (1564–1632)', *Economic History Review,* vol. 9: 3, pp. 433–450.

Batho, G. R. (1960), 'The Library of Henry Percy, 9th Earl of Northumberland (1564–1632)', *The Library,* vol. 15, pp. 246–61.

Batho, G. R. (2000), 'Thomas Harriot and the Northumberland Household', *Thomas Harriot: An Elizabethan Man of Science,* R. Fox (Aldershot, Ashgage Press): pp. 28–47.

Beckwith, I. (1976), 'Captain John Smith: The Yeoman Background', *History Today,* vol. 26: 7, p. 444.

Bemiss, S. M., 1894–1966 (1957), 'John Martin, Ancient Adventurer', *Virginia Magazine of History and Biography*, vol. 65, pp. 209–221.

Bennett, A. F. (1936), 'Bennett', *William and Mary College Quarterly Historical Magazine*, vol. 16: 2, pp. 316–18.

Bennett, E. (1620), *A treatise deuided into three parts, touching the inconueniences, that the importation of tobacco out of Spaine, hath brought into this land viz. 1 In the first is shewed how treasure was vsually brought into this land. 2 In the second, what hath and doth hinder the bringing of it, with other inconueniences. 3 In the third, how to remedie the one, and the other* (London: For John Budge).

Bennett, M. K. (1955), 'The Food Economy of the New England Indians, 1605–75', *The Journal of Political Economy*, vol. 63: 5, pp. 369–97.

Benson, G. (1609), *A sermon preached at Paules Crosse the seauenth of May, M.DC.IX. By George Benson . . .* (Imprinted at London: By H. L[ownes] for Richard Moore, and are to be sold at his shop in S. Dunstans Church-yard).

Bernhard, V. (1992), ' "Men, Women and Children" at Jamestown: Population and Gender in Early Virginia, 1607–1610', *The Journal of Southern History*, vol. 58: 4, pp. 599–618.

Bernhard, V. (1994), 'A Response: The Forest and the Trees: Thomas Camfield and the History of Early Virginia', *The Journal of Southern History*, vol. 60: 4, pp. 663–70.

Berry, H. (1989), 'The First Public Playhouses, Especially the Red Lion', *Shakespeare Quarterly*, vol. 40: 2, pp. 133–48.

Beverley, R. (1722), *The history of Virginia, in four parts* (London, B. and S. Tooke).

Beverley, R., L. B. Wright, et al., eds (1947), *The History and Present State of Virginia* (Chapel Hill, University of North Carolina Press).

Blanton, D. B. (2000), 'Drought as a Factor in The Jamestown Colony, 1607–1612', *Historical Archaeology*, vol. 34: 4.

Blick, J. P. (2000), 'The Quiyoughcohannock Ossuary Ritual and the Feast of the Dead', *Proceedings of the 6th Internet World Congress for Biomedical Sciences* (Ciudad Real, Spain, INABIS).

Boddie, J. B. (1933), 'Edward Bennett of London and Virginia', *William and Mary College Quarterly Historical Magazine*, vol. 13: 2, pp. 117–30.

Boddie, J. B. (1936), 'Genealogical Queries', *William and Mary College Quarterly Historical Magazine*, vol. 16: 3, pp. 493–8.

Bottigheimer, K. S. (1979), 'Kingdom and Colony: Ireland in the Westward Enterprise, 1536–1660', *The Westward Enterprise*, K. R. Andrews, N. P. Canny and P. E. H. Hair, (Liverpool, Liverpool University Press: pp. 45–65).

Boulton, J. (1996), 'Wage Labour in Seventeenth-Century London', *Economic History Review,* vol. 49: 2, pp. 268–90.

Bovill, E. W. (1968), 'The Madre de Dios', *Mariner's Mirror [Great Britain],* vol. 54, pp. 129–52.

Bradford, W. (1981), *Of Plymouth Plantation, 1620–1647* (New York, Random House).

Brinsley, J. (1622), *A consolation for our grammar schooles: or, a faithfull and most comfortable incouragement, for laying of a sure foundation of all good learning in our schooles, and for prosperous building thereupon More specially for all those of the inferiour sort, and all ruder countries and places; namely, for Ireland, Wales, Virginia, with the Sommer Ilands, and for their more speedie attaining of our English tongue by the same labour, that all may speake one and the same language. . . .* (London: Printed by Richard Field for Thomas Man, dwelling in Pater noster Row, at the signe of the Talbot).

Brown, A. (1890), *The genesis of the United States ; a narrative of the movement in England, 1605–1616, which resulted in the plantation of North America by Englishmen, disclosing the contest between England and Spain for the possession of the soil now occupied by the United States of America; set forth through a series of historical manuscripts now first printed, together with a reissue of rare contemporaneous tracts, accompanied by bibliographical memoranda, notes, and brief biographies* (London, W. Heinemann).

Brown, A. (1898), *The First Republic in America: An Account of the Origin of this Nation* (Boston and New York, Houghton, Mifflin).

Bruce, D. H. and T. M. Cranfill (1953), *Barnaby Rich: A Short Biography* (Austin; Edinburgh, University of Texas Press: Thomas Nelson).

Bushnell, E. (1664), *The compleat ship-wright plainly and demonstratively teaching the proportions used by experienced ship-wrights according to their custome of building, both geometrically and arithmetically performed: to which by Edmund Bushnell, ship-wright* (London: Printed by W. Leybourn for George Hurlock, and are to be sold at his shop. . .).

Butler, N., J. Smith, et al. (1882), *The Historye of the Bermudaes or Summer Islands* (London, Printed for the Hakluyt Society).

Camfield, T. M. (1994), 'A Can or Two of Worms: Virginia Bernhard and the Historiography of Early Virginia, 1607–1610', *The Journal of Southern History,* vol. 60: 4, pp. 649–62.

Campion, E., Saint, M. Hanmer, et al. (1633), *Tvvo histories of Ireland. The one written by Edmund Campion, the other by Meredith Hanmer Dr of Divinity* (Dublin: Printed by the Society of Stationers [and London: by Thomas Harper]).

Canny, N. P. (1973), 'The Ideology of English Colonization: From Ireland to America', *William and Mary College Quarterly Historical Magazine*, vol. 30: 4, pp. 575–98.

Casas, B. d. l. and M. M. S. (1583), *The Spanish colonie, or Briefe chronicle of the acts and gestes of the Spaniardes in the West Indies, called the newe world, for the space of xl. yeeres: written in the Castilian tongue by the reuerend Bishop Bartholomew de las Cases or Casaus, a friar of the order of S. Dominicke. And nowe first translated into english, by M.M.S.* (Imprinted at London: [By Thomas Dawson] for William Brome).

Casey, J. (1999), *Early Modern Spain: A Social History* (London; New York, Routledge).

Chamberlain, J., N. E. McClure, et al. (1939), *The Letters of John Chamberlain* (Philadelphia, The American Philosophical Society).

Chapman, G. and I. Jones (1613), *The memorable maske of the two honorable houses or Innes of Court; the Middle Temple, and Lyncolns Inne As it was performd before the King, at White-Hall on Shroue Munday at night; being the 15. of February. 1613. At the princely celebration of the most royall nuptialls of the Palsgraue, and his thrice gratious Princesse Elizabeth. &c. With a description of their whole show; in the manner of their march on horse-backe to the Court from the Maister of the Rolls his house: with all their right noble consorts, and most showfull attendants. Inuented, and fashioned, with the ground, and speciall structure of the whole worke: by our kingdomes most artfull and ingenious architect Innigo Iones. Supplied, aplied, digested, and written, by Geo: Chapman* (At London: Printed by G. Eld, for George Norton, and are to be sould at his shoppe neere Temple-bar).

Chapman, G. and T. M. Parrott (1914), *The Plays and Poems of George Chapman* (London).

Churchyard, T. (1579), *A generall rehearsall of warres, called Churchyardes choise wherein is fiue hundred seuerall seruices of land and sea as seiges, battailes, skirmiches, and encounters. A thousande gentle mennes names, of the beste sorte of warriours. A praise and true honour of soldiours. A proofe of perfite nobilitie. A triall and first erection of heraldes. A discourse of calamitie. And ioyned to the same some tragedies & epitaphes, as many as was necessarie for this firste booke. All which workes are dedicated to the hounourable sir Christopher Hatton knight, . . . Written by Thomas Churchyard Gent. 1579* (Imprinted at London: By [John Kingston for] Edward White, dwellyng at the little north-doore of S. Paules Churche, at the signe of the Gunne).

Colet, C. and A. Munday (1588), *The famous, pleasant, and variable historie, of Palladine of England Discoursing of honorable aduentures, of knightly deedes of armes and chiualrie: enterlaced likewise with the loue of sundrie noble personages, as time and affection limited their desires. . . . Translated out of French by A.M. one of the messengers of her Maiesties Chamber* (At London: Printed by Edward Allde for Iohn Perin, dwelling in Paules Churchyard at the signe of the Angell, and are there to be sould).

Cook, J. (2001), *Dr Simon Forman: A Most Notorious Physician* (London).

Copland, P. and P. Pope (1622), *Virginia's God be thanked, or A sermon of thanksgiving for the happie successe of the affayres in Virginia this last yeare. Preached by Patrick Copland at Bow-Church in Cheapside, before the Honorable Virginia Company, on Thursday, the 18. of Aprill 1622. And now published by the commandement of the said honorable Company. Hereunto are adjoyned some epistles, written first in Latine (and now Englished) in the East Indies by Peter Pope, an Indian youth, borne in the bay of Bengala, who was first taught and converted by the said P.C. And after baptized by Master Iohn Wood, Dr in Divinitie, in a famous assembly before the Right Worshipfull, the East India Company, at S. Denis in Fan-Church streete in London, December 22. 1616* (London: Printed by I[ohn] D[awson] for William Sheffard and Iohn Bellamie, and are to be sold at his shop at the two Grey-hounds in Corne-hill, neere the Royall Exchange).

Cornwallis, C., Sir (1641), *The life and death of our late most incomparable and heroique prince, Henry Prince of Wales A prince (for valour and vertue) fit to be imitated in succeeding times. Written by Sir Charles Cornvvallis knight, treasurer of his Highnesse houshold* (London: printed by Iohn Dawson for Nathanael Butter).

Coryate, T. (1616), *Thomas Coriate traueller for the English vvits: greeting From the court of the Great Mogul, resident at the towne of Asmere, in easterne India* ([London]: Printed by W. Iaggard, and Henry Fetherston).

Crakanthorpe, R. (1609), *A sermon at the solemnizing of the happie inauguration of our most gracious and religious soueraigne King Iames wherein is manifestly proued, that the soueraignty of kings is immediatly from God, and second to no authority on earth whatsoeuer : preached at Paules Crosse, the 24. of March last 1608 / by Richard Crakanthorpe . . .* (London: Printed by VV. Iaggard for Tho. Adams, dwelling in Paules Church-yard, at the signe of the Blew Bell).

Crashaw, W. (1610), *A sermon preached in London before the right honourable the Lord Lavvarre, Lord Gouernour and Captaine Generall of Virginea, and others of his Maiesties Counsell for that kingdome, and the rest of the aduenturers in that plantation At the said Lord Generall his leaue taking of England his natiue countrey, and departure for Virginea, Febr. 21. 1609. By W. Crashaw Bachelar of Diuinitie, and preacher at the Temple. Wherein both the lawfulnesse of that action is maintained, and the necessity thereof is also demonstrated, not so much out of the grounds of policie, as of humanity, equity, and Christianity. Taken from his mouth, and published by direction* (London: Printed [by W. Hall] for William Welby, and are to be sold in Pauls Church-yard at the signe of the Swan).

Craven, W. F. (1930), 'The Earl of Warwick, a Speculator in Piracy', *Hispanic American Historical Review*, vol. 10: 4, pp. 457–79.

Craven, W. F. (1937), 'An Introduction to the History of Bermuda I', *William and Mary College Quarterly Historical Magazine*, vol. 17: 2, pp. 176–215.

Craven, W. F. (1937), 'An Introduction to the History of Bermuda II', *William and Mary Quarterly*, vol. 17: 3, pp. 317–62.

Craven, W. F. (1964), *Dissolution of the Virginia Company: The Failure of a Colonial Experiment* (Gloucester, Massachusetts, P. Smith).

Croft, P. (1991), 'The Religion of Robert Cecil', *Historical Journal*, vol. 34: 4, pp. 773–96.

Cromwell, O. (1655), *A declaration of His Highnes, by the advice of his Council setting forth, on the behalf of this Commonwealth, the justice of their cause against Spain. Friday the 26. of October, 1655. Ordered by His Highness the Lord Protector, and the Council, that this declaration be forthwith printed and published. Hen: Scobel, Clerk of the Council* (Edinburgh: re-printed by Christopher Higgins, in Harts-Close, over against the Trone-Church).

Cronin, M. (2001), *A History of Ireland* (Houndmills, Basingstoke, Hampshire; New York, Palgrave).

Culliford, S. G. (1965), *William Strachey, 1572–1621* (Charlottesville, Virginia, University Press of Virginia).

Davenport, F. G. (1917), *European Treaties Bearing on the History of the United States and its Dependencies to 1648* (Washington DC, Carnegie Institution of Washington).

Davies, J. (1617), *VVits bedlam ——vvhere is had, whipping-cheer, to cure the mad* (London: Printed by G. Eld, and are to be sould by Iames Dauies, at the Red Crosse nere Fleete-streete Conduit).

Davis, R. B. (1955), *George Sandys, Poet-Adventurer: A Study in Anglo-American Culture in the Seventeenth Century* (London; New York, The Bodley Head: Columbia University Press).

De la Warr, T. W., Baron, 1577–1618 (1611), *The relation of the Right Honourable the Lord De-La-Warre, Lord Gouernour and Captaine*

Generall of the colonie, planted in Virginea (London: Printed by VVilliam Hall, for William Welbie, dwelling in Pauls Church-yeard at the signe of the Swan).

Dee, J. (1604), *To the Kings most excellent Maiestie* (London: Printed by E. Short, dwelling on Bred-streete hill neere to the end of old Fish-streete, at the signe of the Starre).

Dee, J. and E. Fenton (1998), *The Diaries of John Dee* (Charlbury, Day Books).

Dekker, T. (1630), *The blacke rod, and the vvhite rod (justice and mercie) striking, and sparing, [brace] London* (London: Printed by B.A. and T.F. for Iohn Covvper).

Dietz, B. (1991), 'The Royal Bounty and English Merchant Shipping in the Sixteenth and Seventeenth Centuries', *Mariner's Mirror [Great Britain]*, vol. 77, pp. 5–20.

Dodd, A. H. (1938), 'A Spy's Report, 1604', *Bulletin of Celtic Studies*, vol. 9, pp. 154–67.

Donne, J. (1623), *Three sermons vpon speciall occasions preached by Iohn Donne* (London: Printed for Thomas Iones, and are to be sold at his shop in the Strand at the Blacke Rauen neere St. Clements Church).

Drayton, M. (1622), *The second part, or a continuance of Poly-Olbion from the eighteenth song Containing all the tracts, riuers, mountaines, and forrests: intermixed with the most remarkable stories, antiquities, wonders, rarities, pleasures, and commodities of the east, and northerne parts of this isle, lying betwixt the two famous riuers of Thames, and Tweed. By Michael Drayton, Esq.* (London: Printed by Augustine Mathewes for Iohn Marriott, Iohn Grismand, and Thomas Dewe).

Drummond, W. (1842), *Notes of Ben Jonson's Conversations with W. D.* (London, Shakespeare Society Publications).

Elliott, J. H. (2002), *Imperial Spain, 1469–1716* (London, Penguin).

Elton, G. R. (1968), 'The Law of Treason in the Early Reformation', *Historical Journal*, vol. 11: 2, pp. 211–36.

Emerson, K. L. (1984), *Wives and Daughters: The Women of Sixteenth-Century England* (Troy, New York, Whitston Publishing Company).

Federal Writers' Project (NY) (1939), *New York city guide; a comprehensive guide to the five boroughs of the metropolis: Manhattan, Brooklyn, the Bronx, Queens, and Richmond* (New York, Random House).

Fitzmaurice, A. (1999), 'The Civic Solution to the Crisis of English Colonization, 1609–1625', *Historical Journal*, vol. 42: 1, pp. 25–51.

Foster, D. W. (1987), 'Master W. H., R. I. P.', *Publications of the Modern Language Association of America*, vol. 102: 1, pp. 42–54.

Fourquevaux, R. d. B. d. P. b. d., P. Ive, et al. (1589), *Instructions for the warres: Amply, learnedly, and politiquely, discoursing the method of militarie discipline. Originally written in French by that rare and worthy generall, Monsieur William de Bellay, Lord of Langey, Knight of the order of Fraunce, and the Kings lieutenant in Thurin. Translated by Paule Iue, Gent* (London, printed [by Thomas Orwin] for Thomas Man, and Tobie Cooke).

Fowler, E. (2000), 'The Ship Adrift', *Critical Views* (London, Reaktion: 37–40).

Fox, John (1576), *Actes and monuments of these latter and perillous dayes* (London, John Daye).

Fuller, T., J. Fuller, et al. (1662), *The history of the worthies of England who for parts and learning have been eminent in the several counties : together with an historical narrative of the native commodities and rarities in each county / endeavoured by Thomas Fuller* (London: Printed by J.G.W.L. and W.G.).

Gallivan, M. D. (2003), *James River Chiefdoms: The Rise of Social Inequality in the Chesapeake* (Lincoln, Nebraska, University of Nebraska Press).

Gayley, C. M. (1917), *Shakespeare and the Founders of Liberty in America* (New York, The Macmillan Company).

Gerard, W. R. (1904), 'The Tapehanek Dialect of Virginia', *American Anthropologist,* vol. 6: 2, pp. 313–30.

Gerard, W. R. (1907), 'Virginia's Indian Contributions to English', *American Anthropologist,* vol. 9: 1, pp. 87–112.

Gleach, F. W. (1997), *Powhatan's World and Colonial Virginia: A Conflict of Cultures* (Lincoln, Nebraska, University of Nebraska Press).

Gookin, W. F. (1949), 'Who was Bartholomew Gosnold?', *William and Mary Quarterly,* vol. 6: 3, pp. 398–415.

Gookin, W. F., 1882–1953, and L. Barbour Philip (1963), *Bartholomew Gosnold, Discoverer and Planter. New England–1602, Virginia–1607. By Warner F. Gookin, B.D., with footnotes and a concluding part by Philip L. Barbour [With a map]* (Hamden, Connecticut; London, Archon Books).

Gray, R., (1609), *A Good Speed to Virginia* (London: Printed by Felix Kyngston for VVilliam Welbie, and are to be sold at his shop at the signe of the Greyhound in Pauls Churchyard).

Greenblatt, S. (2004), *Will in the World: How Shakespeare Became Shakespeare* (London, Jonathan Cape).

Greene, T. and S. B. Flexner (1991), *The Language of the Constitution: A Sourcebook and Guide to the Ideas, Terms, and Vocabulary Used by the Framers of the United States Constitution* (New York; London, Greenwood Press).

Greengrass, M. (1999), 'Hidden Transcripts, Secret Histories And Personal Testimonies of Religious Violence In The French Wars of Religion',

The Massacre in History, M. Levene and P. Roberts (New York: 69–88).

Guasco, M. J. (2005), 'Settling with Slavery: Human Bondage in the Early Anglo-American World', *Envisioning an English Empire: Jamestown and the Making of the North Atlantic World*, R. Appelbaum and J. W. Sweet (Philadelphia, University of Pennsylvania Press): pp. 236–253.

Guybert, P. and I. W. (1639), *The charitable physitian with The charitable apothecary / vvritten in French by Philbert Guibert Esquire physitian regent in Paris, and by him after many severall editions, reviewed, corrected, amended, and augmented ; and now faithfully translated into English, for the benefit of this kingdome, by I.W.* (London: Printed by Thomas Harper, and are to bee sold by Lawrence Chapman at his shop at Chancery lane end, next Holborne).

Hagthorpe, J. (1622), *Diuine meditations, and elegies. By Iohn Hagthorpe Gentleman* (London: Printed by Bernard Alsop).

Haile, E. W. (1998), *Jamestown Narratives: Eyewitness Accounts of the Virginia Colony, the First Decade, 1607–1617* (Champlain, Virginia, RoundHouse).

Hakluyt, R. (1599), *The principal nauigations, voyages, traffiques and discoueries of the English nation made by sea or ouer-land, to the remote and farthest distant quarters of the earth, at any time within the compasse of these 1600. yeres: deuided into three seuerall volumes, according to the positions of the regions, whereunto they were directed. The first volume containeth the worthy discoueries, &c. of the English . . . The second volume comprehendeth the principall nauigations . . . to the south and south-east parts of the world . . . By Richard Hakluyt preacher, and sometime student of Christ-Church in Oxford* (Imprinted at London: By George Bishop, Ralph Newberie, and Robert Barker).

Hakluyt, R., 1552?–1616 (1609), *Virginia richly valued, by the description of the maine land of Florida, her next neighbourout of the foure yeeres continuall trauell and discouerie, for aboue one thousand miles east and west, of Don Ferdinando de Soto, and sixe hundred able men in his companie. Wherin are truly obserued the riches and fertilitie of those parts, abounding with things necessarie, pleasant, and profitable for the life of man: with the natures and dispositions of the inhabitants. Written by a Portugall gentleman of Eluas, emploied in all the action, and translated out of Portugese by Richard Hakluyt.; Relaçam verdadeira dos trabalhos que ho governador dom Fernando de Souto e certos fidalgos portugueses passarom no descobrimento da Frolida.English* (London:

Printed by Felix Kyngston for Matthew Lownes, and are to be sold at the signe of the Bishops head in Pauls Churchyard).

Hall, J., A. Gentili, et al. (1613), *The discouery of a new world or A description of the South Indies Hetherto vnknowne by an English Mercury* ([London]: Imprinted for Ed: Blount. and W. Barrett).

Hammer, P. E. J. (1998), 'A Welshman Abroad: Captain Peter Wynn of Jamestown', *Parergon*, vol. 16: 1, pp. 59–92.

Handover, P. M. (1959), *The Second Cecil: The Rise to Power, 1563–1604, of Sir Robert Cecil, Late First Earl of Salisbury* (London, Eyre & Spottiswoode).

Hantman, J. L. (1990), 'Between Powhatan and Quirank: Reconstructing Monacan Culture and History in the Context of Jamestown', *American Anthropologist*, vol. 92: 3, pp. 676–90.

Hariot, T., 1560–1621 (1588), *A briefe and true report of the new found land of Virginia of the commodities there found and to be raysed, as well marchantable, as others for victuall, building and other necessarie vses for those that are and shalbe the planters there; and of the nature and manners of the naturall inhabitants: discouered by the English colony there seated by Sir Richard Greinuile Knight in the yeere 1585. which remained vnder the gouernment of Rafe Lane Esquier, one of her Maiesties Equieres, during the space of twelue monethes: at the speciall charge and direction of the Honourable Sir Walter Raleigh Knight, Lord Warden of the stanneries; who therein hath beene fauored and authorised by her Maiestie and her letters patents: directed to the aduenturers, fauourers, and welwillers of the action, for the inhabiting and planting there: by Thomas Hariot; seruant to the abouenamed Sir Walter, a member of the Colony, and there imployed in discouering* (London: [By R. Robinson]).

Herford, C. H., B. Jonson, et al. (1986), *Ben Jonson* (Oxford, Clarendon).

Heywood, T. (1612), *An apology for actors Containing three briefe treatises. 1 Their antiquity. 2 Their ancient dignity. 3 The true vse of their quality. Written by Thomas Heywood* (London: Printed by Nicholas Okes).

Hills Wallace, H. (1906), *The History of East Grinstead* (East Grinstead, Farncombe & Co.).

Honan, P. (1999) *Shakespeare: A Life* (Oxford, Oxford University Press).

Hotten, J. C. (1962), *The original lists of persons of quality, emigrants, religious exiles, political rebels, serving men sold for a term of years, apprentices, children stolen, maidens pressed, and others who went from Great Britain to the American plantations, 1600–1700 : with their ages, the localities where they formerly lived in the mother country, the names of the ships in which they embarked, and other interesting particulars, from mss.*

preserved in the State Paper Department of Her Majesty's Public Record Office, England (Baltimore, Genealogical Publishing Company).

Hulme, P. and W. H. Sherman (2000), *The Tempest and its Travels* (London, Reaktion).

Hunter, W. W. S. (1899), *A History of British India* (London).

Husselby, J. and P. Henderson (2002), 'Location, Location, Location! Cecil House in the Strand', *Architectural History*, vol. 45, pp. 159–93.

James I, (1604), *A Counterblaste to Tobacco* (Imprinted at London: By R. B[arker]).

Jardine, L. and A. Stewart (1999), *Hostage to Fortune: The Troubled Life of Francis Bacon* (London, Phoenix Giant).

Jefferson, T., A. A. Lipscomb, et al. (1905), *The Writings of Thomas Jefferson* (Washington, D.C., Issued under the auspices of the Thomas Jefferson Memorial Association of the United States).

Johnson, R., *fl.* 1586–1626 (1609), *Nova Britannia offering most excellent fruites by planting in Virginia: exciting all such as be well affected to further the same* (London: Printed for Samuel Macham, and are to be sold at his shop in Pauls church-yard, at the signe of the bul-head).

Johnson, R., *fl.* 1586–1626 (1612), *The nevv life of Virginea declaring the former successe and present estate of that plantation, being the second part of Noua Britannia. Published by the authoritie of his Maiesties Counsell of Virginea* (London: Imprinted by Felix Kyngston for William Welby, dwelling at the signe of the Swan in Pauls Churchyard).

Jourdain, S., d. 1650 (1613), *A plaine description of the Barmudas, now called Sommer IlandsVVith the manner of their discouerie anno 1609. by the shipwrack and admirable deliuerance of Sir Thomas Gates, and Sir George Sommers, wherein are truly set forth the commodities and profits of that rich, pleasant, and healthfull countrie. With an addition, or more ample relation of diuers other remarkeable matters concerning those ilands since then experienced, lately sent from thence by one of the colonie now there resident; Discovery of the Barmudas, otherwise called the Ile of Divels* (London: Printed by W. Stansby, for W. Welby).

Kamen, H. (2003), *Spain's Road to Empire: The Making of a World Power, 1492–1763* (London, Penguin).

Kelso, W. M. (1995–2001), *Rediscovering Jamestown*, (Richmond, Virginia, APVA).

Kelso, W. M. (2004), *Jamestown Rediscovery 1994–2004* (Richmond, Virginia, APVA).

King, J. C. H. (2000), 'Native American Ethnicity: a View from the British Museum', *Historical Research* vol. 73: 182, pp. 221–38.

Kingsbury, S. M., L. Virginia Company of, et al. (1906), *The Records of the Virginia Company of London* (Washington, Government Printing Office).

Konig, D. T. (1982), ' "Dales' Laws" and the Non Common-Law Origins of Criminal Justice in Virginia', *American Journal of Legal History*, vol. 26, pp. 338–66.

Kuin, R. (1998), 'Querre-Muhau: Sir Philip Sidney and the New World', *Renaissance Quarterly*, vol. 51: 2, pp. 549–85.

Kupperman, K. O. (1982), 'The Puzzle of the American Climate in the Early Colonial Period', *American Historical Review*, vol. 87: 5, pp. 1,262–89.

Kupperman, K. O. (2000), *Indians and English: Facing Off in Early America* (Ithaca, New York; London, Cornell University Press).

Lamb, H. H. (1995), *Climate, History and the Modern World* (London, Routledge).

Lederer, J. and W. Talbot, Sir (1672), *The discoveries of John Lederer in three several marches from Virginia to the west of Carolina and other parts of the continent begun in March, 1669 and ended in September, 1670: together with a general map of the whole territory which he traversed/ collected and translated out of Latine from his discourse and writings, by Sir William Talbot, Baronet* (London: Printed by J.C. for Samuel Heyrick . . .).

Leppard, M. J. (2001), 'Drew Pickesse of Brambletye and Stephen French', *Wealden Iron Research Group Newsletter*, vol. 34, pp. 5–6.

Lescarbot, M. and P. Erondelle, *fl.* 1586–1609 (1609), *Noua Francia: or The description of that part of Nevv France, which is one continent with Virginia Described in the three late voyages and plantation made by Monsieur de Monts, Monsieur du Pont-Graué, and Monsieur de Poutrincourt, into the countries called by the Frenchmen La Cadie, lying to the southwest of Cape Breton. Together with an excellent seuerall treatie of all the commodities of the said countries, and maners of the naturall inhabitants of the same. Translated out of French into English by P.E.; Histoire de la Nouvelle France. Selections* (Londini: [Printed by Eliot's Court Press] impensis Georgii Bishop).

Lewkenor, L., Sir, (1595), *The estate of English fugitiues vnder the king of Spaine and his ministers Containing, besides, a discourse of the sayd Kings manner of gouernment, and the iniustice of many late dishonorable practises by him contriued* (London: Printed [by Thomas Scarlet] for Iohn Drawater, and are to be solde at his shop in Canon lane neere Powles).

London, V. C. O. (1616), *A declaration for the certaine time of dravving the great standing lottery* (Imprinted at London: By Felix Kyngston, for VVilliam VVelby).

Loomie, A. J. (1963), *The Spanish Elizabethans: The English Exiles at the Court of Philip II* (New York, Fordham University Press).

Lorimer, J. (1979), 'The English Contraband Tobacco Trade from Trinidad and Guiana, 1590–1617', *The Westward Enterprise*, K. R. Andrews, N. P. Canny and P. E. H. Hair (Liverpool, Liverpool University Press): pp. 124–50.

MacCulloch, D. (2004), *Reformation: Europe's House Divided, 1490–1700* (London, Penguin).

Malynes, G. (1622), *Consuetudo, vel lex mercatoria, or The ancient law-merchant Diuided into three parts: according to the essentiall parts of trafficke. Necessarie for all statesmen, iudges, magistrates, temporall and ciuile lawyers, mint-men, merchants, marriners, and all others negotiating in all places of the world. By Gerard Malynes merchant* (London: Printed by Adam Islip).

Malynes, G. (1622), *The maintenance of free trade according to the three essentiall parts of traffique; namely, commodities, moneys and exchange of moneys, by bills of exchanges for other countries, or, An answer to a treatise of free trade, or the meanes to make trade flourish, lately published. . . . By Gerard Malynes merchant* (London: Printed by I. L[egatt] for William Sheffard, and are to bee sold at his shop, at the entring in of Popes head Allie out of Lumbard street).

Markham, C. R. (1888), *The Fighting Veres* (London).

Marlowe, C., D. Fuller, et al. (1998), *Tamburlaine the Great, Parts 1 and 2* (Oxford; New York, Clarendon Press: Oxford University Press).

Martin, P. H. and J. Finnis (2003), 'Thomas Thorpe, "W.S.", and the Catholic Intelligencers', *English Literary Renaissance*, vol. 33: 1, pp. 3–43.

Mayes, C. R. (1957), 'The Sale of Peerages in Early Stuart England', *The Journal of Modern History*, vol. 29: 1, pp. 21–37.

McCartney, M. W. (2003), *A Study of the Africans and African Americans on Jamestown Island and at Green Spring, 1619–1803* (Williamsburg, Virginia, Colonial Williamsburg Foundation).

McPherson, B. H. (1998), 'Revisiting the Manor of East Greenwich', *The American Journal of Legal History*, vol. 42: 1, pp. 35–56.

Miller, P. (1948), 'The Religious Impulse in the Founding of Virginia: Religion and Society in the Early Literature', *William and Mary College Quarterly Historical Magazine*, vol. 5: 4, pp. 492–522.

Mook, M. A. (1943), 'The Ethnological Significance of Tindall's Map of Virginia, 1608', *William and Mary College Quarterly Historical Magazine*, vol. 23: 4, pp. 371–408.

Morgan, E. S. (1971), 'The First American Boom: Virginia 1618 to 1630',

William and Mary College Quarterly Historical Magazine, vol. 28: 2, pp. 169–98.

Morgan, E. S. (1975), *American Slavery, American Freedom: The Ordeal of Colonial Virginia* (New York; London, W. W. Norton).

Mornay, P. d., seigneur du Plessis-Marly, P. Sidney, Sir, et al. (1587), *A vvoorke concerning the trewnesse of the Christian religion, written in French: against atheists, Epicures, Paynims, Iewes, Mahumetists, and other infidels. By Philip of Mornay Lord of Plessie Marlie. Begunne to be translated into English by Sir Philip Sidney Knight, and at his request finished by Arthur Golding* (Imprinted at London: [By [John Charlewood and] George Robinson] for Thomas Cadman).

Morris, E. (2002), *Theodore Rex 1901–1909* (London, HarperCollins).

Morris, T. A. (1998), *Europe and England in the Sixteenth Century* (London; New York, Routledge).

Moryson, F. (1617), *An itinerary vvritten by Fynes Moryson Gent. First in the Latine tongue, and then translated by him into English: containing his ten yeeres trauell through the tvvelue dominions of Germany, Bohmerland, Sweitzerland, Netherland, Denmarke, Poland, Jtaly, Turky, France, England, Scotland, and Ireland. Diuided into III parts. The I. part. Containeth a iournall through all the said twelue dominions: shewing particularly the number of miles, the soyle of the country, the situation of cities, the descriptions of them, with all monuments in each place worth the seeing, as also the rates of hiring coaches or horses from place to place, with each daies expences for diet, horse-meate, and the like. The II. part. Containeth the rebellion of Hugh, Earle of Tyrone, and the appeasing thereof: written also in forme of a iournall. The III. part. Containeth a discourse vpon seuerall heads, through all the said seuerall dominions* (At London: Printed by Iohn Beale, dwelling in Aldersgate street).

Mossiker, F. (1996), *Pocahontas: The Life and the Legend* (Cambridge, Massachusetts, Da Capo Press).

N., E. D. (1868), 'William Brewster of the Plymouth Plantation', *Notes and Queries [Great Britain]*, vol. 2: 32, pp. 125–6.

Nelson, A. H. (2003), *Monstrous Adversary: The Life of Edward de Vere, 17th Earl of Oxford* (Liverpool, Liverpool University Press).

Nicholls, M. (1992), 'As Happy a Fortune as I Desire: The Pursuit of Financial Security by the Younger Brothers of Henry Percy, 9th Earl of Northumberland', *Historical Research [Great Britain]*, vol. 65: 158, pp. 296–314.

O'Brien, T. H. (1960), 'The London Livery Companies and the Virginia Company', *Virginia Magazine of History and Biography*, vol. 68, [137]–155.

O'Mara, J. (1982), 'Town Founding in Seventeenth-Century North America: Jamestown in Virginia', *Journal of Historical Geography*, vol. 8: 1, pp. 1–11.

Peckard, P. (1790), *Memoirs of the life of Mr. Nicholas Ferrar* (Cambridge, printed by J. Archdeacon; and sold by J. & J. Merrill, and J. Bowtell; and T. Payne & Son, London).

Peterson, M. L. R. (1988), 'The Sea Venture', *Mariner's Mirror [Great Britain]* vol. 74: 1, pp. 37–48.

Pont, R. (1599), *Against sacrilege three sermons / preached by Maister Robert Pont . . .* (Edinburgh: Printed by Robert Waldegraue).

Pope, F. J. (1911), 'Sir George Somers and his Family', *Proceedings of the Dorset Natural History and Archaeological Society*, vol. 32, pp. 26–32.

Price, D., 1581–1631 (1609), *Sauls prohibition staide. Or The apprehension, and examination of Saule And the inditement of all that persecute Christ, with a reproofe of those that traduce the honourable plantation of Virginia. Preached in a sermon commaunded at Pauls Crosse, vpon Rogation Sunday, being the 28. of May. 1609. By Daniel Price, Chapleine in ordinarie to the Prince, and Master of Artes of Exeter Colledge in Oxford* (London: Printed [by John Windet] for Matthew Law, and are to be sold in Pauls Churchyard, neere vnto Saint Austines Gate, at the signe of the Foxe).

Pritchard, R. E. (2004), 'Shakespeare and Thomas Coryate', *Notes and Queries [Great Britain]*, vol. 51: 3, pp. 295–6.

Prynne, W. (1633), *Histrio-mastix The players scourge, or, actors tragaedie, divided into two parts. Wherein it is largely evidenced, by divers arguments, by the concurring authorities and resolutions of sundry texts of Scripture ... That popular stage-playes ... are sinfull, heathenish, lewde, ungodly spectacles, and most pernicious corruptions; condemned in all ages, as intolerable mischiefes to churches, to republickes, to the manners, mindes, and soules of men. And that the profession of play-poets, of stage-players; together with the penning, acting, and frequenting of stage-playes, are unlawfull, infamous and misbeseeming Christians. All pretences to the contrary are here likewise fully answered; and the unlawfulnes of acting, of beholding academicall enterludes, briefly discussed; besides sundry other particulars concerning dancing, dicing, health-drinking, &c. of which the table will informe you. By William Prynne, an vtter-barrester of Lincolnes Inne* (London: Printed by E[dward] A[llde, Augustine Mathewes, Thomas Cotes] and W[illiam] I[ones] for Michael Sparke, and are to be sold at the Blue Bible, in Greene Arbour, in little Old Bayly).

Purchas, S. (1614), *Purchas his pilgrimage. Or Relations of the vvorld and the religions obserued in all ages and places discouered, from the Creation vnto this present In foure parts. This first containeth a theologicall and geographical historie of Asia, Africa, and America, with the ilands adiacent. . . . Declaring the ancient religions before the Floud . . . With briefe descriptions of the countries, nations, states, discoueries; priuate and publike customes, and the most remarkable rarities of nature, or humane industrie, in the same. The second edition, much enlarged with additions through the whole worke; by Samuel Purchas, minister at Estwood in Essex* (London: Printed by William Stansby for Henrie Fetherstone, and are to be sold at his shop in Pauls Church-yard at the signe of the Rose).

Purchas, S. (1617), *Purchas his pilgrimage, or Relations of the vvorld and the religions obserued in al ages and places discouered, from the Creation vnto this present In foure parts. This first contayneth a theologicall and geographicall historie of Asia, Africa, and America, with the ilands adiacent. Declaring the ancient religions before the Floud . . . With briefe descriptions of the countries, nations, states, discoueries; priuate and publike customes, and the most remarkable rarities of nature, or humane industrie, in the same. The third edition, much enlarged with additions through the whole worke; by Samuel Purchas, parson of St. Martins by Ludgate London* (London: Printed by William Stansby for Henry Fetherstone, and are to be sold at his shop in Pauls Church-yard at the signe of the Rose).

Purchas, S. (1623), *The kings tovvre and triumphant arch of London. A sermon preached at Pauls Crosse, August. 5. 1622. By Samuel Purchas, Bacheler of Diuinitie, and parson of Saint Martins Ludgate, in London* (London: Printed by W. Stansby, and are to be sold by Henrie Fetherstone).

Purchas, S. (1625), *Purchas his pilgrimes In fiue bookes. The first, contayning the voyages and peregrinations made by ancient kings, patriarkes, apostles, philosophers, and others, to and thorow the remoter parts of the knowne world: enquiries also of languages and religions, especially of the moderne diuersified professions of Christianitie. The second, a description of all the circum-nauigations of the globe. The third, nauigations and voyages of English-men, alongst the coasts of Africa ... The fourth, English voyages beyond the East Indies, to the ilands of Iapan, China, Cauchinchina, the Philippinae with others ... The fifth, nauigations, voyages, traffiques, discoueries, of the English nation in the easterne parts of the world ... The first part.* (London: Printed by William Stansby for Henrie Fetherstone, and are to be sold at his shop in Pauls Church-yard at the signe of the Rose).

Quinn, D. B. (1940), *The Voyages and Colonising Enterprises of Sir Humphrey Gilbert* (London, Hakluyt Society).

Quinn, D. B. (1955), *The Roanoke Voyages, 1584-1590: Documents to Illustrate the English Voyages to North America under the Patent Granted to Walter Raleigh in 1584* (London, for the Hakluyt Society).

Quinn, D. B. (1974), *England and the Discovery of America, 1481–1620: From the Bristol Voyages of the Fifteenth Century to the Pilgrim Settlement at Plymouth; the Exploration, Exploitation, and Trial-and-Error Colonization of North America by the English* (London, G. Allen & Unwin Ltd).

Quinn, D. B., A. M. Quinn, et al. (1979), *New American World: A Documentary History of North America to 1612* (London, Macmillan).

Rabb, T. K. (1964), 'Sir Edwin Sandys and the Parliament of 1604', *American Historical Review,* vol. 69: 3, pp. 646–70.

Rabb, T. K. (1966), 'Investment in English Overseas Enterprise, 1575–1630', *Economic History Review,* vol. 19: 1, pp. 70–81.

Rabb, T. K. (1998), *Jacobean Gentleman: Sir Edwin Sandys, 1561-1629* (Princeton, N. J., Princeton University Press).

Raleigh, W., Sir, (1596), *The discouerie of the large, rich, and bevvtiful empire of Guiana with a relation of the great and golden citie of Manoa (which the spanyards call El Dorado) and the prouinces of Emeria, Arromaia, Amapaia, and other countries, with their riuers, adioyning. Performed in the yeare 1595. by Sir W. Ralegh Knight, captaine of her Maiesties Guard, Lo. Warden of the Sannerries [sic], and her Highnesse Lieutenant generall of the countie of Cornewall* (Imprinted at London: By Robert Robinson).

Ransome, D. R. (1992), *The Ferrar Papers, 1590–1790* (Wakefield, UK, Microform Academic Publishers).

Read, C. (1960), *Lord Burghley and Queen Elizabeth* (London, Jonathan Cape).

Renshaw, W. C. (1906), 'Notes from the Act Books of the Archdeaconry Court of Lewes', *Sussex Archaeological Collections: Illustrating the History and Antiquities of the County*, vol. 49.

Review, A. H. (1922), 'Lord Sackville's Papers Respecting Virginia, 1613–1631, I.', *American Historical Review,* vol. 27: 3, pp. 493–538.

Rich, B. (1615), *The honestie of this age proouing by good circumstance, that the world was neuer honest till now / by Barnabe Rych . . .* (Printed at London: For T.A).

Roberts, R. J., A. G. Watson, et al. (1990), *John Dee's Library Catalogue* (London, Bibliographical Society).

Roberts, V. M. (1962), *On Stage: A History of Theatre* (New York, Harper & Row).

Rose, E. (2005), 'The Politics of Pathos: Richard Frethorne's Letters Home', *Envisioning an English Empire: Jamestown and the Making of the North Atlantic World* (R. Appelbaum and J. W. Sweet. Philadelphia, University of Pennsylvania Press): pp. 92–108).

Rountree, H. C. (1989), *The Powhatan Indians of Virginia: Their Traditional Culture* (Norman; London, University of Oklahoma Press).

Salkeld, D. (2004), 'The Bell and The Bel Savage Inns, 1576–1577', *Notes and Queries [Great Britain]*, vol. 51: 3, pp. 242–3.

Sandys, E., Sir, (1605), *A relation of the state of religion and with what hopes and pollicies it hath beene framed, and is maintained in the severall states of these westerne parts of the world* (London: Printed for Simon Waterson dwelling in Paules Churchyard at the signe of the Crowne).

Sandys, E., Sir, (1629), *Europae speculum. Or, A vievv or survey of the state of religion in the vvesterne parts of the world VVherein the Romane religion, and the pregnant policies of the Church of Rome to support the same, are notably displayed: with some other memorable discoueries and memorations, never before till now published according to the authours originall copie* (Hagae-Comitis [i.e. The Hague: Printed for Michael Sparke, London]).

Seaver, P. S. (1985), *Wallington's World: A Puritan Artisan in Seventeenth-Century London* (Stanford, California, Stanford University Press).

Shakespeare, W. and S. Orgel (1987), *The Tempest* (Oxford, Clarendon).

Shakespeare, W., V. M. Vaughan, et al. (1999), *The Tempest* (London, Arden Shakespeare).

Shapiro, I. A. (1950), 'The Mermaid Club', *Modern Language Review*, vol. 45, pp. 6–17.

Shirley, J. W., 1908– (1949), 'George Percy at Jamestown, 1607–1612', *Virginia Magazine of History and Biography*, vol. 57, pp. 227–43.

Sidney, P., (1595), *The Defence of Poesie by Sir Phillip Sidney . . .* (London: Printed for VVilliam Ponsonby).

Simpson, P. (1951), 'The Mermaid Club: An Answer and a Rejoinder', *Modern Language Review*, vol. 46, pp. 58.

Sluiter, E. (1997), 'New Light on the "20. and Odd Negroes" Arriving in Virginia, August 1619', *The William and Mary Quarterly*, vol. 54: 2, pp. 395–8.

Smith, B. (2001), 'A Wealden ironmaster in Jamestown', *Wealden Iron Research Group Newsletter*, vol. 33, pp. 5–6.

Smith, J., P. L. Barbour, et al. (1986), *The Complete Works of Captain John Smith (1580–1631)*, (Chapel Hill, Published for the Institute of Early American History and Culture, Williamsburg, Virginia, by the University of North Carolina Press).

Smith, J. D. (2002), *Managing White Supremacy: Race Politics and Citizenship in Jim Crow Virginia* (Chapel Hill, North Carolina Press).

Spenser, E. (1590), *The faerie qveene disposed into twelue books, fashioning XII. morall vertues.* (London: Printed for William Ponsonbie).

Stahle, D. W., M. K. Cleaveland, et al. (1998), 'The Lost Colony and Jamestown Droughts', *Science,* vol. 280: 5,363, pp. 564–7.

Stephenson, N. W. (1913), 'Some Inner History of the Virginia Company pt1', *William and Mary College Quarterly,* vol. 22: 2, pp. 89–98.

Still, B. (1956), *Mirror for Gotham: New York as Seen by Contemporaries from Dutch days to the present* (Washington Square, NY: University Press).

Stow, J., E. Howes, et al. (1631), *Annales, or, a generall chronicle of England* (Londini, A. Mathewes for Impensis Richardi Meighen).

Strachan, M. (1967), 'The Mermaid Tavern Club, a New Discovery', *History Today,* vol. 17: 8, pp. 533–8.

Strachey, W., 1572? –1621 (1612), *For the colony in Virginea Britannia. Lavves diuine, morall and martiall, &c.* (Printed at London: [By William Stansby] for Walter Burre).

Strong, R. C. (2000), *Henry, Prince of Wales and England's Lost Renaissance* (London, Pimlico).

Stymeist, D. (2002), ' "Strange Wives": Pocahontas in Early Modern Colonial Advertisement', *Mosaic,* vol. 35, p. 109+.

Symonds, W., 1556–1616? (1609), *Virginia. A sermon preached at VVhite-Chappel, in the presence of many, honourable and worshipfull, the aduenturers and planters for Virginia. 25. April. 1609 Published for the benefit and vse of the colony, planted, and to bee planted there, and for the aduancement of their Christian purpose. By William Symonds, preahcer at Saint Sauiors in Southwarke* (London: Printed by I. Windet, for Eleazar Edgar, and William Welby, and are to be sold in Paules Church-yard at the signe of the Windmill).

Taylor, E. G. R. and R. Hakluyt (1935), *The Original Writings and Correspondence of the Two Richard Hakluyts* (London, Hakluyt Society).

Thomas, P. A. (1976), 'Contrastive Subsistence Strategies and Land Use as Factors for Understanding Indian–White Relations in New England', *Ethnohistory,* vol. 23: 1, pp. 1–18.

Thompson, C. (1997), 'The Reaction of the House of Commons in November and December 1621 to the Confinement of Sir Edwin Sandys', *Historical Journal,* vol. 40: 3, pp. 779–86.

Thompson, E. K. (1959), 'English and Spanish Tonnage in 1588' *Mariner's Mirror [Great Britain]* vol. 46, p. 154.

Thorndale, W. (1995), 'The Virginia Census of 1619', *Magazine of Virginia Genealogy,* vol. 13: 3, pp. 155–70.

Thornton, J. (1998), 'The African Experience of the "20. and Odd Negroes" Arriving in Virginia in 1619', *The William and Mary Quarterly,* vol. 55: 3, pp. 421–34.

Thornton, J. K. (1999), *Warfare in Atlantic Africa, 1500–1800* (London, UCL Press).

Totten, G. (2005), 'Zitkala-Sa and the Problem of Regionalism: Nations, Narratives, and Critical Traditions', *The American Indian Quarterly,* vol. 29: 1, pp. 84–123.

Traister, B. H. (2001), *The Notorious Astrological Physician of London: Works and Days of Simon Forman* (Chicago; London, University of Chicago Press).

Trevelyan, R. (2002), *Sir Walter Raleigh* (London, Allen Lane).

Trumbull, J. H. (1870), *The Composition of Indian Geographical Names, Illustrated from the Algonkin Languages.*

Tyler, L. G. (1893), 'Virginia under the Commonwealth', *William and Mary College Quarterly Historical Papers,* vol. 1: 4, pp. 189–96.

Vaughan, A. T. (2000), 'Trinculo's Indian: American Natives in Shakespeare's England', *Critical Views* (London, Reaktion): pp. 49–59.

Vaughan, A. T. (2005), 'Powhatans Abroad: Virginia Indians in England', *Envisioning an English Empire: Jamestown and the Making of the North Atlantic World*, R. Appelbaum and J. W. Sweet (Philadelphia, University of Pennsylvania Press: 49–67).

Virginia Company (1610), *A publication by the counsell of Virginea, touching the plantation there* (Jmprinted at London: By Thomas Haueland for William Welby, and are to be sold at his shop in Pauls Church-yard at the signe of the Swanne).

Virginia Company (1610), *A true and sincere declaration of the purpose and ends of the plantation begun in Virginia of the degrees which it hath receiued; and meanes by which it hath beene aduanced: and the resolution and conclusion of his Maiesties councel of that colony, for the constant and patient prosecution thereof, vntill by the mercies of God it shall retribute a fruitful haruest to the kingdome of heauen, and this common-wealth. Sett forth by the authority of the gouernors and councellors established for that plantation* (At London: Printed [by George Eld] for I. Stepneth, and are to be sold [by W. Burre] at the signe of the Crane in Paules Churchyard).

Virginia Company (1610), *A true declaration of the estate of the colonie in Virginia vvith a confutation of such scandalous reports as haue tended to the disgrace of so worthy an enterprise. Published by aduise and direction of the Councell of Virginia* (London: Printed [by Eliot's Court Press and

William Stansby] for William Barret, and are to be sold [by Edward Blount] at the blacke Beare in Pauls Churchyard).

W., W. (1612), *Londons lotterie with an incouragement to the furtherance thereof for the good of Virginia, and the benefite of this our natiue countrie, wishing good fortune to all that venture in the same[.] To the tune of Lusty Gallant* (Imprinted at London: by W. W[hite]. for Henry Robards, and are to be sold at his shop neere to S[. Dunstons] Church without Aldgate).

Walne, P. (1962), 'The "Running Lottery" of the Virginia Company', *Virginia Magazine of History and Biography*, vol. 70, pp. 30–34.

Waterhouse, E., colonist. and H. Briggs, 1561–1630 (1622), *A declaration of the state of the colony and affaires in Virginia With a relation of the barbarous massacre in the time of peace and league, treacherously executed by the natiue infidels vpon the English, the 22 of March last. Together with the names of those that were then massacred; that their lawfull heyres, by this notice giuen, may take order for the inheriting of their lands and estates in Virginia. And a treatise annexed, written by that learned mathematician Mr. Henry Briggs, of the Northwest passage to the South Sea through the continent of Virginia, and by Fretum Hudson. Also a commemoration of such worthy benefactors as haue contributed their Christian charitie towards the aduancement of the colony. And a note of the charges of necessary prouisions fit for euery man that intends to goe to Virginia. Published by authoritie* (Imprinted at London: By G. Eld, for Robert Mylbourne, and are to be sold at his shop, at the great south doore of Pauls).

Weinreb, B. and C. Hibbert (1987), *The London Encyclopaedia* (London, Papermac).

Whitaker, A., and W. Crashaw (1613), *Good nevves from Virginia Sent to the Counsell and Company of Virginia, resident in England. From Alexander Whitaker, the minister of Henrico in Virginia. Wherein also is a narration of the present state of that countrey, and our colonies there. Perused and published by direction from that Counsell. And a preface prefixed of some matters touching that plantation, very requisite to be made knowne* (At London: Imprinted by Felix Kyngston for William Welby, and are to be sold at his shop in Pauls Church-yard at the signe of the Swanne).

White, J. (1630), *The planters plea· Or The grounds of plantations examined, and vsuall objections answered Together with a manifestation of the causes mooving such as have lately vndertaken a plantation in Nevv-England: for the satisfaction of those that question the lawfulnesse of the action* (London: Printed by– William Iones [, M. Flesher, and J. Dawson]).

Wight, I. o. (1899), 'Isle of Wight County Records', *William and Mary College Quarterly*, vol. 7: 4, pp. 205–315.

Wilding, M. (1999), 'Edward Kelley: A Life', *Cauda Pavonis*, vol. 18: 1 & 2, pp. 1–26.

Williams, P. (1995), *The Later Tudors: England, 1547–1603* (Oxford; New York, Clarendon Press: Oxford University Press).

Wingfield, J. R. (1993), *Virginia's True Founder: Edward-Maria Wingfield and his Times 1550-c.1614* (Athens, GA, Wingfield Family Society).

Woodnoth, A. (1651), *A short collection of the most remarkable passages from the originall to the dissolution of the Virgina company* (London: Prin[t]ed by Richard Cotes for Edward Husband . . .).

Woodward, D. (1981), 'Wage Rates and Living Standards in Pre-Industrial England', *Past and Present*, vol. 91, pp. 28–46.

Wright, I. A. (1920), 'Spanish Policy Toward Virginia, 1606–1612; Jamestown, Ecija, and John Clark of the Mayflower', *The American Historical Review*, vol. 25: 3, pp. 448–79.

Zimmermann, Warren, 'Jingoes, Goo-Goos and the Rise of America's Empire', *The Wilson Quarterly*, vol. 22, Spring 1998.

INDEX